An Introduction to Domain Decomposition Methods

An Introduction to Domain Decomposition Methods

Algorithms, Theory, and Parallel Implementation

Victorita Dolean
University of Strathclyde
Glasgow, United Kingdom

Pierre Jolivet
Toulouse Institute of Computer Science Research
Toulouse, France

Frédéric Nataf
University Pierre and Marie Curie
Paris, France

siam.

Society for Industrial and Applied Mathematics
Philadelphia

10 9 8 7 6 5 4 3 2 1

Publisher	David Marshall
Acquisitions Editor	Elizabeth Greenspan
Developmental Editor	Gina Rinelli
Managing Editor	Kelly Thomas
Production Editor	David Riegelhaupt
Copy Editor	Nicola Howcroft
Production Manager	Donna Witzleben
Production Coordinator	Cally Shrader
Compositor	Techsetters, Inc.
Graphic Designer	Lois Sellers

Library of Congress Cataloging-in-Publication Data

Dolean, Victorita.

An introduction to comain decomposition methods : algorithms, theory, and parallel implementation / Victorita Dolean, University of Strathclyde, Glasgow, United Kingdom, Pierre Jolivet, Frédéric Nataf, Unive-sity Pierre and Marie Curie, Paris, France.

pages cm. – (Other titles in applied mathematics ; 144)

Includes bibliographical references and index.

ISBN 978-1-611974-05-8

1. Decomposition method. 2. Differential equations, Partial. I. Jolivet, Pierre, 1988- II. Nataf, Frédéric (Computer scientist) III. Title.

QA402.2.D65 2015

515'.353–dc23

2015027040

Contents

Preface

The purpose of this text is to offer an overview of the most popular domain decomposition methods for partial differential equations (PDEs). The presentation is kept as much as possible at an elementary level with a special focus on the definitions of these methods both in terms of PDEs and of the sparse matrices arising from their discretizations. We also provide implementations available at www.siam.org/books/ot144 written with open-source finite element software. In addition, we consider a number of methods that have not been presented in other books. We think that this book will give a new perspective and that it will complement those of Smith, Bjørstad, and Gropp [175], Quarteroni and Valli [165], Mathew [138], and Toselli and Widlund [185], as well as the review article [23].

The book is addressed to computational scientists, mathematicians, physicists, and, in general, to people involved in numerical simulation of PDEs. It can also be used as a textbook for advanced undergraduate/first-year graduate students. The mathematical tools needed are basic linear algebra, notions of programming, variational formulation of PDEs, and basic knowledge of finite element discretization.

The value of domain decomposition methods is part of a general need for parallel algorithms for professional and consumer use. We will focus on scientific computing and more specifically on the solution of the algebraic systems arising from the approximation of a PDE.

Domain decomposition methods are a family of methods used to solve problems of linear algebra on parallel machines in the context of simulation. In scientific computing, the first step is to model mathematically a physical phenomenon. This often leads to systems of PDEs such as the Navier–Stokes equations in fluid mechanics, elasticity systems in solid mechanics, Schrödinger equations in quantum mechanics, the Black–Scholes equation in finance, Lighthill–Whitham equations for traffic, etc. Functional analysis is used to study the well-posedness of the PDEs, which is a necessary condition for their possible numerical approximation. Numerical analysis enables one to design stable and consistent discretization schemes. This leads to the discrete equations $F(u) = b \in \mathbb{R}^n$, where n is the number of degrees of freedom of the discretization. If F is linear, calculating u is a problem of linear algebra. If F is nonlinear, a method for solving it is the classical Newton method, which also leads to solving a series of linear systems.

In the past, improving performance of a program, either in speed or in the amount of data processed, was only a matter of waiting for the next generation of processors. Every 18 months, computer performance doubled. As a consequence, linear solver research would take second place to the search for new discretization schemes. But since around 2005 the clock speed has stagnated at 2–3 GHz. Increase in performance is almost entirely due to the increase in the number of cores per processor. All major processor vendors are producing multicore chips, and now every machine is a parallel machine. Waiting for the next-generation machine no longer guarantees better performance of software. To keep doubling, performance parallelism must double, which implies a huge effort in algorithmic development. Scientific computing is only one illustration of this general need in computer science. Visualization, data storage, mesh generation, operating systems, etc., must all be designed with parallelism in mind.

We focus here on parallel linear iterative solvers. Contrary to direct methods, the appealing feature of domain decomposition methods is that they are naturally parallel. We introduce the reader to the main classes of domain decomposition algorithms: Schwarz, Neumann–Neumann/FETI, and optimized Schwarz. For each method we start with the continuous formulation in terms of PDEs for two subdomains. We then give the definition in terms of stiffness matrices and their implementation in a free finite element package in the many-subdomain case. This presentation reflects the dual nature of domain decomposition methods. They are solvers of linear systems, keeping in mind that the matrices arise from the discretization of partial differential operators. As for domain decomposition methods that directly address nonlinearities, we refer the reader to, e.g., [63, 122], [17] or [18] and references therein. As for iterative solvers non related to domain decomposition we refer the reader to [12] or [140] e.g.

In Chapter 1 we start by introducing different versions of Schwarz algorithms at the continuous level, taking as our starting point the methods of H. Schwarz (see [174]): the Jacobi–Schwarz method (JSM), the additive Schwarz method (ASM) and the restricted additive Schwarz (RAS) method. The first natural feature of these algorithms is that they are equivalent to a block Jacobi method (default solver in PETSc [8, 7]) when the overlap is minimal. We move on to the algebraic versions of the Schwarz methods. In order to do this, several concepts are necessary: restriction and prolongation operators, as well as partitions of unity which make the global definition possible. These concepts are explained in detail for different types of discretizations (finite difference or finite element) and spatial dimensions. The convergence of the Schwarz method in the two-subdomain case is illustrated for one-dimensional problems and then for two-dimensional problems by using Fourier analysis. A short paragraph introduces P. L. Lions' algorithm that will be considered in detail in Chapter 2. The last part of the chapter is dedicated to numerical implementation using FreeFem++ [108] for general decompositions into subdomains.

In Chapter 2 we present the optimized Schwarz methods applied to the Helmholtz equation, which models acoustic wave propagation in the frequency domain. We begin with the two-subdomain case. We show the need for the use of interface conditions which are different from Dirichlet or Neumann boundary conditions. The Lions and Després algorithms, which are based on Robin interface conditions, are analyzed together with their implementations. We also show that by taking even more general interface conditions, much better convergence can be achieved at no extra cost when compared to the use of Robin interface conditions. We also consider the many-subdomain case. These algorithms are the method of choice for wave propagation phenomena in the frequency regime. Such situations occur in acoustics, electromagnetics, and elastodynamics.

In Chapter 3 we present the main ideas which justify the use of Krylov methods instead of stationary iterations. Since the Schwarz methods introduced in Chapters 1 and 2 represent fixed-point iterations applied to preconditioned global problems, and consequently do not provide the fastest convergence possible, it is natural to apply Krylov methods instead. This provides the justification for using Schwarz methods as preconditioners rather than solvers. Numerical implementations and results using FreeFem++ close the chapter. Although some aspects of the presentation of some Krylov methods are not standard, readers already familiar with Krylov methods may as well skip this part.

Chapter 4 is devoted to the introduction of two-level methods. In the presence of many subdomains, the performance of Schwarz algorithms, i.e., the iteration number and execution time, will grow linearly with the number of subdomains in one direction. From a parallel computing point of view this translates into a lack of scalability. The latter can be achieved by adding a second level or a coarse space. This is closely related to multigrid methods and deflation methods from numerical linear algebra. The simplest coarse space, that of Nicolaides, is introduced and then implemented in FreeFem++.

In Chapter 5, we show that Nicolaides' coarse space (see above) is a particular case of a more general class of spectral coarse spaces which are generated by vectors resulting from solving some local generalized eigenvalue problems. Then, a theory of these two-level algorithms is presented. First, a general variational setting is introduced, as well as elements from the abstract theory of the two-level additive Schwarz methods (e.g., the concept of stable decomposition). The analysis of spectral and classical coarse spaces goes through some properties and functional analysis results. These results are valid for scalar elliptic PDEs. This chapter is more technical than the others and is not necessary to the rest of the book.

Chapter 6 is devoted to the Neumann–Neumann and FETI algorithms. We start with the two-subdomain case for the Poisson problem. Then, we consider the formulation in terms of stiffness matrices and stress the duality of these methods. We also establish a connection with block factorization of the stiffness matrix of the original problem. We then show that in the many-subdomain case Neumann–Neumann and FETI are no longer strictly equivalent. For the sake of simplicity, we give a FreeFem++ implementation of only the Neumann–Neumann algorithm. The reader is then ready to delve into the abundant literature devoted to the use of these methods for solving complex mechanical problems.

In Chapter 7, we return to two-level methods. This time, a quite recent adaptive abstract coarse space, together with most classical two-level methods, is presented in a different light, under a common framework. Moreover, their convergence properties are proven in an abstract setting, provided that the assumptions of the fictitious space lemma are satisfied. The new coarse space construction is based on solving GENeralized Eigenvalue problems in the Overlap (GenEO). The construction is provable in the sense that the condition number is given in terms of an explicit formula where the constants that appear are the maximal number of neighbors of a subdomain and a threshold prescribed by the user. The latter can be applied to a broader class of elliptic equations, which include systems of PDEs such as those for linear elasticity, even with highly heterogeneous coefficients. From sections 7.1 to 7.6, we give all the materials necessary to build and analyze two-level methods for additive Schwarz methods. In section 7.7, we build a coarse space for the one-level optimized Schwarz methods of Chapter 2. It is based on introducing the SORAS algorithm and two related generalized eigenvalue problems. The resulting algorithm is named SORAS-GenEO-2. Section 7.8 is devoted to endowing the one-level Neumann–Neumann algorithm of Chapter 6 with a GenEO-type coarse space.

In Chapter 8 we introduce the parallel computational framework used in the parallel version of the free finite element package FreeFem++, which is currently linked with HPDDM, a framework for high-performance domain decomposition methods, available at `https://github.com/hpddm/hpddm`. For the sake of simplicity we restrict ourselves to the two-level Schwarz methods. Numerical simulations of very large scale problems on high-performance computers show the weak and strong scalabilities of the Schwarz methods for 2D and 3D Darcy and elasticity problems with highly heterogeneous coefficients with billions of degrees of freedom. A self-contained FreeFem++ parallel script is given.

In Figure 1, we give a dependency graph of the various chapters. For instance, in order to understand Chapter 4 it is necessary to be familiar with both Chapters 1 *and* 3. From

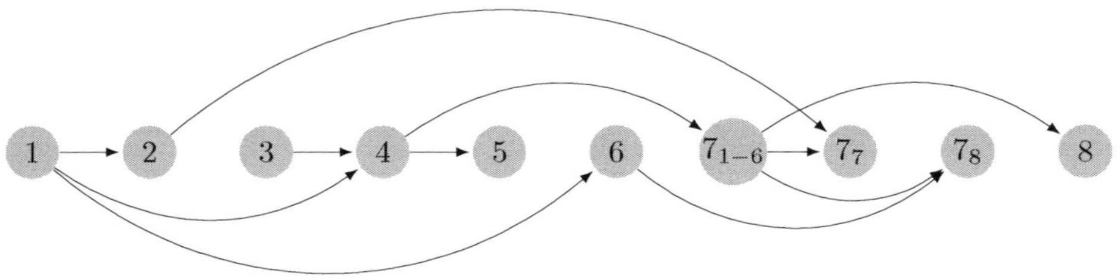

Figure 1. *Dependency graph of chapters.*

this graph, the reader is able to choose their own way of reading the book. We suggest some possible partial readings. A reader interested in having only a quick and partial view, and who is already familiar with Krylov methods, may very well read only Chapter 1 followed by Chapter 4. For newcomers to Krylov methods, a reading of Chapter 3 must be intercalated between Chapter 1 and Chapter 4.

For a quick view on all Schwarz methods without entering into the technical details of coarse spaces, one could consider beginning with Chapter 1, followed by Chapter 2 and then Chapter 3 on the use of Schwarz methods as preconditioners, to finish with Chapter 4 on classical coarse spaces.

For the more advanced reader, Chapters 5 and 7 provide the technical framework for the analysis and construction of more sophisticated coarse spaces. And last but not least, Chapter 8 provides the key to parallel implementation and illustrates the previously introduced methods with large-scale numerical results.

Acknowledgments

We have been developing the material for this book for five years. Over the years we have benefited from feedback and suggestions from many people, including our own graduate students, students in our courses, and our colleagues at Laboratory J.-L. Lions and elsewhere. This printing benefited greatly from the SIAM editorial staff's expertise, as well as from the anonymous reviewers' valuable comments and suggestions. We would like to single out for special acknowledgments the following people: Frédéric Hecht (the father of FreeFem++) for his expertise in scientific computing and for initiating our work on parallel domain decomposition methods with FreeFem++; Olivier Pironneau for his motivational comments and positive response on writing this book; Jed Brown for his thorough knowledge of PETSc, and for sitting down with us and tuning such a powerful software adequately; and Olof Widlund for our fruitful discussions together and for his leading role in domain decomposition methods.

Chapter 1

Schwarz methods

1.1 ▪ Three continuous Schwarz algorithms

Hermann Schwarz was a German analyst of the 19th century. He was interested in proving the existence and uniqueness of the Poisson problem. At that time, there were no Sobolev spaces and no Lax–Milgram theorem. The only available tool was the Fourier transform, limited by its very nature to simple geometries. In order to consider more general situations, Schwarz devised an iterative algorithm for solving a Poisson problem set on a union of simple geometries; see [174]. For a historical presentation of these kinds of methods, see [88].

Let the domain Ω be the union of a disk and a rectangle; see Figure 1.1. Consider the Poisson problem which consists in finding $u : \Omega \to \mathbb{R}$ such that

$$\begin{aligned} -\Delta(u) &= f \quad \text{in } \Omega, \\ u &= 0 \quad \text{on } \partial\Omega. \end{aligned} \tag{1.1}$$

Definition 1.1 (Original Schwarz algorithm). *The Schwarz algorithm is an iterative method based on solving subproblems alternatively in domains Ω_1 and Ω_2. It updates $(u_1^n, u_2^n) \to (u_1^{n+1}, u_2^{n+1})$ by*

$$\begin{aligned} -\Delta(u_1^{n+1}) &= f && \text{in } \Omega_1, \\ u_1^{n+1} &= 0 && \text{on } \partial\Omega_1 \cap \partial\Omega, \\ u_1^{n+1} &= u_2^n && \text{on } \partial\Omega_1 \cap \overline{\Omega_2}; \end{aligned} \quad \text{then,} \quad \begin{aligned} -\Delta(u_2^{n+1}) &= f && \text{in } \Omega_2, \\ u_2^{n+1} &= 0 && \text{on } \partial\Omega_2 \cap \partial\Omega, \\ u_2^{n+1} &= u_1^{n+1} && \text{on } \partial\Omega_2 \cap \overline{\Omega_1}. \end{aligned} \tag{1.2}$$

Schwarz proved the convergence of the algorithm and thus the well-posedness of the Poisson problem in complex geometries.

With the advent of digital computers, this method also acquired a practical interest as an iterative linear solver. Subsequently, parallel computers became available and a small modification of the algorithm [129] makes it suited to these architectures. Its convergence can be proved using the maximum principle [128].

Definition 1.2 (Jacobi Schwarz algorithm). *The Jacobi Schwarz algorithm is an iterative method which solves concurrently in all subdomains, $i = 1, 2$:*

$$\begin{aligned} -\Delta(u_i^{n+1}) &= f && \text{in } \Omega_i, \\ u_i^{n+1} &= 0 && \text{on } \partial\Omega_i \cap \partial\Omega, \\ u_i^{n+1} &= u_{3-i}^n && \text{on } \partial\Omega_i \cap \overline{\Omega}_{3-i}. \end{aligned} \tag{1.3}$$

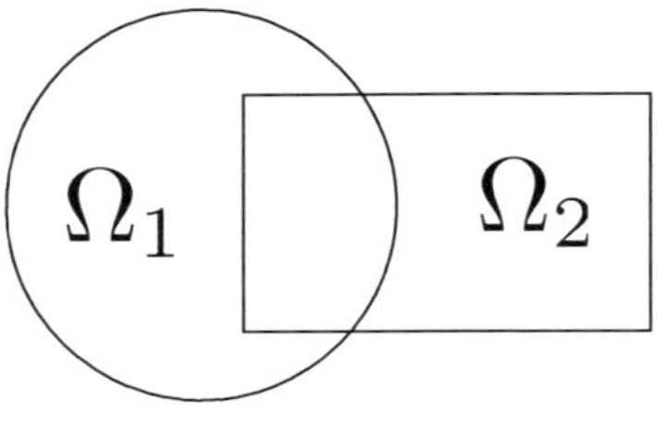

Figure 1.1. *A complex domain made from the union of two simple geometries.*

It is easy to see that if the algorithm converges, the solutions u_i^∞, $i = 1,2$, in the intersection of the subdomains take the same values. Indeed, in the overlap $\Omega_{12} := \Omega_1 \cap \Omega_2$, let $e^\infty := u_1^\infty - u_2^\infty$. By the last line of (1.3), we know that $e^\infty = 0$ on $\partial\Omega_{12}$. By linearity of the Poisson equation, we also have that e^∞ is harmonic. Thus, e^∞ solves the homogeneous well-posed boundary value problem (BVP)

$$\begin{aligned} -\Delta(e^\infty) &= 0 \quad \text{in} \quad \Omega_{12}, \\ e^\infty &= 0 \quad \text{on} \quad \partial\Omega_{12}, \end{aligned}$$

and thus $e^\infty = 0$.

Algorithms (1.2) and (1.3) act on the local functions $(u_i)_{i=1,2}$. In order to write algorithms that act on global functions we need extension operators and a partition of unity.

Definition 1.3 (Extension operators and a partition of unity). *Let the extension operator E_i be such that $E_i(w_i) : \Omega \to \mathbb{R}$ is the extension of a function $w_i : \Omega_i \mapsto \mathbb{R}$, by zero outside Ω_i. We also define the partition of unity functions $\chi_i : \Omega_i \to \mathbb{R}$, $\chi_i \geq 0$, and $\chi_i(x) = 0$ for $x \in \partial\Omega_i \setminus \partial\Omega$ and such that*

$$w = \sum_{i=1}^{2} E_i(\chi_i\, w_{|\Omega_i}) \tag{1.4}$$

for any function $w : \Omega \mapsto \mathbb{R}$.

There are two ways to write related algorithms that act on global functions. They are given in Definitions 1.4 and 1.7.

Definition 1.4 (First global Schwarz iteration). *Let u^n be an approximation of the solution to the Poisson problem* (1.1). *u^{n+1} is computed by first solving local subproblems*

$$\begin{aligned} -\Delta(w_i^{n+1}) &= f \quad \text{in} \quad \Omega_i, \; w_i^{n+1} = u^n \text{ on } \quad \partial\Omega_i \cap \overline{\Omega}_{3-i}, \\ w_i^{n+1} &= 0 \quad \text{on} \quad \partial\Omega_i \cap \partial\Omega \end{aligned} \tag{1.5}$$

and then gluing them together using the partition of unity functions:

$$u^{n+1} := \sum_{i=1}^{2} E_i(\chi_i\, w_i^{n+1}). \tag{1.6}$$

We can prove the following property.

Lemma 1.5. *Algorithm* (1.5)–(1.6) *which iterates on u^n and algorithm* (1.3) *which iterates on (u_1^n, u_2^n) are equivalent.*

Proof. Starting from initial guesses which satisfy $u^0 = \sum_{i=1}^2 E_i(\chi_i\, u_i^0)$, we prove by induction that

$$u^n = \sum_{i=1}^2 E_i(\chi_i\, u_i^n) \tag{1.7}$$

holds for all $n \geq 0$. Assume the property holds at step n of the algorithm. Then, using the fact that $\chi_2 = 1$ and $\chi_1 = 0$ on $\partial\Omega_1 \cap \bar{\Omega}_2$, see Figure 2.6 we have by definition that w_1^{n+1} is a solution to BVP (1.3)

$$\begin{aligned} -\Delta(w_1^{n+1}) &= f && \text{in} \quad \Omega_1, \\ w_1^{n+1} &= 0 && \text{on} \quad \partial\Omega_1 \cap \partial\Omega, \\ w_1^{n+1} = u^n = \sum_{i=1}^2 E_i(\chi_i\, u_i^n) &= u_2^n && \text{on} \quad \partial\Omega_1 \cap \overline{\Omega}_2, \end{aligned} \tag{1.8}$$

and thus $w_1^{n+1} = u_1^{n+1}$. The proof is the same for $w_2^{n+1} = u_2^{n+1}$. Finally, we have, using (1.6),

$$u^{n+1} = \sum_{i=1}^2 E_i(\chi_i\, w_i^{n+1}) = \sum_{i=1}^2 E_i(\chi_i\, u_i^{n+1}). \qquad \square$$

This result can be seen as a continuous version of the algebraic formulation established in [71].

We introduce in Algorithm 1.1 another formulation of algorithm (1.5)–(1.6) in terms of the continuous residual $r^n := f + \Delta u^n$. This way, we get closer to the algebraic definition of domain decomposition methods. Algorithm 1.1 is called the RAS (restricted additive Schwarz) algorithm.

Lemma 1.6 (Equivalence between the Schwarz algorithm and RAS). *The algorithm defined by* (1.12), (1.13), *and* (1.14) *is called the continuous RAS algorithm. It is equivalent to the Schwarz algorithm* (1.3).

Proof. Here, we have to prove the equality

$$u^n = E_1(\chi_1 u_1^n) + E_2(\chi_2 u_2^n),$$

where $u_{1,2}^n$ is given by (1.3) and u^n is given by (1.12)–(1.14). We assume that the property holds for the initial guesses

$$u^0 = E_1(\chi_1 u_1^0) + E_2(\chi_2 u_2^0)$$

and proceed by induction assuming the property holds at step n of the algorithm, i.e., $u^n = E_1(\chi_1 u_1^n) + E_2(\chi_2 u_2^n)$. From (1.14) we have

$$u^{n+1} = E_1(\chi_1(u^n + v_1^n)) + E_2(\chi_2(u^n + v_2^n)). \tag{1.9}$$

We prove now that $u^n_{|\Omega_1} + v_1^n = u_1^{n+1}$ by proving that $u^n_{|\Omega_1} + v_1^n$ satisfies (1.3) as u_1^{n+1} does. We first note that, using (1.13)–(1.12), we have

$$\begin{aligned} -\Delta(u^n + v_1^n) &= -\Delta(u^n) + r^n = -\Delta(u^n) + f + \Delta(u^n) = f \quad \text{in } \Omega_1, \\ u^n + v_1^n &= u^n \quad \text{on } \partial\Omega_1 \cap \overline{\Omega_2}. \end{aligned} \tag{1.10}$$

It remains to prove that

$$u^n = u_2^n \text{ on } \partial\Omega_1 \cap \overline{\Omega_2}.$$

By the induction hypothesis we have $u^n = E_1(\chi_1 u_1^n) + E_2(\chi_2 u_2^n)$. On $\partial\Omega_1 \cap \overline{\Omega_2}$, we have $\chi_1 \equiv 0$ and thus $\chi_2 \equiv 1$. So, on $\partial\Omega_1 \cap \overline{\Omega_2}$ we have

$$u^n = \chi_1 u_1^n + \chi_2 u_2^n = u_2^n\,. \tag{1.11}$$

Finally from (1.10) and (1.11) we can conclude that $u^n_{|\Omega_1} + v_1^n$ satisfies problem (1.3) and is thus equal to u_1^{n+1}. The same holds for domain Ω_2, $u^n_{|\Omega_2} + v_2^n = u_2^{n+1}$. Then (1.9) reads as

$$u^{n+1} = E_1(\chi_1 u_1^{n+1}) + E_2(\chi_2 u_2^{n+1}),$$

which ends the proof of the equivalence between the Schwarz algorithm and the continuous RAS algorithm (1.12)–(1.14).

Algorithm 1.1 RAS algorithm at the continuous level

1. Compute the residual $r^n : \Omega \to \mathbb{R}$:

$$r^n := f + \Delta(u^n). \tag{1.12}$$

2. For $i = 1, 2$ solve for a local correction v_i^n:

$$-\Delta(v_i^n) = r^n \quad \text{in} \quad \Omega_i\,,\ v_i^n = 0 \quad \text{on} \quad \partial\Omega_i\,. \tag{1.13}$$

3. Compute an average of the local corrections and update u^n:

$$u^{n+1} = u^n + E_1(\chi_1 v_1^n) + E_2(\chi_2 v_2^n), \tag{1.14}$$

where $(\chi_i)_{i=1,2}$ and $(E_i)_{i=1,2}$ define a partition of unity as defined in (1.4) in section 1.1.

Another global variant of the parallel Schwarz algorithm (1.3) consists in replacing formula (1.6) by a simpler one not based on the partition of unity.

Definition 1.7 (Second global Schwarz iteration). *Let u^n be an approximation of the solution to the Poisson problem* (1.1). *u^{n+1} is computed by solving first local subproblems* (1.5) *and then gluing them together without the use of the partition of unity:*

$$u^{n+1} := \sum_{i=1}^{2} E_i(w_i^{n+1}). \tag{1.15}$$

It is easy to check that this algorithm is equivalent to Algorithm 1.2 which is called the ASM (additive Schwarz method) algorithm.

To sum up, starting from the original Schwarz algorithm (1.2) that is sequential, we thus have three continuous algorithms that are essentially parallel:

- Algorithm (1.3): Jacobi–Schwarz method (JSM)
- Algorithm (1.12)–(1.14): Restricted additive Schwarz (RAS)
- Algorithm (1.16)–(1.18): Additive Schwarz method (ASM)

Algorithm 1.2 ASM algorithm at the continuous level

1. Compute the residual $r^n : \Omega \to \mathbb{R}$:
$$r^n := f + \Delta(u^n). \tag{1.16}$$
2. For $i = 1, 2$ solve for a local correction v_i^n:
$$-\Delta(v_i^n) = r^n \text{ in } \Omega_i\,, \quad v_i^n = 0 \text{ on } \partial\Omega_i\,. \tag{1.17}$$
3. Update u^n:
$$u^{n+1} = u^n + E_1(v_1^n) + E_2(v_2^n). \tag{1.18}$$

The discrete version of the first algorithm is seldom implemented since it involves duplication of unknowns. The discrete version of the second algorithm is the *restricted additive Schwarz* method (RAS; see [20, 21]). The discrete version of the third algorithm is the *additive Schwarz method* (ASM), for which many theoretical results have been derived; see [185] and references therein. The latter term was first introduced by Dryja and Widlund in [66] for a variant of the algorithm first introduced at continuous level in [139].

1.2 ▪ Connection with the block Jacobi algorithm

In the previous section we remarked that the three methods illustrate different points of view of the Schwarz iteration. The continuous aspect emphasized the interest of the overlap (see section 1.5), which is hidden in the discrete formulation. When going to the discrete level, we will see that the Schwarz algorithm is, from a linear algebra point of view, a variation of a block Jacobi algorithm.

We first recall the definition of a block Jacobi algorithm and then establish a connection with the Schwarz algorithms. Let us consider a linear system

$$A\mathbf{U} = \mathbf{F} \tag{1.19}$$

with a matrix A of size $m \times m$, a right-hand side $\mathbf{F} \in \mathbb{R}^m$, and a solution $\mathbf{U} \in \mathbb{R}^m$, where m is a non-zero integer. The set of indices $\{1, \ldots, m\}$ is partitioned into two sets

$$\mathcal{N}_1 := \{1, \ldots, m_s\} \text{ and } \mathcal{N}_2 := \{m_s + 1, \ldots, m\}\,.$$

Let $\mathbf{U}_1 := (U_k)_{k \in \mathcal{N}_1} := \mathbf{U}_{|\mathcal{N}_1}$, $\mathbf{U}_2 := (U_k)_{k \in \mathcal{N}_2} := \mathbf{U}_{|\mathcal{N}_2}$ and similarly $\mathbf{F}_1 := \mathbf{F}_{|\mathcal{N}_1}$, $F_2 := F_{|\mathcal{N}_2}$. The linear system has the following block form:

$$\begin{pmatrix} A_{11} & A_{12} \\ A_{21} & A_{22} \end{pmatrix} \begin{pmatrix} \mathbf{U}_1 \\ \mathbf{U}_2 \end{pmatrix} = \begin{pmatrix} \mathbf{F}_1 \\ \mathbf{F}_2 \end{pmatrix},$$

where $A_{ij} := A_{|\mathcal{N}_i \times \mathcal{N}_j}$, $1 \le i, j \le 2$.

Definition 1.8 (Jacobi algorithm). *Let D be the diagonal of A. The Jacobi algorithm reads as*

$$D\mathbf{U}^{n+1} = D\mathbf{U}^n + (\mathbf{F} - A\mathbf{U}^n),$$

or equivalently,

$$\mathbf{U}^{n+1} = \mathbf{U}^n + D^{-1}(\mathbf{F} - A\mathbf{U}^n) = \mathbf{U}^n + D^{-1}\mathbf{r}^n,$$

where $\mathbf{r}^n = F - A\mathbf{U}^n$ *is the residual of the equation.*

We now define the block Jacobi algorithm.

Definition 1.9 (Block Jacobi algorithm). *The block Jacobi algorithm reads as*

$$\begin{pmatrix} A_{11} & 0 \\ 0 & A_{22} \end{pmatrix} \begin{pmatrix} \mathbf{U}_1^{n+1} \\ \mathbf{U}_2^{n+1} \end{pmatrix} = \begin{pmatrix} A_{11} & 0 \\ 0 & A_{22} \end{pmatrix} \begin{pmatrix} \mathbf{U}_1^n \\ \mathbf{U}_2^n \end{pmatrix} + \begin{pmatrix} F_1 \\ F_2 \end{pmatrix} - A \begin{pmatrix} \mathbf{U}_1^n \\ \mathbf{U}_2^n \end{pmatrix}, \tag{1.20}$$

or equivalently,

$$\begin{pmatrix} A_{11} & 0 \\ 0 & A_{22} \end{pmatrix} \begin{pmatrix} \mathbf{U}_1^{n+1} \\ \mathbf{U}_2^{n+1} \end{pmatrix} = \begin{pmatrix} F_1 - A_{12}\mathbf{U}_2^n \\ F_2 - A_{21}\mathbf{U}_1^n \end{pmatrix}. \tag{1.21}$$

In order to have a more compact form of the previous algorithm, we introduce R_1 as the restriction operator from $\mathcal{N}$ into $\mathcal{N}_1$ and similarly R_2 as the restriction operator from $\mathcal{N}$ into $\mathcal{N}_2$. The transpose operators R_i^T are extension operators from $\mathcal{N}_i$ into $\mathcal{N}$. Note that $A_{ii} = R_i A R_i^T$.

Lemma 1.10 (Compact form of the block Jacobi algorithm). *The algorithm* (1.21) *can be rewritten as*

$$\boxed{\mathbf{U}^{n+1} = \mathbf{U}^n + \left(R_1^T(R_1AR_1^T)^{-1}R_1 + R_2^T(R_2AR_2^T)^{-1}R_2\right)\mathbf{r}^n.} \tag{1.22}$$

Proof. Let $\mathbf{U}^n = (\mathbf{U}_1^{n\,T}, \mathbf{U}_2^{n\,T})^T$. Algorithm (1.21) becomes

$$\begin{pmatrix} A_{11} & 0 \\ 0 & A_{22} \end{pmatrix} \mathbf{U}^{n+1} = F - \begin{pmatrix} 0 & A_{12} \\ A_{21} & 0 \end{pmatrix} \mathbf{U}^n. \tag{1.23}$$

On the other hand, (1.20) can be rewritten equivalently as

$$\begin{pmatrix} \mathbf{U}_1^{n+1} \\ \mathbf{U}_2^{n+1} \end{pmatrix} = \begin{pmatrix} \mathbf{U}_1^n \\ \mathbf{U}_2^n \end{pmatrix} + \begin{pmatrix} A_{11} & 0 \\ 0 & A_{22} \end{pmatrix}^{-1} \begin{pmatrix} \mathbf{r}_1^n \\ \mathbf{r}_2^n \end{pmatrix} \Leftrightarrow \mathbf{U}^{n+1} = \mathbf{U}^n + \begin{pmatrix} A_{11}^{-1} & 0 \\ 0 & A_{22}^{-1} \end{pmatrix} \mathbf{r}^n, \tag{1.24}$$

where $\mathbf{r}_i^n := \mathbf{r}^n_{|\mathcal{N}_i}$, $i = 1,2$. By taking into account that

$$\begin{pmatrix} A_{11}^{-1} & 0 \\ 0 & 0 \end{pmatrix} = R_1^T A_{11}^{-1} R_1 = R_1^T(R_1AR_1^T)^{-1}R_1$$

and

$$\begin{pmatrix} 0 & 0 \\ 0 & A_{22}^{-1} \end{pmatrix} = R_2^T A_{22}^{-1} R_2 = R_2^T(R_2AR_2^T)^{-1}R_2,$$

the conclusion follows easily. □

In order to establish a connection with the Schwarz algorithms, consider the following BVP on $\Omega := (0,1)$: Find u such that

$$\begin{aligned} -\Delta u = f \quad & \text{in} \quad \Omega, \\ u(0) = u(1) \quad & = \quad 0. \end{aligned}$$

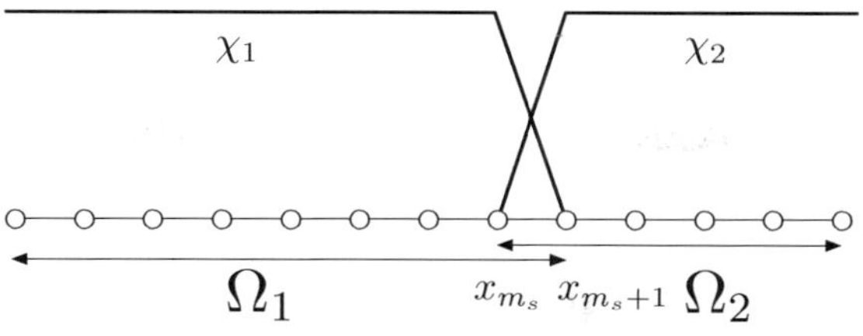

Figure 1.2. *Domain decomposition with minimal overlap and a partition of unity.*

We discretize it by a three-point finite difference scheme on the grid $x_j := j\,h$, $1 \le j \le m$, where $h := 1/(m+1)$. Let $u_j \simeq u(x_j)$, $f_j := f(x_j)$, $1 \le j \le m$, and $\mathbf{U} = (u_j)_{1\le j\le m}$, and let $\mathbf{F} = (f_j)_{1\le j\le m}$ satisfy (1.19) where A is the tridiagonal matrix $A_{j\,j} := 2/h^2$ and $A_{j\,j+1} = A_{j+1\,j} := -1/h^2$.

Let domains $\Omega_1 := (0, (m_s+1)\,h)$ and $\Omega_2 := (m_s\,h, 1)$ define an overlapping decomposition with a minimal overlap of width h. The discretization of (1.5) for domain Ω_1 reads as

$$\begin{cases} -\dfrac{u_{1,j-1}^{n+1} - 2u_{1,j}^{n+1} + u_{1,j+1}^{n+1}}{h^2} = f_j, \; 1 \le j \le m_s, \\ u_{1,0}^{n+1} = 0, \\ u_{1,m_s+1}^{n+1} = u_{2,m_s+1}^{n}. \end{cases}$$

Solving for $\mathbf{U}_1^{n+1} = (u_{1,j}^{n+1})_{1\le j\le m_s}$ corresponds to solving a Dirichlet boundary value problem in subdomain Ω_1 with Dirichlet data taken from the other subdomain at the previous step. With the notations introduced previously, $\mathbf{U}_1^{n+1}$ satisfies

$$A_{11}\mathbf{U}_1^{n+1} + A_{12}\mathbf{U}_2^{n} = \mathbf{F}_1.$$

Similarly, we have

$$A_{22}\mathbf{U}_2^{n+1} + A_{21}\mathbf{U}_1^{n} = \mathbf{F}_2.$$

These two equations are equivalent to (1.21) and represent the discretization of the JSM (1.3).

The discrete counterpart of the extension operator E_1 (resp., E_2) is defined by $E_1(\mathbf{U}_1) = (\mathbf{U}_1^T, 0)^T$ (resp., $E_2(\mathbf{U}_2) = (0, \mathbf{U}_2^T)^T$). The discretization of the ASM (1.15) is then given by (1.23). *Thus when the overlap is minimal, the ASM reduces to the block Jacobi algorithm.*

Let χ_i, $i = 1,2$, be the piecewise linear functions that define a partition of unity on the domain decomposition; see Figure 1.2. In this very simple configuration,

$$\chi_1(x) = \begin{cases} 1 & \text{if } 0 \le x \le x_{m_s}, \\ \dfrac{x_{m_s+1} - x}{h} & \text{if } x_{m_s} \le x \le x_{m_s+1} \end{cases}$$

and

$$\chi_2(x) = \begin{cases} \dfrac{x - x_{m_s}}{h} & \text{if } x_{m_s} \le x \le x_{m_s+1}, \\ 1 & \text{if } x_{m_s+1} \le x \le 1. \end{cases}$$

Functions χ_i, $i = 1,2$, define a partition of unity in the sense of (1.4). Since the overlap is minimal, the discretization of (1.6) is equivalent to that of (1.15). Thus RAS reduces, in this case, to the ASM.

Remark 1.2.1. *In conclusion, when the overlap is minimal the discrete counterparts of the three Schwarz methods of section* 1.1 *are equivalent to the same block Jacobi algorithm. Notice here a counterintuitive feature: a nonoverlapping decomposition of the set of indices* $\mathcal{N}$ *corresponds to a geometric decomposition of the domain* Ω *with minimal overlap.*

1.3 ▪ Algebraic algorithms: Discrete partition of unity

Our goal is to introduce in the general case the algebraic counterparts of the RAS and ASM algorithms defined in section 1.1. The simplest way to do so is to write the iterative method in terms of residuals, as done in (1.22). In order to do this, we need to settle some necessary elements in this text. One of them is the proper definition of the partition of unity.

At the continuous level (partial differential equations), the main ingredients of the partition of unity are

- an open domain Ω and an overlapping decomposition into N open subsets $\Omega = \cup_{i=1}^N \Omega_i$;
- a function $u : \Omega \to \mathbb{R}$;
- the extension operator E_i of a function $\Omega_i \to \mathbb{R}$ to a function $\Omega \to \mathbb{R}$ equal to zero in $\Omega \backslash \Omega_i$;
- the partition of unity functions χ_i, $1 \le i \le N$, introduced in formula (1.4) which verify for all functions $u : \Omega \to \mathbb{R}$:

$$u = \sum_{i=1}^{2} E_i(\chi_i \, u_{|\Omega_i}).$$

We can give a similar definition at the discrete level.

Definition 1.11 (Algebraic partition of unity). *At the discrete level, the main ingredients of the partition of unity are*

- *a set of indices of degrees of freedom* $\mathcal{N}$ *and a decomposition into* N *subsets* $\mathcal{N} = \cup_{i=1}^N \mathcal{N}_i$;
- *a vector* $\mathbf{U} \in \mathbb{R}^{\#\mathcal{N}}$;
- *the restriction of a vector* $\mathbf{U} \in \mathbb{R}^{\#\mathcal{N}}$ *to a subdomain* Ω_i, $1 \le i \le N$, *which can be expressed as* $R_i \, \mathbf{U}$ *where* R_i *is a rectangular* $\#\mathcal{N}_i \times \#\mathcal{N}$ *Boolean matrix (the extension operator will be the transpose matrix* R_i^T*);*
- *the partition of unity "functions" at the discrete level which correspond to diagonal matrices of size* $\#\mathcal{N}_i \times \#\mathcal{N}_i$ *with non-negative entries such that for all vectors* $\mathbf{U} \in \mathbb{R}^{\#\mathcal{N}}$

$$U = \sum_{i=1}^{N} R_i^T D_i R_i \, U,$$

or in other words

$$\boxed{I_d = \sum_{i=1}^{N} R_i^T D_i R_i,} \tag{1.25}$$

where $I_d \in \mathbb{R}^{\#\mathcal{N} \times \#\mathcal{N}}$ *is the identity matrix.*

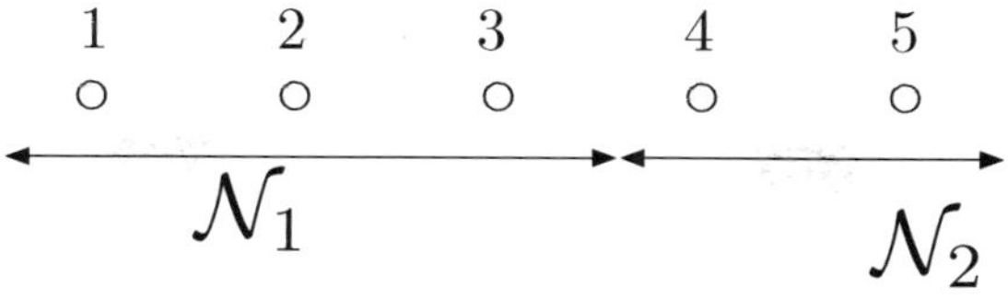

Figure 1.3. *Algebraic partition of the set of indices.*

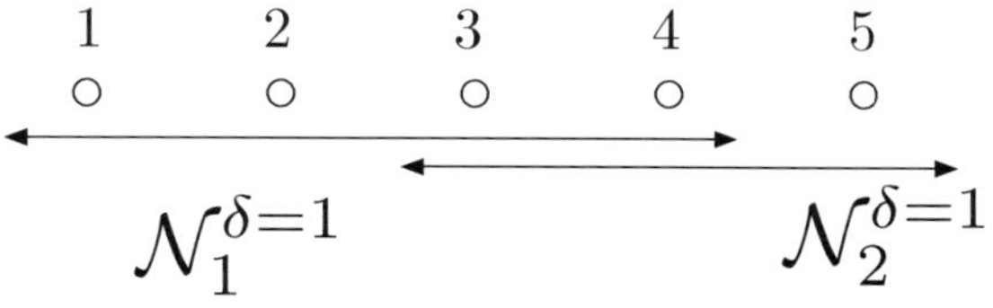

Figure 1.4. *Algebraic decomposition of the set of indices into overlapping subsets.*

As pointed out in Remark 1.2.1, an overlapping decomposition of a domain Ω might correspond to a partition of the set of indices.

In the following, we will give some simple examples in which all the ingredients of Definition 1.11 are detailed and we will check that (1.25) is verified in those cases.

1.3.1 ▪ Two-subdomain case in one dimension

1D algebraic setting

We start from the 1D example of section 1.2 with $n=5$, $n_s=3$ so that the set of indices $\mathcal{N} := \{1,\dots,5\}$ is partitioned into two sets; see Figure 1.3:

$$\mathcal{N}_1 := \{1,2,3\} \text{ and } \mathcal{N}_2 := \{4,5\}.$$

Then, matrix R_1 is of size 3×5 and matrix R_2 is of size 2×5:

$$R_1 = \begin{pmatrix} 1 & 0 & 0 & 0 & 0 \\ 0 & 1 & 0 & 0 & 0 \\ 0 & 0 & 1 & 0 & 0 \end{pmatrix} \text{ and } R_2 = \begin{pmatrix} 0 & 0 & 0 & 1 & 0 \\ 0 & 0 & 0 & 0 & 1 \end{pmatrix},$$

and

$$R_1^T = \begin{pmatrix} 1 & 0 & 0 \\ 0 & 1 & 0 \\ 0 & 0 & 1 \\ 0 & 0 & 0 \\ 0 & 0 & 0 \end{pmatrix} \text{ and } R_2^T = \begin{pmatrix} 0 & 0 \\ 0 & 0 \\ 0 & 0 \\ 1 & 0 \\ 0 & 1 \end{pmatrix}.$$

We also have

$$D_1 = \begin{pmatrix} 1 & 0 & 0 \\ 0 & 1 & 0 \\ 0 & 0 & 1 \end{pmatrix} \text{ and } D_2 = \begin{pmatrix} 1 & 0 \\ 0 & 1 \end{pmatrix}.$$

It is clear that relation (1.25) holds.

Consider now the case where each subset is extended with a neighboring point; see Figure 1.4:

$$\mathcal{N}_1^{\delta=1} := \{1,2,3,4\} \text{ and } \mathcal{N}_2^{\delta=1} := \{3,4,5\}.$$

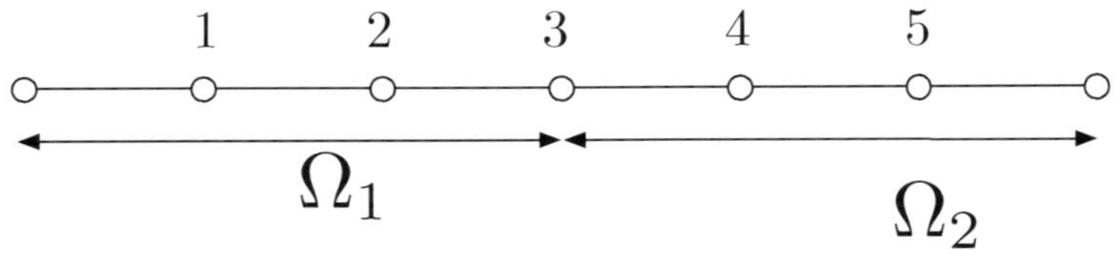

Figure 1.5. *Finite element partition of the mesh.*

Then, matrices R_1 and R_2 are

$$R_1 = \begin{pmatrix} 1 & 0 & 0 & 0 & 0 \\ 0 & 1 & 0 & 0 & 0 \\ 0 & 0 & 1 & 0 & 0 \\ 0 & 0 & 0 & 1 & 0 \end{pmatrix} \text{ and } R_2 = \begin{pmatrix} 0 & 0 & 1 & 0 & 0 \\ 0 & 0 & 0 & 1 & 0 \\ 0 & 0 & 0 & 0 & 1 \end{pmatrix}.$$

The simplest choices for the partition of unity matrices are

$$D_1 = \begin{pmatrix} 1 & 0 & 0 & 0 \\ 0 & 1 & 0 & 0 \\ 0 & 0 & 1 & 0 \\ 0 & 0 & 0 & 0 \end{pmatrix} \text{ and } D_2 = \begin{pmatrix} 0 & 0 & 0 \\ 0 & 1 & 0 \\ 0 & 0 & 1 \end{pmatrix}$$

or

$$D_1 = \begin{pmatrix} 1 & 0 & 0 & 0 \\ 0 & 1 & 0 & 0 \\ 0 & 0 & 1/2 & 0 \\ 0 & 0 & 0 & 1/2 \end{pmatrix} \text{ and } D_2 = \begin{pmatrix} 1/2 & 0 & 0 \\ 0 & 1/2 & 0 \\ 0 & 0 & 1 \end{pmatrix}.$$

Again, it is clear that relation (1.25) holds.

1D finite element decomposition

We still want to consider the 1D example with a decomposition into two subdomains, but this time in a finite element setting. A *partition* of the 1D mesh in Figure 1.5 corresponds to an *overlapping* decomposition of the set of indices

$$\mathcal{N}_1 := \{1,2,3\} \text{ and } \mathcal{N}_2 := \{3,4,5\}.$$

Then, matrices R_1 and R_2 are

$$R_1 = \begin{pmatrix} 1 & 0 & 0 & 0 & 0 \\ 0 & 1 & 0 & 0 & 0 \\ 0 & 0 & 1 & 0 & 0 \end{pmatrix} \text{ and } R_2 = \begin{pmatrix} 0 & 0 & 1 & 0 & 0 \\ 0 & 0 & 0 & 1 & 0 \\ 0 & 0 & 0 & 0 & 1 \end{pmatrix}.$$

In order to satisfy relation (1.25), the simplest choice for the partition of unity matrices is

$$D_1 = \begin{pmatrix} 1 & 0 & 0 \\ 0 & 1 & 0 \\ 0 & 0 & 1/2 \end{pmatrix} \text{ and } D_2 = \begin{pmatrix} 1/2 & 0 & 0 \\ 0 & 1 & 0 \\ 0 & 0 & 1 \end{pmatrix}.$$

Consider now the situation where we add a vertex to each subdomain; see Figure 1.6. Accordingly, the set of indices is decomposed as

$$\mathcal{N}_1^{\delta=1} := \{1,2,3,4\} \text{ and } \mathcal{N}_2^{\delta=1} := \{2,3,4,5\}.$$

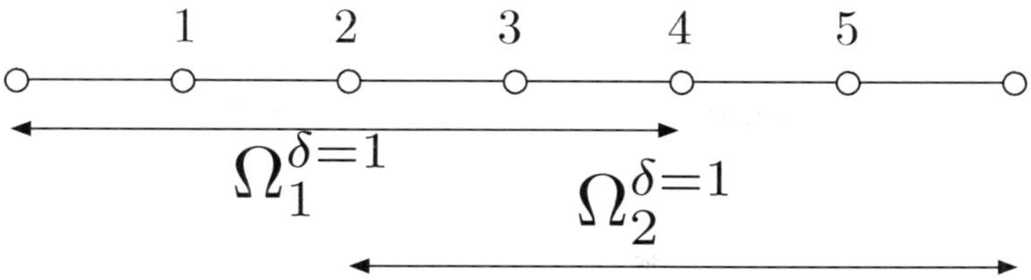

Figure 1.6. *Finite element decomposition of the mesh into overlapping subdomains.*

Then, matrices R_1 and R_2 are

$$R_1 = \begin{pmatrix} 1 & 0 & 0 & 0 & 0 \\ 0 & 1 & 0 & 0 & 0 \\ 0 & 0 & 1 & 0 & 0 \\ 0 & 0 & 0 & 1 & 0 \end{pmatrix} \text{ and } R_2 = \begin{pmatrix} 0 & 1 & 0 & 0 & 0 \\ 0 & 0 & 1 & 0 & 0 \\ 0 & 0 & 0 & 1 & 0 \\ 0 & 0 & 0 & 0 & 1 \end{pmatrix}.$$

In order to satisfy relation (1.25), the simplest choice for the partition of unity matrices is

$$D_1 = \begin{pmatrix} 1 & 0 & 0 & 0 \\ 0 & 1 & 0 & 0 \\ 0 & 0 & 1/2 & 0 \\ 0 & 0 & 0 & 0 \end{pmatrix} \text{ and } D_2 = \begin{pmatrix} 0 & 0 & 0 & 0 \\ 0 & 1/2 & 0 & 0 \\ 0 & 0 & 1 & 0 \\ 0 & 0 & 0 & 1 \end{pmatrix}.$$

Another possible choice that will satisfy relation (1.25) as well is

$$D_1 = \begin{pmatrix} 1 & 0 & 0 & 0 \\ 0 & 1/2 & 0 & 0 \\ 0 & 0 & 1/2 & 0 \\ 0 & 0 & 0 & 1/2 \end{pmatrix} \text{ and } D_2 = \begin{pmatrix} 1/2 & 0 & 0 & 0 \\ 0 & 1/2 & 0 & 0 \\ 0 & 0 & 1/2 & 0 \\ 0 & 0 & 0 & 1 \end{pmatrix}.$$

1.3.2 ▪ Multidimensional problems and many subdomains

In the general case, the set of indices $\mathcal{N}$ can be partitioned by an automatic graph partitioner such as METIS [120] or SCOTCH [27]. From the input matrix A, a connectivity graph is created. Two indices $i, j \in \mathcal{N}$ are connected if the matrix coefficient $A_{ij} \neq 0$. Usually, even if matrix A is not symmetric, the connectivity graph is symmetrized. Then algorithms that find good partitioning of the vertices even for highly unstructured graphs are used. This distribution must be done so that the number of elements assigned to each processor is roughly the same, and the number of adjacent elements assigned to different processors is minimized (graph cuts). The goal of the first condition is to balance the computations among the processors. The goal of the second condition is to minimize the communication resulting from the placement of adjacent elements to different processors.

Multidimensional algebraic setting

Let us consider a partition into N subsets (see Figure 1.7):

$$\mathcal{N} := \bigcup_{i=1}^{N} \mathcal{N}_i, \quad \mathcal{N}_i \cap \mathcal{N}_j = \emptyset \text{ for } i \neq j. \tag{1.26}$$

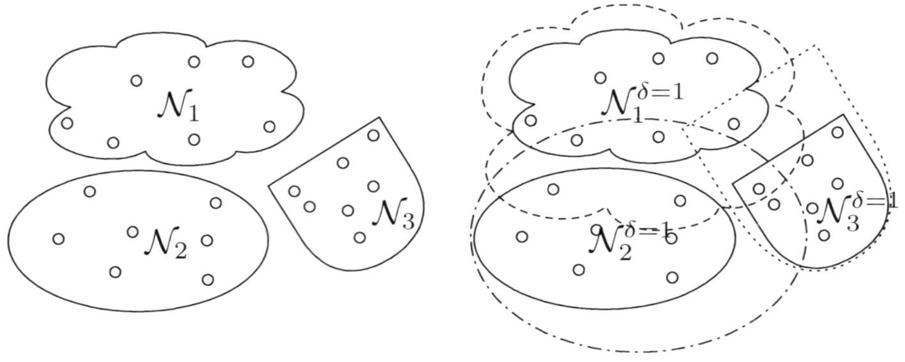

Figure 1.7. *Partition and overlapping decomposition of the set of indices.*

Let R_i be the restriction matrix from set $\mathcal{N}$ to the subset $\mathcal{N}_i$ and let D_i be the identity matrix of size $\#\mathcal{N}_i \times \#\mathcal{N}_i$, $1 \le i \le N$. Then, relation (1.25) is satisfied.

Consider now the case where each subset $\mathcal{N}_i$ is extended with its direct neighbors to form $\mathcal{N}_i^{\delta=1}$; see Figure 1.7. Let R_i be the restriction matrix from set $\mathcal{N}$ to the subset $\mathcal{N}_i^{\delta=1}$ and let D_i be a diagonal matrix of size $\#\mathcal{N}_i^{\delta=1} \times \#\mathcal{N}_i^{\delta=1}$, $1 \le i \le N$. For the choice of the coefficients of D_i there are two main options. The simplest one is to define it as a Boolean matrix:

$$(D_i)_{jj} := \begin{cases} 1 \text{ if } j \in \mathcal{N}_i, \\ 0 \text{ if } j \in \mathcal{N}_i^{\delta=1} \backslash \mathcal{N}_i. \end{cases}$$

Then, relation (1.25) is satisfied. Another option is to introduce for all $j \in \mathcal{N}$ the set of subsets having j as an element:

$$\mathcal{M}_j := \{1 \le i \le N \,|\, j \in \mathcal{N}_i^{\delta=1}\}.$$

Then, define

$$(D_i)_{jj} := 1/\#\mathcal{M}_j \text{ for } j \in \mathcal{N}_i^{\delta=1}.$$

Then, relation (1.25) is satisfied.

Multidimensional finite element decomposition

Partitioning a set of indices is well adapted to an algebraic framework. In a finite element setting, the computational domain is the union of elements of the finite element mesh $\mathcal{T}_h$. A geometric partition of the computational domain is natural. Here again, graph partitioning can be used by first modeling the finite element mesh with a graph and then partitioning the elements into N parts $(\mathcal{T}_{i,h})_{1 \le i \le N}$; see Figure 1.8. By adding layers of elements to each part, it is possible to create overlapping subdomains resolved by the finite element meshes:

$$\Omega_i = \bigcup_{\tau \in \mathcal{T}_{i,h}} \tau \quad \text{for } 1 \le i \le N. \tag{1.27}$$

Let $\{\phi_k\}_{k \in \mathcal{N}}$ be a basis of the finite element space. We define

$$\mathcal{N}_i := \{k \in \mathcal{N} : \text{meas}(\text{supp}(\phi_k) \cap \Omega_i) > 0\} \text{ for } 1 \le i \le N. \tag{1.28}$$

For each degree of freedom $k \in \mathcal{N}$, let

$$\mu_k := \#\{j : 1 \le j \le N \text{ and } k \in \mathcal{N}_j\}.$$

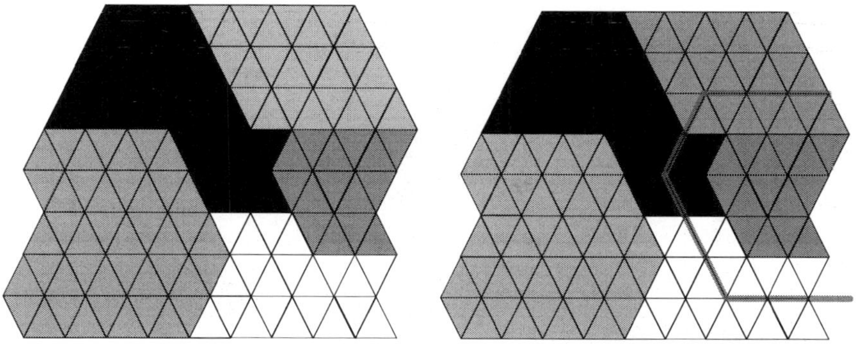

Figure 1.8. *Left: Finite element partition. Right: One-layer extension of the right subdomain.*

Let R_i be the restriction matrix from set $\mathcal{N}$ to the subset $\mathcal{N}_i$ and let D_i be a diagonal matrix of size $\#\mathcal{N}_i \times \#\mathcal{N}_i$, $1 \le i \le N$, such that

$$(D_i)_{kk} := 1/\mu_k, \, k \in \mathcal{N}_i.$$

Then, relation (1.25) is satisfied.

1.4 ▪ Iterative Schwarz methods: RAS and ASM

In a similar way as was done for the block Jacobi algorithm in (1.22), we can define RAS (the counterpart of algorithm (1.12)–(1.14)) and ASM algorithms (the counterpart of algorithm (1.16)–(1.18)).

Definition 1.12 (RAS algorithm). *The iterative RAS algorithm is the preconditioned fixed-point iteration defined by*

$$\mathbf{U}^{n+1} = \mathbf{U}^n + M_{RAS}^{-1}\mathbf{r}^n, \, \mathbf{r}^n := \mathbf{F} - A\,\mathbf{U}^n,$$

where the matrix

$$\boxed{M_{RAS}^{-1} := \sum_{i=1}^{N} R_i^T D_i \left(R_i A R_i^T\right)^{-1} R_i} \tag{1.29}$$

is called the RAS preconditioner.

Definition 1.13 (ASM algorithm). *The iterative ASM algorithm is the preconditioned fixed-point iteration defined by*

$$\mathbf{U}^{n+1} = \mathbf{U}^n + M_{ASM}^{-1}\mathbf{r}^n, \, \mathbf{r}^n := \mathbf{F} - A\,\mathbf{U}^n,$$

where the matrix

$$\boxed{M_{ASM}^{-1} := \sum_{i=1}^{N} R_i^T \left(R_i A R_i^T\right)^{-1} R_i} \tag{1.30}$$

is called the ASM preconditioner.

1.5 ▪ Convergence analysis

In order to have an idea about the convergence of these methods, we perform a simple yet revealing analysis. In section 1.5.1 we consider a 1D domain decomposed into two subdomains. This shows that the size of the overlap between the subdomains is key to the convergence of the method. In section 1.5.2 an analysis of the multidimensional case is carried out by a Fourier analysis. It reveals that the high-frequency component of the error is very quickly damped thanks to the overlap, whereas the low-frequency part will demand special treatment; see Chapter 4 on coarse spaces and two-level methods.

1.5.1 ▪ 1D case: A geometrical analysis

In the 1D case, the original sequential Schwarz method (1.2) can be analyzed easily. Let $L > 0$ and the domain $\Omega = (0, L)$ be decomposed into two subdomains $\Omega_1 := (0, L_1)$ and $\Omega_2 := (l_2, L)$ with $l_2 \le L_1$. By linearity of the equation and of the algorithm the error $e_i^n := u_i^n - u_{|\Omega_i}$, $i = 1, 2$, satisfies

$$\begin{array}{lcl} -\dfrac{d^2 e_1^{n+1}}{dx^2} = 0 \quad \text{in } (0, L_1), & & -\dfrac{d^2 e_2^{n+1}}{dx^2} = 0 \quad \text{in } (l_2, L), \\ e_1^{n+1}(0) = 0, & \text{then,} & e_2^{n+1}(l_2) = e_1^{n+1}(l_2), \\ e_1^{n+1}(L_1) = e_2^n(L_1); & & e_2^{n+1}(L) = 0. \end{array} \tag{1.31}$$

Thus the errors are affine functions in each subdomain:

$$e_1^{n+1}(x) = e_2^n(L_1)\frac{x}{L_1} \text{ and } e_2^{n+1}(x) = e_1^{n+1}(l_2)\frac{L - x}{L - l_2}.$$

Thus, we have

$$e_2^{n+1}(L_1) = e_1^{n+1}(l_2)\frac{L - L_1}{L - l_2} = e_2^n(L_1)\frac{l_2}{L_1}\frac{L - L_1}{L - l_2}.$$

Let $\delta := L_1 - l_2$ denote the size of the overlap; we have

$$e_2^{n+1}(L_1) = \frac{l_2}{l_2 + \delta}\frac{L - l_2 - \delta}{L - l_2} e_2^n(L_1) = \frac{1 - \delta/(L - l_2)}{1 + \delta/l_2} e_2^n(L_1).$$

We see that the following quantity is the convergence factor of the algorithm:

$$\boxed{\rho_1 = \frac{1 - \delta/(L - l_2)}{1 + \delta/l_2}.}$$

It is clear that $\delta > 0$ is a sufficient and necessary condition to have convergence. The convergence becomes faster as the ratio of the size of the overlap over the size of the subdomain gets bigger. A geometric illustration of the convergence history can be found in Figure 1.9.

1.5.2 ▪ 2D case: Fourier analysis for two subdomains

For the sake of simplicity we consider the plane $\mathbb{R}^2$ decomposed into two half-planes $\Omega_1 = (-\infty, \delta) \times \mathbb{R}$ and $\Omega_2 = (0, \infty) \times \mathbb{R}$ with an overlap of size $\delta > 0$. We choose as an example a symmetric positive definite problem ($\eta > 0$)

$$\begin{array}{c} (\eta - \Delta)(u) = f \quad \text{in} \quad \mathbb{R}^2, \\ u \text{ is bounded at infinity}. \end{array}$$

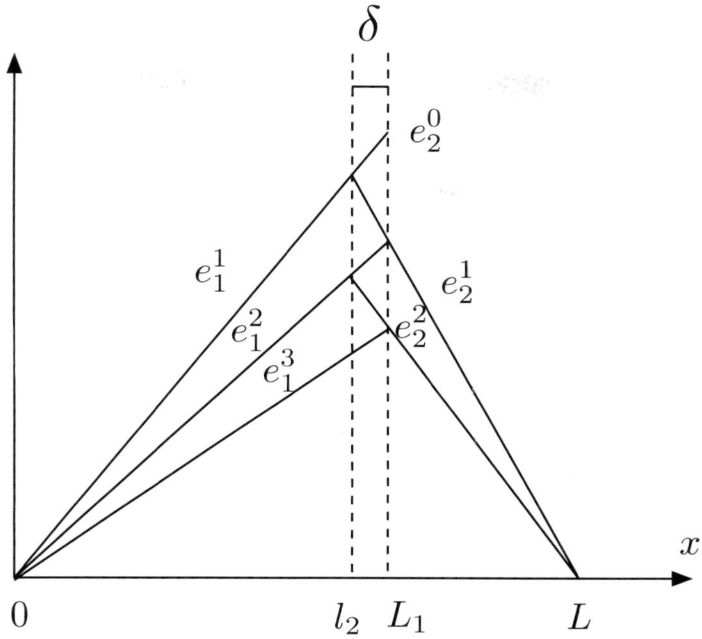

Figure 1.9. *Convergence of the Schwarz method.*

The Jacobi–Schwarz method for this problem is the following iteration:

$$\begin{aligned} (\eta-\Delta)(u_1^{n+1}) &= f(x,y), \quad (x,y)\in\Omega_1, \\ u_1^{n+1}(\delta,y) &= u_2^n(\delta,y), \quad y\in\mathbb{R}, \end{aligned} \tag{1.32}$$

and

$$\begin{aligned} (\eta-\Delta)(u_2^{n+1}) &= f(x,y), \quad (x,y)\in\Omega_2, \\ u_2^{n+1}(0,y) &= u_1^n(0,y), \quad y\in\mathbb{R}, \end{aligned} \tag{1.33}$$

with the local solutions u_j^{n+1}, $j=1,2$, bounded at infinity.

In order to compute the convergence factor, we introduce the errors

$$e_i^n := u_i^n - u|_{\Omega_i}, \; i=1,2.$$

By linearity, the errors satisfy the above algorithm with $f=0$:

$$\begin{aligned} (\eta-\Delta)(e_1^{n+1}) &= 0, \quad (x,y)\in\Omega_1, \\ e_1^{n+1}(\delta,y) &= e_2^n(\delta,y), \quad y\in\mathbb{R}, \end{aligned} \tag{1.34}$$

and

$$\begin{aligned} (\eta-\Delta)(e_2^{n+1}) &= 0, \quad (x,y)\in\Omega_2, \\ e_2^{n+1}(0,y) &= e_1^n(0,y), \quad y\in\mathbb{R}, \end{aligned} \tag{1.35}$$

with e_i^{n+1} bounded at infinity.

By taking the partial Fourier transform of the first line of (1.34) in the y-direction we get

$$\left(\eta-\frac{\partial^2}{\partial x^2}+k^2\right)(\hat{e}_1^{n+1}(x,k))=0 \quad \text{in} \quad \Omega_1.$$

For a given Fourier variable k, this is an ordinary differential equation (ODE) whose solution is sought in the form

$$\hat{e}_1^{n+1}(x,k)=\sum_j \gamma_j(k)\exp(\lambda_j(k)x).$$

A simple computation gives

$$\lambda_1(k)=\lambda^+(k),\ \lambda_2(k)=\lambda^-(k),\ \text{with}\ \lambda^{\pm}(k)=\pm\sqrt{\eta+k^2}.$$

Therefore we have

$$\hat{e}_1^{n+1}(x,k)=\gamma_+^{n+1}(k)\exp(\lambda^+(k)x)+\gamma_-^{n+1}(k)\exp(\lambda^-(k)x).$$

Since the solution must be bounded at $x=-\infty$, this implies that $\gamma_-^{n+1}(k)\equiv 0$. Thus we have

$$\hat{e}_1^{n+1}(x,k)=\gamma_+^{n+1}(k)\exp(\lambda^+(k)x)$$

or equivalently, by changing the value of the coefficient γ_+,

$$\hat{e}_1^{n+1}(x,k)=\gamma_1^{n+1}(k)\exp(\lambda^+(k)(x-\delta)).$$

Similarly, in domain Ω_2 we have

$$\hat{e}_2^{n+1}(x,k)=\gamma_2^{n+1}(k)\exp(\lambda^-(k)x)$$

with $\gamma_{1,2}^{n+1}$ to be determined. From the interface conditions we get

$$\gamma_1^{n+1}(k)=\gamma_2^{n}(k)\exp(\lambda^-(k)\delta)$$

and

$$\gamma_2^{n+1}(k)=\gamma_1^{n}(k)\exp(-\lambda^+(k)\delta).$$

Combining these two and denoting $\lambda(k)=\lambda^+(k)=-\lambda^-(k)$, we get for $i=1,2$,

$$\gamma_i^{n+1}(k)=\rho(k;\alpha,\delta)^2\,\gamma_i^{n-1}(k)$$

with ρ the convergence factor given by

$$\boxed{\rho(k;\alpha,\delta)=\exp(-\lambda(k)\delta),\ \lambda(k)=\sqrt{\eta+k^2}.} \tag{1.36}$$

A graphical representation can be found in Figure 1.10 for some values of the overlap. This formula deserves a few remarks.

Remark 1.5.1. *We have the following properties:*

- *For all $k\in\mathbb{R}$ and $\delta>0$, $\rho(k)\le\exp(-\sqrt{\eta}\,\delta)<1$ so that $\gamma_i^n(k)\to 0$ uniformly as n goes to infinity.*
- *$\rho\to 0$ as k tends to infinity; high-frequency modes of the error converge very fast.*
- *When there is no overlap ($\delta=0$), $\rho=1$ and there is stagnation of the method.*

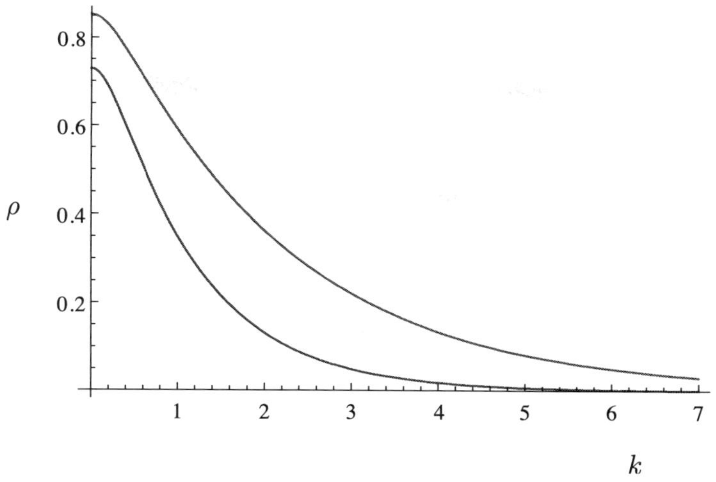

Figure 1.10. *Convergence rate of the Schwarz method for* $\eta = 0.1$, $\delta = 0.5$ *(top curve), or* $\delta = 1$ *(bottom curve).*

1.6 ▪ More sophisticated Schwarz methods: P. L. Lions' algorithm

During the last few decades, more sophisticated Schwarz methods have been designed, namely the optimized Schwarz methods. These methods are based on a classical domain decomposition, but they use more effective transmission conditions than the classical Dirichlet conditions at the interfaces between subdomains. The first of these more effective transmission conditions were first introduced by P. L. Lions [130]. For elliptic problems, we have seen that Schwarz algorithms work only for overlapping domain decompositions and their performance in terms of iteration counts depends on the width of the overlap. The algorithm introduced by P. L. Lions [130] can be applied to both overlapping and nonoverlapping subdomains. It is based on improving Schwarz methods by replacing the Dirichlet interface conditions with Robin interface conditions.

Let α be a positive number. The modified algorithm reads as

$$\begin{aligned} -\Delta(u_1^{n+1}) &= f && \text{in } \Omega_1, \\ u_1^{n+1} &= 0 && \text{on } \partial\Omega_1 \cap \partial\Omega, \\ \left(\frac{\partial}{\partial \mathbf{n}_1} + \alpha\right)(u_1^{n+1}) &= \left(\frac{\partial}{\partial \mathbf{n}_1} + \alpha\right)(u_2^n) && \text{on } \partial\Omega_1 \cap \overline{\Omega_2} \end{aligned} \tag{1.37}$$

and

$$\begin{aligned} -\Delta(u_2^{n+1}) &= f && \text{in } \Omega_2, \\ u_2^{n+1} &= 0 && \text{on } \partial\Omega_2 \cap \partial\Omega, \\ \left(\frac{\partial}{\partial \mathbf{n}_2} + \alpha\right)(u_2^{n+1}) &= \left(\frac{\partial}{\partial \mathbf{n}_2} + \alpha\right)(u_1^n) && \text{on } \partial\Omega_2 \cap \overline{\Omega_1}, \end{aligned} \tag{1.38}$$

where $\mathbf{n}_1$ and $\mathbf{n}_2$ are the outward normals on the boundary of the subdomains; see Figure 1.11.

This algorithm was extended to the Helmholtz problem by Després [44]. It is also possible to consider interface conditions other than Robin conditions (a.k.a. Fourier conditions) and optimize their choice with respect to the convergence factor. All these ideas will be presented in detail in Chapter 2.

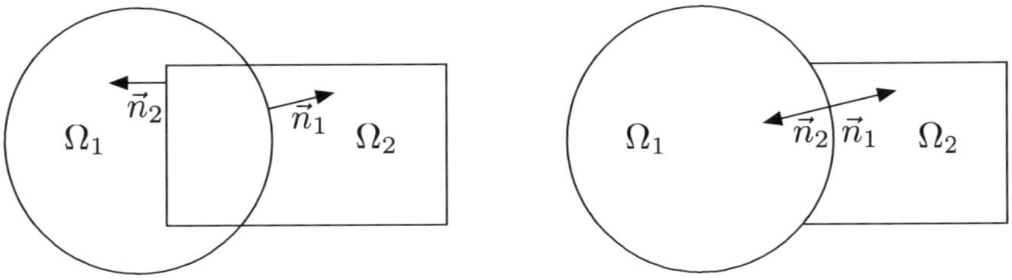

Figure 1.11. *Outward normals for overlapping and nonoverlapping subdomains for P. L. Lions' algorithm.*

1.7 ▪ Schwarz methods using FreeFem++

The aim of this section is to illustrate numerically the previously defined Schwarz methods applied to second-order elliptic boundary value problems (e.g., the Laplace equation and the system of linear elasticity). In order to do this we will use the open-source finite element software FreeFem++ [108], developed at the Laboratoire Jacques-Louis Lions at the Université Pierre et Marie Curie.

1.7.1 ▪ A very short introduction to FreeFem++

FreeFem++ provides a very simple and natural way of solving a great variety of variational problems by finite element-type methods including discontinuous Galerkin discretizations. It is also possible to have access to the underlying linear algebra such as the stiffness or mass matrices. In this section we will provide only a minimal number of elements of this software, which are necessary for understanding the programs in the next section; see also `http://www.cmap.polytechnique.fr/spip.php?article239`. A very detailed documentation of FreeFem++ is available from the official website `http://www.freefem.org/ff++` at the following address: `http://www.freefem.org/ff++/ftp/freefem++doc`. The standard implementation includes lots of very useful examples that make up a tutorial by themselves. It is also possible to use the integrated environment FreeFem++-cs [124] which provides an intuitive graphical interface to FreeFem++ users.

To begin with, suppose we want to solve a very simple homogeneous Dirichlet boundary value problem for a Laplacian defined on a unit square $\Omega =]0,1[^2$:

$$\begin{cases} -\Delta u = f & \text{in} \quad \Omega, \\ u = 0 & \text{on} \quad \partial\Omega. \end{cases} \tag{1.39}$$

The variational formulation of this problem reads as follows:

$$\text{Find } u \in H_0^1(\Omega) := \{w \in H^1(\Omega) \ : \ w = 0 \text{ on } \partial\Omega\} \text{ such that}$$

$$\int_\Omega \nabla u . \nabla v dx - \int_\Omega f \, v \, dx = 0 \quad \forall v \in H_0^1(\Omega).$$

The above variational formulation is first replaced by

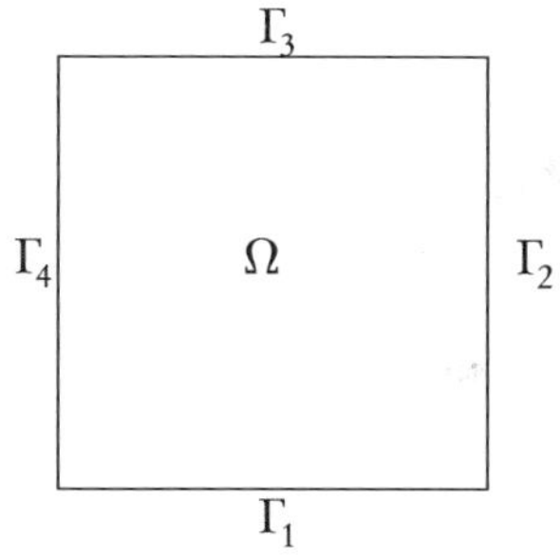

Figure 1.12. *Numbering of square borders in FreeFem++.*

Find $u \in H^1(\Omega)$ such that

$$\int_\Omega \nabla u . \nabla v dx - \int_\Omega f v dx = 0 \quad \forall v \in H^1(\Omega).$$

Then the finite element approximation leads to a system of the type

$$\sum_{j=1}^{M} A_{ij} u_j - F_j = 0, i = 1, \ldots, M, A_{ij} = \int_\Omega \nabla \phi_i . \nabla \phi_j dx, F_i = \int_\Omega \phi_i dx,$$

where $(\phi_i)_{1 \leq i \leq M}$ are the finite element functions. Note that the discretized system corresponds to a Neumann problem. Dirichlet conditions of the type $u = g$ are then implemented by penalty, namely by setting

$$A_{ii} = 10^{30}, F_i = 10^{30} \cdot g_i$$

if i is a boundary degree of freedom. The penalty number 10^{30} is called TGV,[1] and it is possible to change this value. The keyword on imposes the Dirichlet boundary condition through this penalty term.

The following FreeFem++ script solves this problem in a few lines. The text after // are comments ignored by the FreeFem++ interpreter. Each new variable must be declared with its type (here `int` designs integers).

```
// Number of mesh points in x— and y—directions
int Nbnoeuds=10;
```

Listing 1.1. `./FreefemCommon/survival.edp`

The function `square` returns a structured `mesh` of a square: the first two arguments are the number of mesh points according to x- and y-directions, and the third one is a parametrization of Ω for x and y varying between 0 and 1 (here it is the identity). The sides of the square are labeled from 1 to 4 counterclockwise (see Figure 1.12).

```
//Mesh definition
mesh Th=square(Nbnoeuds,Nbnoeuds,[x,y]);
```

Listing 1.2. `./FreefemCommon/survival.edp`

[1]Très Grande Valeur (Terrifically Great Value) = Very big value in French.

We define the function representing the right-hand side using the keyword `func`:

```
// Functions of x and y
func f=x*y;
func g=1.;
```

Listing 1.3. `./FreefemCommon/survival.edp`

and the $P1$ finite element space `Vh` over the mesh `Th` using the keyword `fespace`:

```
// Finite element space on the mesh Th
fespace Vh(Th,P1);
//uh and vh are of type Vh
Vh uh,vh;
```

Listing 1.4. `./FreefemCommon/survival.edp`

The functions u_h and v_h belong to the $P1$ finite element space V_h which is an approximation of $H^1(\Omega)$. Note here that if one wants to use $P2$ instead of $P1$ finite elements, it is enough to replace `P1` by `P2` in the definition of `Vh`.

```
// variational problem definition
problem heat(uh,vh,solver=LU)=
        int2d(Th)(dx(uh)*dx(vh)+dy(uh)*dy(vh))
        -int2d(Th)(f*vh)
        +on(1,2,3,4,uh=0);
```

Listing 1.5. `./FreefemCommon/survival.edp`

The keyword `problem` allows the definition of a variational problem, here called `heat`, which can be expressed mathematically as follows:

Find $u_h \in V_h$ such that

$$\int_\Omega \nabla u_h . \nabla v_h dx - \int_\Omega f v_h dx = 0 \quad \forall v_h \in V_h .$$

Afterwards, for the Dirichlet boundary condition the penalization is imposed using TGV which usually is equal to 10^{30}.

Note that the keyword `problem` defines problem (1.39) without solving it. The parameter `solver` sets the method that will be used to solve the resulting linear system, here a Gauss elimination. In order to effectively solve the finite element problem, we need the command

```
//Solving the problem
heat;
// Plotting the result
plot(uh,wait=1);
```

Listing 1.6. `./FreefemCommon/survival.edp`

The FreeFem++ script can be saved with your preferred text editor (e.g., under the name `heat.edp`). In order to execute the script FreeFem++, it is enough to write the shell command `FreeFem++ heat.edp`. The result will be displayed in a graphic window.

One can easily modify the script in order to solve the same kinds of problems but with mixed Neumann and Robin boundary conditions such as

$$\begin{cases} -\Delta u + u = f \text{ in } \Omega, \\ \dfrac{\partial u}{\partial n} = 0 \text{ on } \Gamma_1, \\ u = 0 \text{ on } \Gamma_2, \\ \frac{\partial u}{\partial n} + \alpha u = g \text{ on } \Gamma_3 \cup \Gamma_4, \end{cases} \tag{1.40}$$

where f and g are arbitrary functions and α is positive real.

The new variational formulation consists in determining $u_h \in V_h$ such that

$$\int_\Omega \nabla u_h . \nabla v_h dx + \int_{\Gamma_3 \cup \Gamma_4} \alpha u_h v_h - \int_{\Gamma_3 \cup \Gamma_4} g v_h - \int_\Omega f v_h dx = 0$$

for all $v_h \in V_h$. Here again the Dirichlet boundary condition will be penalized. The FreeFem++ definition of the problem reads as follows:

```
// Changing boundary conditions to Neumann or Robin
real alpha =1.;
problem heatRobin(uh,vh)=
  int2d(Th)(dx(uh)*dx(vh)+dy(uh)*dy(vh))
  +int1d(Th,3,4)(alpha*uh*vh)
  -int1d(Th,3,4)(g*vh)
  -int2d(Th)(f*vh)
  +on(2,uh=0);
```

Listing 1.7. `./FreefemCommon/survival.edp`

In the variational formulation of (1.40) the extra boundary integral on $\Gamma_3 \cup \Gamma_4$ is defined by the keyword `int1d(Th,3,4)(function to integrate)`.

The keyword `varf` allows the definition of a variational formulation:

```
// Using linear algebra package
varf varheatRobin(uh,vh)=
  int2d(Th)(dx(uh)*dx(vh)+dy(uh)*dy(vh))
  +int1d(Th,3,4)(alpha*uh*vh)
  -int1d(Th,3,4)(g*vh)
  -int2d(Th)(f*vh)
  +on(2,uh=0);
```

Listing 1.8. `./FreefemCommon/survival.edp`

If one wants to use some form of linear algebra package to solve the linear system resulting from the finite element discretization, the program below shows how one can retrieve the stiffness matrix and the vector associated to the right-hand side of the variational formulation. As a general rule, this procedure can be very useful if one wants to use other solvers such as domain decomposition methods. Here, the linear system is solved by UMFPACK [38].

```
// Retrieving the stiffness matrix
matrix Aglobal; // sparse matrix
Aglobal = varheatRobin(Vh,Vh,solver=UMFPACK); // stiffness matrix
                                    // UMFPACK direct solver
// Retrieving the right-hand side
Vh rhsglobal;
rhsglobal[] = varheatRobin(0,Vh); //right-hand side vector of dofs
```

```
// Solving the problem by a sparse LU solver
uh[] = Aglobal^-1*rhsglobal[];
```

Listing 1.9. ./FreefemCommon/survival.edp

Here `rhsglobal` is a finite element function and the associated vector of degrees of freedom is denoted by `rhsglobal[]`.

1.7.2 ▪ Setting the domain decomposition problem

According to the description of the Schwarz algorithms in the previous sections, we need a certain number of data structures which will be built in what follows. The file `data.edp` contains the declaration of these structures as well as the definition of the global problem to be solved.

```
load "metis" // mesh partitioner
load "medit" // OpenGL-based scientific visualization software
int nn=2,mm=2;      // number of the domains in each direction
int npart= nn*mm;      // total number of domains
int nloc = 20; // local no of dofs per domain in one direction
bool withmetis = 1; // =1 (Metis decomp) =0 (uniform decomp)
int sizeovr = 1; // size of the geometric overlap between subdomains, algebraic overlap is 2 sizeovr+1
real allong = real(nn)/real(mm); // aspect ratio of the global domain
// Mesh of a rectangular domain
mesh Th=square(nn*nloc,mm*nloc,[x*allong,y]);// radial mesh
    [(1.+x*allong)*cos(pi*y),(1.+x*allong)*sin(pi*y)]);//
fespace Vh(Th,P1);
fespace Ph(Th,P0);
Ph part; // piecewise constant function
int[int] lpart(Ph.ndof); // giving the decomposition
// Domain decomposition data structures
mesh[int] aTh(npart); // sequence of subdomain meshes
matrix[int] Rih(npart); // local restriction operators
matrix[int] Dih(npart); // partition of unity operators
int[int] Ndeg(npart); // number of dofs for each mesh
real[int] AreaThi(npart); // area of each subdomain
matrix[int] aA(npart),aN(npart); // local matrices
Vh[int] Z(npart); // coarse space, see Chapter 4
// Definition of the problem to solve
// Delta (u) = f, u = 1 on the global boundary
//int[int] chlab=[1,1 ,2,1 ,3,1 ,4,1 ];
//Th=change(Th,refe=chlab); // all label borders are set to 1
macro Grad(u) [dx(u),dy(u)] // EOM
func f = 1; // right-hand side
func g = 0 ; // Dirichlet data
func kappa = 1.; // viscosity
func eta = 0;
Vh rhsglobal,uglob; // rhs and solution of the global problem
varf vaglobal(u,v) = int2d(Th)(eta*u*v+kappa*Grad(u)'*Grad(v))
                                        +on(1,u=g) + int2d(Th)(f*v);
matrix Aglobal;
// Iterative solver parameters
real tol=1e-6; // tolerance for the iterative method
int maxit=300; // maximum number of iterations
```

Listing 1.10. ./FreefemCommon/data.edp

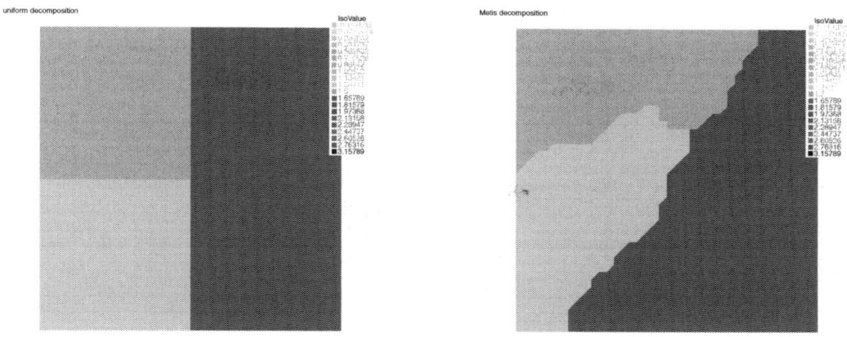

Figure 1.13. *Uniform and METIS decomposition.*

Afterwards, we have to define a piecewise constant function `part` which takes integer values. The isovalues of this function implicitly define a nonoverlapping partition of the domain. We have a coloring of the subdomains.

Suppose we want the decomposition of a rectangle Ω into `nn`×`mm` domains with approximately `nloc` points in one direction, or a more general partitioning method, using, for example, METIS [120] or SCOTCH [27]. In order to perform one of these decompositions, we make use of one of the routines `decompunif` or `decompMetis` defined in the script `decomp.idp`, which will return a vector defined on the mesh that can be recast into the piecewise function `part` we are looking for.

```
if (withmetis)
    {
      metisdual(lpart,Th,npart); // FreeFem++ interface to Metis
      for(int i=0;i<lpart.n;++i)
          part[][i]=lpart[i];
    }
  else
    {
    Ph xx=x,yy=y;
    part= int(xx/allong*nn)*mm + int(yy*mm);
    }
if (verbosity > 1)
   plot(part,wait=1,fill=1,value=1);
```

Listing 1.11. `./FreefemCommon/decomp.idp`

The isovalues of these two `part` functions correspond to respectively uniform or METIS nonoverlapping decompositions as shown in Figure 1.13.

Using the function `part` defined as above as an argument in the routine `SubdomainsPartitionUnity`, we get for each subdomain labeled `i` the overlapping mesh `aTh[i]`.

```
func bool SubdomainsPartitionUnity(mesh & Th, real[int] & partdof, int sizeoverlaps, mesh[int] & aTh,
     ↳ matrix[int] & Rih, matrix[int] & Dih, int[int] & Ndeg, real[int] & AreaThi)
{
 int npart=partdof.max+1;
 mesh Thi=Th; // freefem's trick, formal definition
 fespace Vhi(Thi,P1); // freefem's trick, formal definition
```

```
  Vhi[int] pun(npart); // local fem functions
  Vh sun=0, unssd=0;
  Ph part;
  part[]=partdof;
  for(int i=0;i<npart;++i)
    {
      Ph suppi= abs(part−i)<0.1; // boolean 1 in the subdomain 0 elsewhere
      AddLayers(Th,suppi[],sizeoverlaps,unssd[]); // ovr partitions by adding layers
      Thi=aTh[i]=trunc(Th,suppi>0,label=10,split=1);// ovr mesh interfaces label 10
      Rih[i]=interpolate(Vhi,Vh,inside=1); // Restriction operator : Vh −> Vhi
      pun[i][]=Rih[i]*unssd[];
      pun[i][] = 1.;// a garder par la suite
      sun[] += Rih[i]'*pun[i][];
      Ndeg[i] = Vhi.ndof;
      AreaThi[i] = int2d(Thi)(1.);
    }
  for(int i=0;i<npart;++i)
    {
      Thi=aTh[i];
      pun[i]= pun[i]/sun;
      Dih[i]=pun[i][]; //diagonal matrix built from a vector
      if(verbosity > 1)
          plot(pun[i],wait=1,value=1,fill=1,dim=3);
    }
  return true;
}
```

Listing 1.12. `./FreefemCommon/createPartition.idp`

Note that in the `CreatePartition.idp` script, the function `AddLayers` is defined below:

```
func bool AddLayers(mesh & Th,real[int] &ssd,int n,real[int] &unssd)
{
  // builds a continuous function uussd (P1) and modifies ssd :
  // IN: ssd in the characteristic function on the input subdomain.
  // OUT: ssd is a boolean function, unssd is a smooth function
  // ssd = 1 if unssd >0; add n layer of element and unssd = 0 ouside of this layer
  Ph s;
  assert(ssd.n==Ph.ndof);
  assert(unssd.n==Vh.ndof);
  unssd=0;
  s[]= ssd;
  Vh u;
  varf vM(uu,v)=int2d(Th,qforder=1)(uu*v/area);
  matrix M=vM(Ph,Vh);
  for(int i=0;i<n;++i)
    {
      u[]= M*s[];
      u = u>.1;
      unssd+= u[];
      s[]= M'*u[];
      s = s >0.1;
    }
  unssd /= (n);
  u[]=unssd;
  ssd=s[];
  return true;
}
```

Listing 1.13. `./FreefemCommon/createPartition.idp`

These last two functions are tricky. The reader does not need to understand their behavior in order to use them. They are given here for the sake of completeness. The restriction/interpolation operators `Rih[i]` from the local finite element space `Vh[i]` to the global one `Vh` and the diagonal local matrices `Dih[i]` are thus created.

The program `testdecomp.edp` (see below) shows such an example by checking that the partition of unity is correct.

```
load "medit"

verbosity=2;
include "dataGENEO.edp"
include "decomp.idp"
include "createPartition.idp"
SubdomainsPartitionUnity(Th,part[],sizeovr,aTh,Rih,Dih,Ndeg,AreaThi);
// check the partition of unity
Vh sum=0,fctone=1;
for(int i=0; i < npart;i++)
{
  Vh localone;
    real[int] bi = Rih[i]*fctone[]; // restriction to the local domain
    real[int] di = Dih[i]*bi;
    localone[] = Rih[i]'*di;
    sum[] +=localone[] ;
    plot(localone,fill=1,value=1, dim=3,wait=1);
}
plot(sum,fill=1,value=1, dim=3,wait=1);
```

Listing 1.14. `./FreefemCommon/testdecomp.edp`

Suppose we now want to do the same thing for a three-dimensional case.

```
load "msh3"
func mesh3 Cube(int[int] & NN,real[int,int] &BB ,int[int,int] & L)
// basic functions to build regular mesh of a cube
// int[int] NN=[nx,ny,nz]; the number of seg in the three directions
// real [int,int] BB=[[xmin,xmax],[ymin,ymax],[zmin,zmax]]; bounding bax
// int [int,int] L=[[1,2],[3,4],[5,6]]; label of the 6 faces left, right, front, back, down, up
{
  // first build the 6 faces of the cube.
  real x0=BB(0,0),x1=BB(0,1);
  real y0=BB(1,0),y1=BB(1,1);
  real z0=BB(2,0),z1=BB(2,1);
  int nx=NN[0],ny=NN[1],nz=NN[2];
  mesh Thx = square(nx,ny,[x0+(x1-x0)*x,y0+(y1-y0)*y]);

  int[int] rup=[0,L(2,1)], rdown=[0,L(2,0)],
    rmid=[1,L(1,0), 2,L(0,1), 3, L(1,1), 4, L(0,0) ];
  mesh3 Th=buildlayers(Thx,nz, zbound=[z0,z1],
            labelmid=rmid, labelup = rup, labeldown = rdown);
  return Th;
}
```

Listing 1.15. `./FreefemCommon/cube.idp`

We would like to build a cube or a parallelepiped defined by calling the function `Cube` defined in the script `cube.idp` and then splitting it into several domains. Again we need a certain number of data structures which will be declared in the file `data3d.edp`.

```
load "metis"
load "medit"
int nn=2,mm=2,ll=2; // number of the domains in each direction
int npart= nn*mm*ll; // total number of domains
int nloc = 11; // local no of dofs per domain in one direction
bool withmetis = 1; // =1 (Metis decomp) =0 (uniform decomp)
int sizeovr = 2; // size of the overlap
real allongx, allongz;
allongx = real(nn)/real(mm);
allongz = real(ll)/real(mm);
// Build the mesh
include "cube.idp"
int[int] NN=[nn*nloc,mm*nloc,ll*nloc];
real [int,int] BB=[[0,allongx],[0,1],[0,allongz]]; // bounding box
int [int,int] L=[[1,1],[1,1],[1,1]]; // the label of the 6 faces
mesh3 Th=Cube(NN,BB,L); // left, right, front, back, down, right
fespace Vh(Th,P1);
fespace Ph(Th,P0);
Ph part; // piecewise constant function
int[int] lpart(Ph.ndof); // giving the decomposition
// domain decomposition data structures
mesh3[int] aTh(npart); // sequence of ovr. meshes
matrix[int] Rih(npart); // local restriction operators
matrix[int] Dih(npart); // partition of unity operators
int[int] Ndeg(npart); // number of dofs for each mesh
real[int] VolumeThi(npart); // volume of each subdomain
matrix[int] aA(npart); // local Dirichlet matrices
Vh[int] Z(npart); // coarse space
// Definition of the problem to solve
// Delta (u) = f, u = 1 on the global boundary
Vh intern;
intern = (x>0) && (x<allongx) && (y>0) && (y<1) && (z>0) && (z<allongz);
Vh bord = 1-intern;
macro Grad(u) [dx(u),dy(u),dz(u)] // EOM
func f = 1; // right-hand side
func g = 1; // Dirichlet data
Vh rhsglobal,uglob; // rhs and solution of the global problem
varf vaglobal(u,v) = int3d(Th)(Grad(u)'*Grad(v))
                              +on(1,u=g) + int3d(Th)(f*v);
matrix Aglobal;
// Iterative solver
real tol=1e-10; // tolerance for the iterative method
int maxit=200; // maximum number of iterations
```

Listing 1.16. `./FreefemCommon/data3d.edp`

Then we need to define a piecewise constant function `part` which takes integer values. The isovalues of this function implicitly define a nonoverlapping partition of the domain. Suppose we want the decomposition of a rectangle Ω into `nn`$\times$`mm` $\times$`ll` domains with approximately `nloc` points in one direction, or a more general partitioning method. We will then make use of one of the decomposition routines which will return a vector defined on the mesh that can be recast into the piecewise function `part` we are looking for.

```
if (withmetis)
   {
     metisdual(lpart,Th,npart);
     for(int i=0;i<lpart.n;++i)
         part[][i]=lpart[i];
   }
  else
```

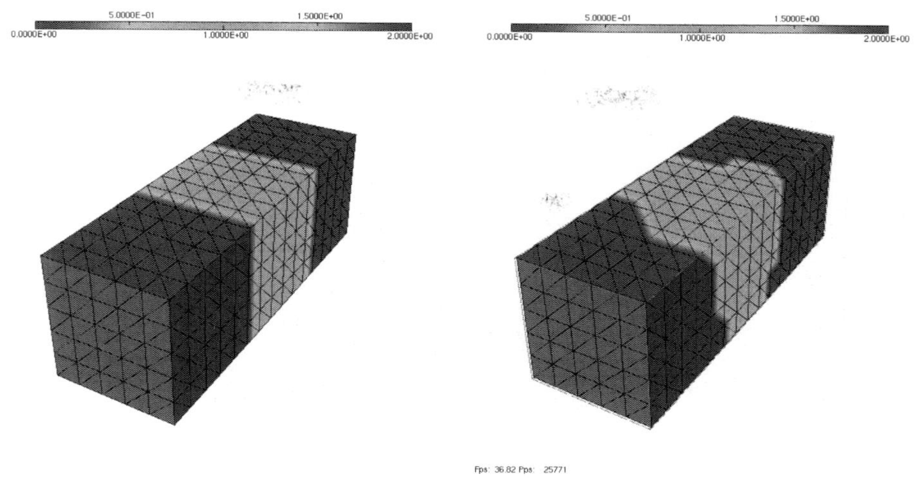

Figure 1.14. *Uniform and METIS decomposition.*

```
{
Ph xx=x,yy=y, zz=z;
part= int(xx/allongx*nn)*mm*ll + int(zz/allongz*ll)*mm+int(y*mm);
}
```

Listing 1.17. `./FreefemCommon/decomp3d.idp`

The isovalues of the function `part` correspond to nonoverlapping decompositions, as shown in Figure 1.14.

Using the function `part` defined as above, function `SubdomainsPartitionUnity3` builds for each subdomain labeled `i` the overlapping meshes `aTh[i]` by making use of the function `AddLayers3` in the `CreatePartition3d.idp`.

```
func bool SubdomainsPartitionUnity3(mesh3 & Th, real[int] & partdof, int sizeoverlaps, mesh3[int] & aTh, matrix[int] & Rih, matrix[int] & Dih, int[int] & Ndeg, real[int] & VolumeThi)
{
  int npart=partdof.max+1;
  mesh3 Thi=Th; // freefem's trick, formal definition
  fespace Vhi(Thi,P1); // freefem's trick, formal definition
  Vhi[int] pun(npart); // local fem functions
  Vh sun=0, unssd=0;
  Ph part;
  part[]=partdof;
  for(int i=0;i<npart;++i)
    {
      // boolean function 1 in the subdomain 0 elswhere
      Ph suppi= abs(part-i)<0.1;
      AddLayers3(Th,suppi[],sizeoverlaps,unssd[]); // overlapping partitions by adding layers
      Thi=aTh[i]=trunc(Th,suppi>0,label=10,split=1); // overlapping mesh, interfaces have label 10
      Rih[i]=interpolate(Vhi,Vh,inside=1); // Restriction operator : Vh -> Vhi
      pun[i][]=Rih[i]*unssd[];
      sun[] += Rih[i]'*pun[i][];
      Ndeg[i] = Vhi.ndof;
      VolumeThi[i] = int3d(Thi)(1.);
    }
  for(int i=0;i<npart;++i)
    {
      Thi=aTh[i];
```

```
        pun[i]= pun[i]/sun;
        Dih[i]=pun[i][];//diagonal matrix built from a vector
    }
  return true;
}
```

Listing 1.18. ./FreefemCommon/createPartition3d.idp

```
func bool AddLayers3(mesh3 & Th,real[int] &ssd,int n,real[int] &unssd)
{
  // builds a continuous function uussd (P1) and modifies ssd :
  // IN: ssd in the characteristic function on the input subdomain.
  // OUT: ssd is a boolean function, unssd is a smooth function
  // ssd = 1 if unssd >0; add n layer of element and unssd = 0 ouside of this layer
  Ph s;
  assert(ssd.n==Ph.ndof);
  assert(unssd.n==Vh.ndof);
  unssd=0;
  s[]= ssd;
  Vh u;
  varf vM(uu,v)=int3d(Th,qforder=1)(uu*v/volume);
  matrix M=vM(Ph,Vh);
  for(int i=0;i<n;++i)
    {
      u[]= M*s[];
      u = u>.1;
      unssd+= u[];
      s[]= M'*u[];
      s = s >0.1;
    }
  unssd /= (n);
  u[]=unssd;
  ssd=s[];
  return true;
}
```

Listing 1.19. ./FreefemCommon/createPartition3d.idp

As in the 2D case, these last two functions are tricky. The reader does not need to understand their behavior in order to use them. They are given here for sake of completeness.

The restriction/interpolation operators `Rih[i]` from the local finite element space `Vh[i]` to the global one `Vh` and the diagonal local matrices `Dih[i]` are thus created. The program `testdecomp3d.edp` shows such an example by checking that the partition of unity is correct.

```
include "data3d.edp"
include "decomp3d.idp"
include "createPartition3d.idp"
medit("part", Th, part, order = 1);
SubdomainsPartitionUnity3(Th,part[],sizeovr,aTh,Rih,Dih,Ndeg,VolumeThi);
// check the partition of unity
Vh sum=0,fctone=1;
for(int i=0; i < npart;i++)
{
  Vh localone;
    real[int] bi = Rih[i]*fctone[]; // restriction to the local domain
    real[int] di = Dih[i]*bi;
```

```
    localone[] = Rih[i]'*di;
    sum[] +=localone[] ;
    medit("loc",Th, localone, order = 1);
    medit("subdomains",aTh[i]);
}
medit("sum",Th, sum, order = 1);
```

Listing 1.20. ./FreefemCommon/testdecomp3d.edp

1.7.3 ▪ Schwarz algorithms as solvers

We are now in a position to code Schwarz solvers. In program schwarz-solver.edp (see below) the RAS method (see (1.29)) is implemented as a solver. First we need to split the domains into subdomains.

```
verbosity=1;
include "../../FreefemCommon/dataGENEO.edp"
include "../../FreefemCommon/decomp.idp"
include "../../FreefemCommon/createPartition.idp"
SubdomainsPartitionUnity(Th,part[],sizeovr,aTh,Rih,Dih,Ndeg,AreaThi);
```

Listing 1.21. ./SCHWARZ/FreefemProgram/schwarz-solver.edp

Then we need to define the global data from the variational formulation.

```
Aglobal = vaglobal(Vh,Vh,solver = UMFPACK); // global matrix
rhsglobal[] = vaglobal(0,Vh); // global rhs
uglob[] = Aglobal^-1*rhsglobal[];
plot(uglob,value=1,fill=1,wait=1,cmm="Solution by a direct method",dim=3);
```

Listing 1.22. ./SCHWARZ/FreefemProgram/schwarz-solver.edp

Afterwards, we build the local problem matrices

```
for(int i = 0;i<npart;++i)
  {
    cout << " Domain :" << i << "/" << npart << endl;
    matrix aT = Aglobal*Rih[i]';
    aA[i] = Rih[i]*aT;
    set(aA[i],solver = UMFPACK);// direct solvers
  }
```

Listing 1.23. ./SCHWARZ/FreefemProgram/schwarz-solver.edp

and finally the Schwarz iteration

```
ofstream filei("Conv.m");
Vh un = 0; // initial guess
Vh rn = rhsglobal;
for(int iter = 0;iter<maxit;++iter)
  {
    real err = 0, res;
    Vh er = 0;
    for(int i = 0;i<npart;++i)
      {
        real[int] bi = Rih[i]*rn[]; // restriction to the local domain
```

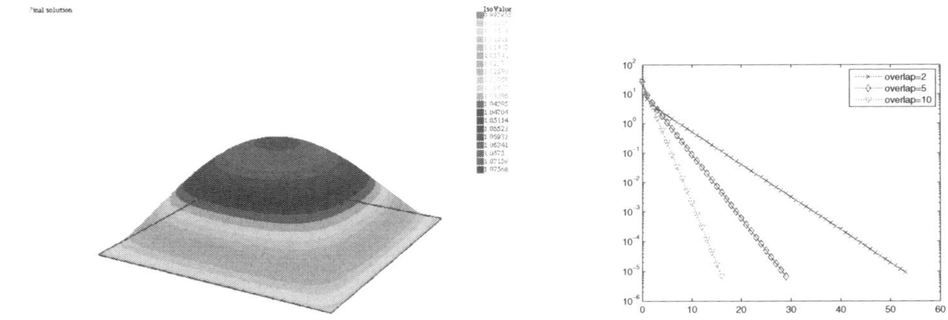

Figure 1.15. *Solution and RAS convergence as a solver for different overlaps.*

```
            real[int] ui = aA[i] ^-1 * bi; // local solve
            bi = Dih[i]*ui;
            // bi = ui; // uncomment this line to test the ASM as a solver
            er[] += Rih[i]'*bi;
        }
    un[] += er[]; // build new iterate
    rn[] = Aglobal*un[]; // computes global residual
    rn[] = rn[] - rhsglobal[];
    rn[] *= -1;
    err = sqrt(er[]'*er[]);
    res = sqrt(rn[]'*rn[]);
    cout << "Iteration: " << iter << " Correction = " << err << " Residual = " << res << endl;
    plot(un,wait=1,value=1,fill=1,dim=3,cmm="Approximate solution at step " + iter);
    int j = iter+1;
    // Store the error and the residual in MATLAB/Scilab/Octave form
    filei << "Convhist("+j+",:)=[" << err << " " << res <<"];" << endl;
    if(err < tol) break;
  }
plot(un,wait=1,value=1,fill=1,dim=3,cmm="Final solution");
```

Listing 1.24. `./SCHWARZ/FreefemProgram/schwarz-solver.edp`

The convergence history of the algorithm is stored in a MATLAB® file (also compatible with Scilab or Octave) `Conv.m`, in the form of a two-column matrix containing the error evolution as well as the residual one.

The convergence graph of the RAS solver for various overlaps is shown in Figure 1.15.

Remark 1.7.1. *Previous tests have shown a very easy way of using the RAS iterative algorithm and some straightforward conclusions from this.*

- *The convergence of RAS, which is not very fast even in a simple configuration of four subdomains, improves when the overlap gets bigger.*
- *Note that it is very easy to test the ASM (see* (1.30)*) when using it as a solver. It is sufficient to uncomment the line* `bi = ui;`.
- *Running the program shows that the ASM does not converge. For this reason, the ASM method is always used as a preconditioner for a Krylov method such as CG, GMRES, or BiCGSTAB, see Chapter* 3.
- *In the three-dimensional case, the only part that changes is the decomposition into subdomains. The other parts of the algorithm are identical.*

```
include "../../FreefemCommon/data3d.edp"
include "../../FreefemCommon/decomp3d.idp"
include "../../FreefemCommon/createPartition3d.idp"
SubdomainsPartitionUnity3(Th,part[],sizeovr,aTh,Rih,Dih,Ndeg,VolumeThi);
```

Listing 1.25. ./SCHWARZ/FreefemProgram/schwarz-solver3d.edp

1.7.4 ▪ Systems of PDEs: The example of linear elasticity

Suppose we now want to solve another kind of problem, such as the system of linear elasticity. A few changes will be necessary.

```
load "metis"
load "medit"
int nn=3,mm=3;      // number of the domains in each direction
int npart= nn*mm;      // total number of domains
int nloc = 20; // local no of dofs per domain in one direction
bool withmetis = 1; // =1 (Metis decomp) =0 (uniform decomp)
int sizeovr = 2; // size of the overlap
real allong = real(nn)/real(mm); // aspect ratio of the global domain
func E = 2*10^11; // Young modulus and Poisson ratio
func sigma = 0.3;
func lambda = E*sigma/((1+sigma)*(1-2*sigma)); // Lame coefficients
func mu = E/(2*(1+sigma));
real sqrt2=sqrt(2.);
func eta = 1.0e-6;
// Mesh of a rectangular domain
mesh Th=square(nn*nloc,mm*nloc,[x*allong,y]);
fespace Vh(Th,[P1,P1]); // vector fem space
fespace Uh(Th,P1); // scalar fem space
fespace Ph(Th,P0);
Ph part; // piecewise constant function
int[int] lpart(Ph.ndof); // giving the decomposition
// Domain decomposition data structures
mesh[int] aTh(npart); // sequence of overlapping meshes
matrix[int] Rih(npart); // local restriction operators
matrix[int] Dih(npart); // partition of unity operators
int[int] Ndeg(npart); // number of dofs for each mesh
real[int] AreaThi(npart); // area of each subdomain
matrix[int] aA(npart); // local Dirichlet matrices
// Definition of the problem to solve
int[int] chlab=[1,11 ,2,2 ,3,33 ,4,1 ]; //Dirichlet conditions for label = 1
Th=change(Th,refe=chlab);
macro Grad(u) [dx(u),dy(u)] // EOM
macro epsilon(u,v) [dx(u),dy(v),(dy(u)+dx(v))/sqrt2] // EOM
macro div(u,v) ( dx(u)+dy(v) ) // EOM
func uboundary = (0.25 - (y-0.5)^2);
varf vaBC([u,v],[uu,vv]) = on(1, u = uboundary, v=0) + on(11, u = 0, v=0) + on(33, u=0,v=0);
// global problem
Vh [rhsglobal,rrhsglobal], [uglob,uuglob];
macro Elasticity(u,v,uu,vv) eta*(u*uu+v*vv) + lambda*(div(u,v)*div(uu,vv))+2.*mu*(
    epsilon(u,v)'*epsilon(uu,vv) ) // EOM
varf vaglobal([u,v],[uu,vv]) = int2d(Th)(Elasticity(u,v,uu,vv)) + vaBC; // on(1,u=uboundary,v=0)
matrix Aglobal;
// Iterative solver parameters
real tol=1e-6; // tolerance for the iterative method
int maxit=200; // maximum number of iterations
```

Listing 1.26. ./FreefemCommon/dataElast.edp

First of all, the file `dataElast.edp` now contains the declarations and data. The definition of the partition is done as before using `decomp.idp`. The `SubdomainsPartitionUnityVec` is the vector adaptation of `SubdomainsPartitionUnity` and will provide the same type of result.

```
func bool SubdomainsPartitionUnityVec(mesh & Th, real[int] & partdof, int sizeoverlaps, mesh[int] &
    aTh, matrix[int] & Rih, matrix[int] & Dih, int[int] & Ndeg, real[int] & AreaThi)
{
  int npart=partdof.max+1;
  mesh Thi=Th; // freefem's trick, formal definition
  fespace Vhi(Thi,[P1,P1]); // freefem's trick, formal definition
  Vhi[int] [pun,ppun](npart); // local fem functions
  Vh [unssd,uunssd], [sun,ssun]=[0,0];
  Uh Ussd = 0;
  Ph part;
  int[int] U2Vc=[0,1]; // no component change
  part[]=partdof;
  for(int i=0;i<npart;++i)
    {
      Ph suppi= abs(part-i)<0.1; // boolean 1 in the subdomain 0 elsewhere
      AddLayers(Th,suppi[],sizeoverlaps,Ussd[]); // ovr partitions by adding layers
      [unssd,uunssd] =[Ussd,Ussd];
      Thi=aTh[i]=trunc(Th,suppi>0,label=10,split=1); // ovr mesh interfaces label 10
      Rih[i]=interpolate(Vhi,Vh,inside=1,U2Vc=U2Vc); // Restriction operator : Vh -> Vhi
      pun[i][]=Rih[i]*unssd[];
      sun[] += Rih[i]'*pun[i][];
      Ndeg[i] = Vhi.ndof;
      AreaThi[i] = int2d(Thi)(1.);
    }
  for(int i=0;i<npart;++i)
    {
      Thi=aTh[i];
      [pun[i],ppun[i]] = [pun[i]/sun, ppun[i]/sun];
      Dih[i]=pun[i][]; //diagonal matrix built from a vector
    }
  return true;
}
```

Listing 1.27. `./FreefemCommon/createPartitionVec.idp`

Note that in the `CreatePartitionVec.idp` script, the function `AddLayers` is the following:

```
func bool AddLayers(mesh & Th,real[int] &ssd,int n,real[int] &unssd)
{
  // builds a continuous function uussd (P1) and modifies ssd :
  // IN: ssd in the characteristic function on the input subdomain.
  // OUT: ssd is a boolean function, unssd is a smooth function
  // ssd = 1 if unssd >0; add n layer of element and unssd = 0 ouside of this layer
  Ph s;
  Uh u;
  assert(ssd.n==Ph.ndof);
  assert(unssd.n==Uh.ndof);
  unssd=0;
  s[]= ssd;
  varf vM(uu,v)=int2d(Th,qforder=1)(uu*v/area);
  matrix M=vM(Ph,Uh);
  for(int i=0;i<n;++i)
    {
      u[]= M*s[];
```

```
        u = u>.1;
        unssd+= u[];
        s[]= M'*u[];
        s = s >0.1;
    }
  unssd /= (n);
  u[]=unssd;
  ssd=s[];
  return true;
}
```

Listing 1.28. `./FreefemCommon/createPartitionVec.idp`

The restriction/interpolation operators `Rih[i]` from the local finite element space `Vh[i]` to the global one `Vh` and the diagonal local matrices `Dih[i]` are thus created.

We are now in a position to code Schwarz solvers. In the program `schwarz-solver-elast.edp` (see below) the RAS method (see (1.29)) is implemented as a solver following the same guidelines as for the case of the Laplace equation. First we need to split the domains into subdomains.

```
include "../../FreefemCommon/dataElast.edp"
include "../../FreefemCommon/decomp.idp"
include "../../FreefemCommon/createPartitionVec.idp"
SubdomainsPartitionUnityVec(Th,part[],sizeovr,aTh,Rih,Dih,Ndeg,AreaThi);
```

Listing 1.29. `./SCHWARZ/FreefemProgram/schwarz-solver-elast.edp`

Then we need to define the global data from the variational formulation.

```
Aglobal = vaglobal(Vh,Vh,solver = UMFPACK); // global matrix
rhsglobal[] = vaglobal(0,Vh); // global rhs
uglob[] = Aglobal^-1*rhsglobal[];
real coeff2 = 1;
mesh Thmv=movemesh(Th,[x+coeff2*uglob,y+coeff2*uuglob]);
medit("Thmv", Thmv);
medit("uex", Th, uglob, Th, uuglob, order=1);
```

Listing 1.30. `./SCHWARZ/FreefemProgram/schwarz-solver-elast.edp`

Afterwards, the local problem matrices are built in the same way as before and finally the Schwarz iteration can be computed:

```
ofstream filei("Conv.m");
Vh [un,uun] = [0,0]; // initial guess
Vh [rn,rrn] = [rhsglobal,rrhsglobal];
for(int iter = 0;iter<maxit;++iter)
  {
    real err = 0, res;
    Vh [er,eer] = [0,0];
    for(int i = 0;i<npart;++i)
      {
        real[int] bi = Rih[i]*rn[]; // restriction to the local domain
        real[int] ui = aA[i] ^-1* bi; // local solve
        bi = Dih[i]*ui;
        // bi = ui; // uncomment this line to test the ASM as a solver
        er[] += Rih[i]'*bi;
      }
```

```
    un[] += er[]; // build new iterate
    rn[] = Aglobal*un[]; // computes global residual
    rn[] = rn[] - rhsglobal[];
    rn[] *= -1;
    err = sqrt(er[]'*er[]);
    res = sqrt(rn[]'*rn[]);
    cout << "Iteration: " << iter << " Correction = " << err << " Residual = " << res << endl;
    int j = iter+1;
    // Store the error and the residual in MATLAB/Scilab/Octave form
    filei << "Convhist("+j+",:)=[" << err << " " << res <<"];" << endl;
    if(res < tol) break;
  }
mesh Thm=movemesh(Th,[x+coeff2*un,y+coeff2*uun]);
medit("Thm", Thm);
medit("uh", Th, un, Th, uun, order=1);
```

Listing 1.31. `./SCHWARZ/FreefemProgram/schwarz-solver-elast.edp`

Chapter 2

Optimized Schwarz methods

During the last few decades, a new class of nonoverlapping and overlapping Schwarz methods was developed for PDEs, namely the optimized Schwarz methods (OSM). These methods are based on a classical domain decomposition, but they use more effective transmission conditions than the classical Dirichlet conditions at the interfaces between subdomains.

We first introduce P. L. Lions' domain decomposition method [130] in section 2.1. Instead of solving a Dirichlet or Neumann boundary value problem (BVP) in the subdomains as in Schwarz or FETI/BNN algorithms, a Robin (a.k.a. mixed) BVP is solved. We compute the convergence factor of the method and give a general convergence proof. This algorithm was extended to the Helmholtz problem by Després [44]; see section 2.2. This extension is all the more important since the original Schwarz method is not convergent in the case of wave propagation phenomena in the frequency regime. Implementation issues are discussed in section 2.3. In particular, when subdomains overlap, OSM can be implemented simply by replacing the Dirichlet solves in the ASM with Robin solves. This justifies the "S" in OSM. Then, in sections 2.4 and 2.5, we show that it is possible to consider interface conditions other than Robin conditions and optimize their choice with respect to the convergence factor. This explains the "O" in OSM. In the final section, section 2.6, we explain a FreeFem++ implementation.

2.1 ▪ P. L. Lions' algorithm

For elliptic problems, Schwarz algorithms work only for overlapping domain decompositions, and their performance in terms of iteration counts depends on the width of the overlap. Substructuring algorithms such as BNN (balancing Neumann–Neumann [132]) or FETI (finite element tearing and interconnecting [80]) are defined for nonoverlapping domain decompositions but not for overlapping subdomains. The algorithm introduced by P. L. Lions [130] can be applied to both overlapping and nonoverlapping subdomains. It is based on improving Schwarz methods by replacing the Dirichlet interface conditions by Robin interface conditions. Let α be a positive number. The modified algorithm reads as

$$\begin{aligned} -\Delta(u_1^{n+1}) &= f && \text{in} \quad \Omega_1, \\ u_1^{n+1} &= 0 && \text{on} \quad \partial\Omega_1 \cap \partial\Omega, \\ \left(\frac{\partial}{\partial \mathbf{n}_1} + \alpha\right)(u_1^{n+1}) &= \left(\frac{\partial}{\partial \mathbf{n}_1} + \alpha\right)(u_2^n) && \text{on} \quad \partial\Omega_1 \cap \overline{\Omega_2} \end{aligned} \tag{2.1}$$

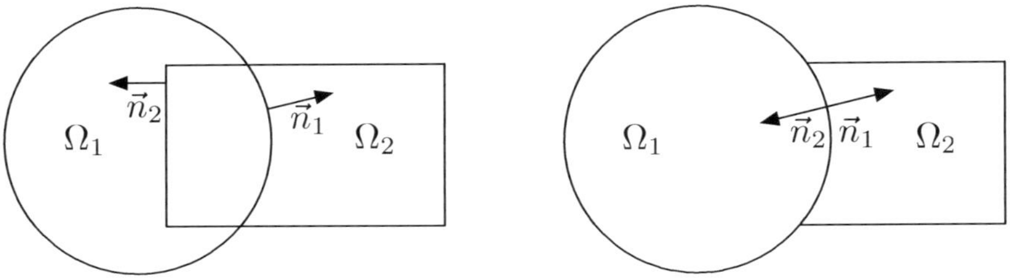

Figure 2.1. *Outward normals for overlapping and nonoverlapping subdomains for P. L. Lions' algorithm.*

and

$$\begin{aligned} -\Delta(u_2^{n+1}) &= f && \text{in} \quad \Omega_2, \\ u_2^{n+1} &= 0 && \text{on} \quad \partial\Omega_2 \cap \partial\Omega, \\ \left(\frac{\partial}{\partial \mathbf{n}_2} + \alpha\right)(u_2^{n+1}) &= \left(\frac{\partial}{\partial \mathbf{n}_2} + \alpha\right)(u_1^n) && \text{on} \quad \partial\Omega_2 \cap \overline{\Omega_1}, \end{aligned} \tag{2.2}$$

where $\mathbf{n}_1$ and $\mathbf{n}_2$ are the outward normals on the boundary of the subdomains; see Figure 2.1.

We use the Fourier transform to analyze the benefit of the Robin interface conditions in a simple case. The domain $\Omega = \mathbb{R}^2$ is decomposed into two half-planes $\Omega_1 = (-\infty, \delta) \times \mathbb{R}$ and $\Omega_2 = (0, \infty) \times \mathbb{R}$ with $\delta \geq 0$. We consider the example of a symmetric positive definite problem

$$(\eta - \Delta)(u) = f \quad \text{in} \quad \mathbb{R}^2,$$

with η being a positive constant. The algorithm can be written as

$$\begin{aligned} (\eta - \Delta)(u_1^{n+1}) &= f(x,y), \quad (x,y) \in \Omega_1, \\ u_1^{n+1} & \quad \text{is bounded at infinity}, \\ \left(\frac{\partial}{\partial \mathbf{n}_1} + \alpha\right)(u_1^{n+1})(\delta, y) &= \left(\frac{\partial}{\partial \mathbf{n}_1} + \alpha\right)(u_2^n)(\delta, y), \quad y \in \mathbb{R}, \end{aligned} \tag{2.3}$$

and

$$\begin{aligned} (\eta - \Delta)(u_2^{n+1}) &= f(x,y), \quad (x,y) \in \Omega_2, \\ u_2^{n+1} & \quad \text{is bounded at infinity}, \\ \left(\frac{\partial}{\partial \mathbf{n}_2} + \alpha\right)(u_2^{n+1})(0, y) &= \left(\frac{\partial}{\partial \mathbf{n}_2} + \alpha\right)(u_1^n)(0, y), \quad y \in \mathbb{R}. \end{aligned} \tag{2.4}$$

2.1.1 ▪ Computation of the convergence factor

In order to compute the convergence factor of the algorithm given by (2.3) and (2.4) we introduce the errors $e_i^n = u_i^n - u|_{\Omega_i}$, $i = 1, 2$. By linearity, the errors satisfy (2.3) and (2.4) with $f = 0$:

$$\begin{aligned} (\eta - \Delta)(e_1^{n+1}) &= 0, \quad (x,y) \in \Omega_1, \\ e_1^{n+1} & \quad \text{is bounded at infinity}, \\ \left(\frac{\partial}{\partial \mathbf{n}_1} + \alpha\right)(e_1^{n+1})(\delta, y) &= \left(\frac{\partial}{\partial \mathbf{n}_1} + \alpha\right)(e_2^n)(\delta, y), \quad y \in \mathbb{R}, \end{aligned} \tag{2.5}$$

and

$$\begin{aligned} (\eta-\Delta)(e_2^{n+1}) &= 0, \quad (x,y)\in\Omega_2, \\ e_2^{n+1} & \text{ is bounded at infinity}, \\ \left(\frac{\partial}{\partial \mathbf{n}_2}+\alpha\right)(e_2^{n+1})(0,y) &= \left(\frac{\partial}{\partial \mathbf{n}_2}+\alpha\right)(e_1^{n})(0,y), \quad y\in\mathbb{R}. \end{aligned} \tag{2.6}$$

By taking the partial Fourier transform of the first line of (2.5) in the y-direction we get

$$\left(\eta-\frac{\partial^2}{\partial x^2}+k^2\right)(\hat{e}_1^{n+1}(x,k))=0 \quad \text{for} \quad x<\delta \text{ and } k\in\mathbb{R}.$$

For a given k, this is an ODE whose solution is sought in the form $\sum_j \gamma_j(k)\exp(\lambda_j(k)x)$. A simple calculation shows that lambda is necessarily given by

$$\lambda^{\pm}(k)=\pm\sqrt{\eta+k^2}.$$

Therefore we have

$$\hat{e}_1^{n+1}(x,k)=\gamma_+^{n+1}(k)\exp(\lambda^+(k)x)+\gamma_-^{n+1}(k)\exp(\lambda^-(k)x).$$

From the second line of (2.5), the solution must be bounded at $x=-\infty$. This implies that $\gamma_-^{n+1}(k)\equiv 0$. Thus we have

$$\hat{e}_1^{n+1}(x,k)=\gamma_+^{n+1}(k)\exp(\lambda^+(k)x),$$

or equivalently, by changing the value of the coefficient γ_+,

$$\hat{e}_1^{n+1}(x,k)=\gamma_1^{n+1}(k)\exp(\lambda^+(k)(x-\delta))$$

and similarly for subdomain 2,

$$\hat{e}_2^{n+1}(x,k)=\gamma_2^{n+1}(k)\exp(\lambda^-(k)x)$$

with $\gamma_{1,2}^{n+1}$ to be determined. From the interface conditions we get

$$\gamma_1^{n+1}(k)(\lambda^+ +\alpha)=\gamma_2^{n}(k)(\lambda^- +\alpha)\exp(\lambda^-(k)\delta)$$

and

$$\gamma_2^{n+1}(k)(-\lambda^- +\alpha)=\gamma_1^{n}(k)(-\lambda^+ +\alpha)\exp(-\lambda^+(k)\delta).$$

Combining these two and denoting $\lambda(k)=\lambda^+(k)=-\lambda^-(k)$, we get for $i=1,2$,

$$\gamma_i^{n+1}(k)=\rho(k,\delta;\alpha)^2\gamma_i^{n-1}(k)$$

with

$$\boxed{\rho(k,\delta;\alpha)=\left|\frac{\lambda(k)-\alpha}{\lambda(k)+\alpha}\right|\exp(-\lambda(k)\delta),} \tag{2.7}$$

where $\lambda(k)=\sqrt{\eta+k^2}$ and $\alpha>0$.

Remark 2.1.1. *Formula* (2.7) *deserves a few remarks.*

- *With or without overlap, for all $k\in\mathbb{R}$, $\rho(k,\delta;\alpha)<1$ so that $\gamma_i^n(k)\to 0$ as n goes to infinity. The method is convergent.*

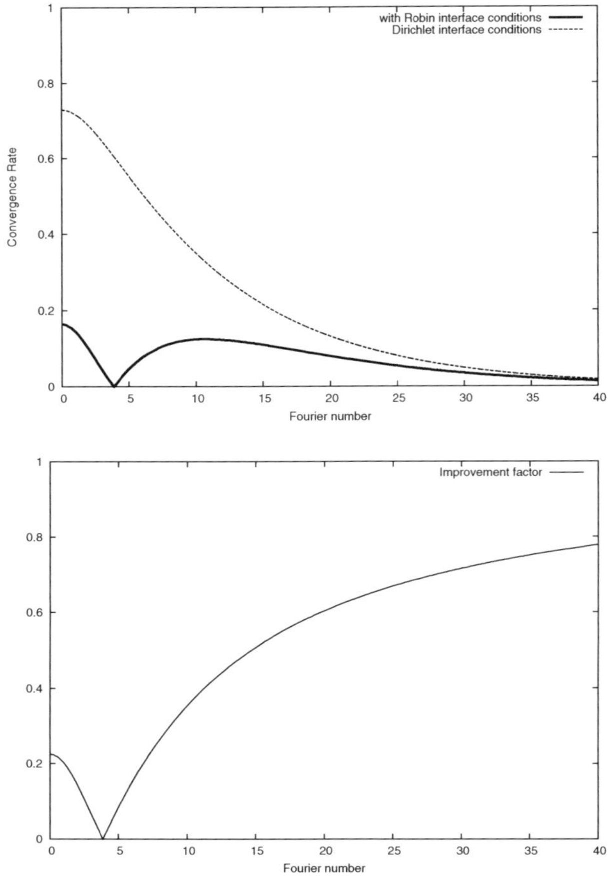

Figure 2.2. *Top: Convergence factor (dotted line: Dirichlet ICs; solid line: Robin ICs) vs. Fourier number. Bottom: Improvement factor due to Robin interface conditions* (2.8) *vs. Fourier number.*

- *When domains overlap* ($\delta > 0$), $\rho(k,\delta;\alpha)$ *is uniformly bounded from above by a constant smaller than* 1, $\rho(k,\delta;\alpha) < \exp(-\sqrt{\eta}\,\delta) < 1$, *and* $\rho \to 0$ *as* k *tends to infinity. Convergence is geometric.*
- *When there is no overlap* ($\delta = 0$), $\rho \to 1$ *as* k *tends to infinity.*
- *Let* $k_c \in \mathbb{R}$. *By taking* $\alpha = \lambda(k_c)$, *we have* $\rho(k_c) = 0$.
- *For the original Schwarz method* (1.3) *the convergence factor is* $\exp(-\lambda(k)\delta)$, *and for* $\delta = 0$ *we see once again that there is no convergence. Replacing the Dirichlet interface conditions (ICs) by Robin conditions enhances the convergence of the* k*th component of the error to zero by an improvement factor*

$$\rho(k,0;\alpha) = \left|\frac{\lambda(k)-\alpha}{\lambda(k)+\alpha}\right| < 1. \tag{2.8}$$

2.1.2 ▪ General convergence proof

The convergence proof given by P. L. Lions in the elliptic case [130] was extended by Després [44] to the Helmholtz equation and then to time-harmonic Maxwell equations [46].

A general presentation for these equations is given in [35]. We treat here the elliptic case with ICs which are more general than Robin ICs since they involve second-order tangential derivatives along the interfaces. Our proof is formal since no functional analysis framework is given. Nevertheless, this general convergence result shows the robustness of this approach in the nonoverlapping case.

Let Ω be an open set. We consider the following problem:

$$\begin{aligned} \eta(\mathbf{x})u - \nabla\cdot(\kappa(\mathbf{x})\nabla u) &= f \quad \text{in } \Omega,\\ u &= 0 \quad \text{on } \partial\Omega, \end{aligned}$$

where the functions $\mathbf{x} \mapsto \eta(\mathbf{x}), \kappa(\mathbf{x})$ are positive functions.

The domain is decomposed into N nonoverlapping subdomains $(\Omega_i)_{1\le i\le N}$,

$$\bar{\Omega} = \cup_{i=1}^{N} \bar{\Omega}_i \text{ and } \Omega_i \cap \Omega_j = \emptyset \quad \text{for } i \neq j.$$

Let Γ_{ij} denote the interface $\Gamma_{ij} = \partial\Omega_i \cap \partial\Omega_j$, $i \neq j$. For two disjoint subdomains,we have $\Gamma_{ij} = \emptyset$.

For the sake of simplicity in the writing of the interface conditions, we consider the two-dimensional case ($\Omega \subset \mathbb{R}^2$), although the proof is valid in arbitrary dimensions. The interface conditions include second-order tangential derivatives and have the form

$$\boxed{\mathscr{B}_{ij} := \kappa(\mathbf{x})\frac{\partial}{\partial \mathbf{n}_i} + \Lambda_{ij},\ \Lambda_{ij} := \alpha_{ij}(\mathbf{x}) - \frac{\partial}{\partial \tau_i}\left(\beta_{ij}(\mathbf{x})\frac{\partial}{\partial \tau_i}\right),} \tag{2.9}$$

where α_{ij} and β_{ij} are functions from Γ_{ij} into $\mathbb{R}$ that verify

$$\begin{aligned} \alpha_{ij}(\mathbf{x}) &= \alpha_{ji}(\mathbf{x}) \ge \alpha_0 > 0,\\ \beta_{ij}(\mathbf{x}) &= \beta_{ji}(\mathbf{x}) \ge 0,\\ \beta_{ij}(\mathbf{x}) &= 0 \text{ on } \partial\Gamma_{ij}. \end{aligned}$$

In this case the operators $\Lambda_{ij} = \Lambda_{ji}$ defined in (2.9) have the following remarkable properties:

- Λ_{ij} is symmetric positive definite (SPD).
- Λ_{ij} has an invertible SPD square root, denoted by $\Lambda_{ij}^{1/2}$, whose inverse is denoted by $\Lambda_{ij}^{-1/2}$, which is SPD as well.

The algorithm consists in computing the iterates u_i^{n+1} of the local solutions from the previous iterates u_i^n by solving

$$\begin{aligned} \eta(\mathbf{x})u_i^{n+1} - \nabla\times(\kappa(\mathbf{x})\nabla u_i^{n+1}) &= f && \text{in } \Omega_i,\\ u_i^{n+1} &= 0 && \text{on } \partial\Omega\cap\partial\Omega_i,\\ \left(\kappa(\mathbf{x})\frac{\partial}{\partial \mathbf{n}_i} + \Lambda_{ij}\right)u_i^{n+1} &= \left(-\kappa(\mathbf{x})\frac{\partial}{\partial \mathbf{n}_j} + \Lambda_{ij}\right)u_j^n && \text{on } \Gamma_{ij}. \end{aligned} \tag{2.10}$$

The interface condition in the algorithm (2.10) can be rewritten as

$$\Lambda_{ij}^{-1/2}\left(\kappa(\mathbf{x})\frac{\partial u_i}{\partial \mathbf{n}_i}\right) + \Lambda_{ij}^{1/2}(u_i) = -\Lambda_{ij}^{-1/2}\left(\kappa(\mathbf{x})\frac{\partial u_j}{\partial \mathbf{n}_j}\right) + \Lambda_{ij}^{1/2}(u_j) \text{ on } \Gamma_{ij}. \tag{2.11}$$

The convergence proof of the algorithm (2.10) follows the arguments given in [35] and is based on a preliminary result which is an energy estimate.

Lemma 2.1 (Energy estimate). *Let u_i denote a function that satisfies*

$$\begin{aligned} \eta(\mathbf{x})u_i - \nabla\cdot(\kappa(\mathbf{x})\nabla u_i) &= 0 \quad \text{in} \quad \Omega_i,\\ u_i &= 0 \quad \text{on} \quad \partial\Omega_i \cap \partial\Omega. \end{aligned}$$

Then,

$$\begin{aligned} \int_{\Omega_i} \eta(\mathbf{x})|u_i|^2 + \kappa(\mathbf{x})|\nabla u_i|^2 + \frac{1}{4}\sum_{j\neq i}\int_{\Gamma_{ij}} \left(\Lambda_{ij}^{-1/2}[\kappa(\mathbf{x})\frac{\partial u_i}{\partial \mathbf{n}_i} - \Lambda_{ij}(u_i)]\right)^2 \\ = \frac{1}{4}\sum_{j\neq i}\int_{\Gamma_{ij}} \left(\Lambda_{ij}^{-1/2}[\kappa(\mathbf{x})\frac{\partial u_i}{\partial \mathbf{n}_i} + \Lambda_{ij}(u_i)]\right)^2. \end{aligned}$$

Proof. From the variational formulation of the local problem

$$\eta(\mathbf{x})u_i - \nabla\cdot(\kappa(\mathbf{x})\nabla u_i) = 0 \quad \text{in}\ \Omega_i,$$

we get by choosing the test function equal to u_i and by integrating by parts

$$\begin{aligned} \int_{\Omega_i} \eta(\mathbf{x})|u_i|^2 + \kappa(\mathbf{x})|\nabla u_i|^2 = \int_{\partial\Omega_i} \kappa(\mathbf{x})\frac{\partial u_i}{\partial \mathbf{n}_i}\, u_i &= \sum_{j\neq i}\int_{\Gamma_{ij}} \kappa(\mathbf{x})\frac{\partial u_i}{\partial \mathbf{n}_i}\, u_i \\ &= \sum_{j\neq i}\int_{\Gamma_{ij}} \kappa(\mathbf{x})\frac{\partial u_i}{\partial \mathbf{n}_i}\, \Lambda_{ij}^{-1/2}\Lambda_{ij}^{1/2}(u_i) \\ &= \sum_{j\neq i}\int_{\Gamma_{ij}} \Lambda_{ij}^{-1/2}\left(\kappa(\mathbf{x})\frac{\partial u_i}{\partial \mathbf{n}_i}\right)\, \Lambda_{ij}^{1/2}(u_i). \end{aligned}$$

From the identity $ab = \frac{1}{4}\left((a+b)^2 - (a-b)^2\right)$, we infer

$$\begin{aligned} \int_{\Omega_i} \eta(\mathbf{x})|u_i|^2 + \kappa(\mathbf{x})|\nabla u_i|^2 + \sum_{j\neq i}\frac{1}{4}\int_{\Gamma_{ij}} \left(\Lambda_{ij}^{-1/2}\left(\kappa(\mathbf{x})\frac{\partial u_i}{\partial \mathbf{n}_i}\right) - \Lambda_{ij}^{1/2}(u_i)\right)^2 \\ = \sum_{j\neq i}\frac{1}{4}\int_{\Gamma_{ij}} \left(\Lambda_{ij}^{-1/2}\left(\kappa(\mathbf{x})\frac{\partial u_i}{\partial \mathbf{n}_i}\right) + \Lambda_{ij}^{1/2}(u_i)\right)^2 \end{aligned}$$

which leads to the conclusion. □

We are ready to prove the following convergence result.

Theorem 2.2. *The algorithm* (2.10) *converges in H^1, that is,*

$$\lim_{n\to\infty} ||u_i^n - u_{|\Omega_i}||_{H^1(\Omega_i)} = 0 \textit{ for } i = 1,\ldots,N.$$

Proof. In order to prove the convergence of the algorithm we prove that $e_i^n = u_i^n - u_{\Omega_i}$ converges to zero. By linearity of the equations and of the algorithm, it is clear that the error e_i^{n+1} satisfies

$$\begin{aligned} \eta(\mathbf{x})e_i^{n+1} - \nabla\cdot(\kappa(\mathbf{x})\nabla e_i^{n+1}) &= 0 && \text{in} \quad \Omega_i,\\ e_i^{n+1} &= 0 && \text{on} \quad \partial\Omega\cap\partial\Omega_i,\\ \Lambda_{ij}^{-1/2}\left(\kappa(\mathbf{x})\frac{\partial e_i^{n+1}}{\partial \mathbf{n}_i}\right) + \Lambda_{ij}^{1/2}(e_i^{n+1}) &= -\Lambda_{ij}^{-1/2}\left(\kappa(\mathbf{x})\frac{\partial e_j^n}{\partial \mathbf{n}_j}\right) + \Lambda_{ij}^{1/2}(e_j^n) && \text{on} \quad \Gamma_{ij}. \end{aligned}$$

We apply the energy estimate to e_i^{n+1}, and taking into account the interface condition (2.11) and noticing that by assumption we have $\Lambda_{ij} = \Lambda_{ji}$, we get

$$\int_{\Omega_i} \eta(\mathbf{x})|e_i^{n+1}|^2 + \kappa(\mathbf{x})|\nabla e_i^{n+1}|^2 = \sum_{j\neq i} \frac{1}{4} \int_{\Gamma_{ij}} \left(\Lambda_{ji}^{-1/2}\left(-\kappa(\mathbf{x})\frac{\partial e_j^n}{\partial \mathbf{n}_j}\right) + \Lambda_{ji}^{1/2}(e_j^n)\right)^2 - \left(\Lambda_{ij}^{-1/2}\left(\kappa(\mathbf{x})\frac{\partial e_i^{n+1}}{\partial \mathbf{n}_i}\right) - \Lambda_{ij}^{1/2}(e_i^{n+1})\right)^2. \tag{2.12}$$

We introduce the following notation:

$$\mathscr{E}_i^{n+1} := \int_{\Omega_i} \eta(\mathbf{x})|e_i^{n+1}|^2 + \kappa(\mathbf{x})|\nabla e_i^{n+1}|^2,$$

$$\mathscr{C}_{ij}^{n+1} := \frac{1}{4}\int_{\Gamma_{ij}} \left(\Lambda_{ij}^{-1/2}\left(\kappa(\mathbf{x})\frac{\partial e_i^{n+1}}{\partial \mathbf{n}_i}\right) - \Lambda_{ij}^{1/2}(e_i^{n+1})\right)^2.$$

With this notation, estimate (2.12) can be rewritten as

$$\mathscr{E}_i^{n+1} + \sum_{j\neq i} \mathscr{C}_{ij}^{n+1} = \sum_{j\neq i} \mathscr{C}_{ji}^n.$$

After summation over the subdomains, we have

$$\sum_{i=1}^N \mathscr{E}_i^{n+1} + \sum_{i,j(j\neq i)} \mathscr{C}_{ij}^{n+1} = \sum_{i,j(j\neq i)} \mathscr{C}_{ji}^n = \sum_{i,j(j\neq i)} \mathscr{C}_{ij}^n.$$

Let us denote

$$\mathscr{E}^{n+1} := \sum_{i=1}^N \mathscr{E}_i^{n+1} \text{ and } \mathscr{C}^n := \sum_{i,j(j\neq i)} \mathscr{C}_{ji}^n.$$

We have

$$\mathscr{E}^{n+1} + \mathscr{C}^{n+1} = \mathscr{C}^n.$$

Hence, by summation over n, we get

$$\sum_{n=0}^{\infty} \mathscr{E}^{n+1} \leq \mathscr{C}^0.$$

This series is convergent only if $\mathscr{E}^n \to 0$ as $n \to \infty$, which proves that for all subdomains $1 \leq i \leq N$ we necessarily have $e_i^n \to 0$ in the H^1 norm as n tends to infinity. □

The same kind of proof holds for the Maxwell system [46] and for the convection-diffusion equation [145].

2.2 ▪ Helmholtz problems

So far, we have considered domain decomposition methods for Poisson or elasticity problems which yield symmetric positive definite matrices. With a suitable coarse space, Schwarz or BNN/FETI-type methods are very powerful methods whose scalability can be proven. But the numerical simulation of propagation phenomena via Helmholtz equations, Maxwell systems, and elasticity systems in the frequency domain yield matrices which are symmetric but not positive. The extension of P. L. Lions' algorithm to the Helmholtz problem in [44, section 2.2.2] is the first iterative method with proven convergence for this indefinite operator.

2.2.1 ▪ Convergence issues of the Schwarz method for the Helmholtz equation

The Helmholtz equation is derived from the acoustic wave equation

$$\frac{1}{c^2}\frac{\partial^2 u}{\partial t^2} - \Delta u = f(t,x,y),$$

which models, for instance, pressure variation with a source term f and a sound velocity c. When the source is time-periodic, it makes sense to look for time-periodic solutions as well. With some abuse of notation, let $f = f(x,y)\exp(i\omega t)$ $(i^2 = -1)$ be a harmonic source term of frequency ω. We seek the solution to the acoustic wave equation in the form $u = u(x,y)\exp(i\omega t)$. Then, u must be a solution to the Helmholtz equation

$$\mathscr{L}(u) := \left(-\frac{\omega^2}{c^2} - \Delta\right)(u) = f(x,y), \quad x,y \in \Omega,$$

whose solution is in general complex valued.

An extra difficulty comes from the nonpositivity of the Helmholtz operator due to the negative sign of the term of order zero. More precisely, a Schwarz algorithm for solving the Helmholtz equation involves the decomposition of domain Ω into N overlapping subdomains $(\Omega_j)_{1\le j\le N}$, solving the Dirichlet problems in the subdomains

$$\begin{array}{rll} (-\frac{\omega^2}{c^2} - \Delta)(u_j^{n+1}) = f(x,y) & \text{in} & \Omega_j,\ 1 \le j \le N, \\ u_j^{n+1} = u^n & \text{on} & \partial\Omega_j \end{array} \tag{2.13}$$

and then using a partition of unity $(\xi_j)_{1\le j\le N}$ related to $(\Omega_j)_{1\le j\le N}$:

$$u^{n+1} := \sum_{j=1}^{N} \xi_j\, u_j^{n+1}.$$

The problem is that this algorithm is not necessarily well-posed. Indeed, if ω^2/c^2 is an eigenvalue of the Poisson operator in a subdomain Ω_k, there exists a nonzero function $v_k : \Omega_k \to \mathbb{R}$ such that

$$\begin{array}{rll} -\Delta v_k = \frac{\omega^2}{c^2} v_k & \text{in} & \Omega_k, \\ v_k = 0 & \text{on} & \partial\Omega_k. \end{array}$$

Then, problem (2.13) is ill-posed in subdomain Ω_k. Either there is no solution, or if there is one, it is possible to add to it any function proportional to the eigenvalue v_k and the result still satisfies (2.13). We have here what is known as a Fredholm alternative.

Moreover, even when the local problems are well-posed, a bad convergence is to be expected. We present the analysis in the case of the plane divided into two subdomains, although it is very similar to the elliptic case considered in section 1.5.2. We introduce the wave number

$$\tilde{\omega} := \frac{\omega}{c}.$$

Consider that the domain is $\Omega = \mathbb{R}^2$ with the Sommerfeld radiation condition at infinity,

$$\lim_{r\to\infty} \sqrt{r}\left(\frac{\partial u}{\partial r} + i\tilde{\omega} u\right) = 0,$$

where $r = \sqrt{x^2+y^2}$. We decompose it into two subdomains with or without overlap, $\delta \geq 0$, $\Omega_1 = (-\infty, \delta) \times \mathbb{R}$ and $\Omega_2 = (0, \infty) \times \mathbb{R}$, and consider the Schwarz algorithm

$$\begin{aligned} -\Delta u_1^{n+1} - \tilde{\omega}^2 u_1^{n+1} &= f(x,y), \quad x,y \in \Omega_1, \\ u_1^{n+1}(\delta, y) &= u_2^n(\delta, y) \quad \forall y \in \mathbb{R}, \\ \lim_{r\to\infty} \sqrt{r}\left(\frac{\partial u_1^{n+1}}{\partial r} + i\tilde{\omega} u_1^{n+1}\right) &= 0 \end{aligned} \tag{2.14}$$

and

$$\begin{aligned} -\Delta u_2^{n+1} - \tilde{\omega}^2 u_2^{n+1} &= f(x,y), \quad x,y \in \Omega_2, \\ u_2^{n+1}(0, y) &= u_1^n(0, y) \quad \forall y \in \mathbb{R}, \\ \lim_{r\to\infty} \sqrt{r}\left(\frac{\partial u_2^{n+1}}{\partial r} + i\tilde{\omega} u_2^{n+1}\right) &= 0. \end{aligned} \tag{2.15}$$

For the convergence analysis it suffices by linearity to consider the case $f(x,y) = 0$ and to analyze convergence to the zero solution. Let the Fourier transform in the y-direction be denoted by

$$\hat{u}(x,k) := (\mathscr{F}u)(x,k) = \int_{\mathbb{R}} u(x,y)e^{-iky}\,dy.$$

Taking a Fourier transform of algorithms (2.14) and (2.15) in the y-direction we obtain

$$\left.\begin{aligned} -\frac{\partial^2 \hat{u}_1^{n+1}}{\partial x^2} - (\tilde{\omega}^2 - k^2)\hat{u}_1^{n+1} &= 0, \\ \hat{u}_1^{n+1}(\delta, k) &= \hat{u}_2^n(\delta, k), \end{aligned}\right. \quad x < \delta,\ k \in \mathbb{R}, \tag{2.16}$$

and

$$\left.\begin{aligned} -\frac{\partial^2 \hat{u}_2^{n+1}}{\partial x^2} - (\tilde{\omega}^2 - k^2)\hat{u}_2^{n+1} &= 0, \\ \hat{u}_2^{n+1}(0, k) &= \hat{u}_1^n(0, k), \end{aligned}\right. \quad x > 0,\ k \in \mathbb{R}. \tag{2.17}$$

The general solutions of these ODEs are

$$\hat{u}_j^{n+1} = A_j e^{\lambda(k)x} + B_j e^{-\lambda(k)x}, \quad j = 1, 2,$$

where $\lambda(k)$ denotes the root of the characteristic equation $\lambda^2 + (\omega^2 - k^2) = 0$ with positive real or imaginary part,

$$\lambda(k) = \begin{cases} \sqrt{k^2 - \tilde{\omega}^2} & \text{for } |k| \geq \tilde{\omega}, \\ i\sqrt{\tilde{\omega}^2 - k^2} & \text{for } |k| < \tilde{\omega}. \end{cases} \tag{2.18}$$

Note that the main difference with the elliptic case is that $\lambda(k)$ is a complex valued function. It takes real values for the evanescent modes $|k| \geq \tilde{\omega}$ and purely imaginary values for propagative modes $|k| < \tilde{\omega}$. Since the Sommerfeld radiation condition excludes growing solutions as well as incoming modes at infinity, we obtain the local solutions

$$\begin{aligned} \hat{u}_1^{n+1}(x,k) &= \hat{u}_1^{n+1}(\delta, k)e^{\lambda(k)(x-\delta)}, \\ \hat{u}_2^{n+1}(x,k) &= \hat{u}_2^{n+1}(0, k)e^{-\lambda(k)x}. \end{aligned} \tag{2.19}$$

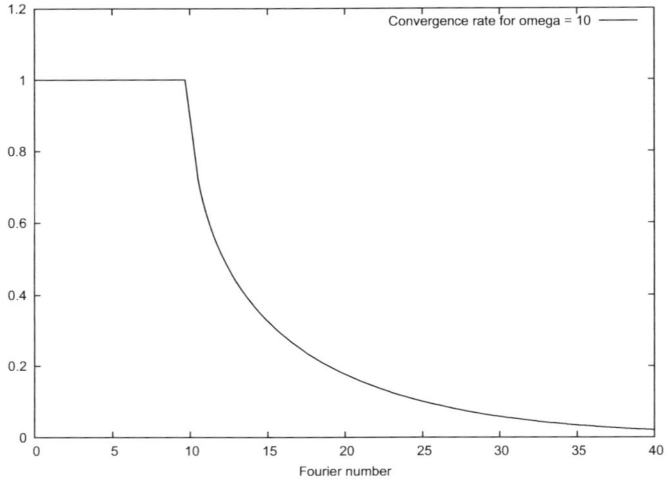

Figure 2.3. *Convergence factor (2.21) of the Schwarz method for the Helmholtz equation vs. the Fourier number.*

Then, it is easy to check that

$$\hat{u}_1^{n+1}(\delta,k) = \exp(-2\lambda(k)\,\delta)\,\hat{u}_1^{n-1}(\delta,k).$$

Defining the convergence factor ρ by

$$\rho(k,\delta) := \left|\frac{\hat{u}_1^{n+1}(\delta,k)}{\hat{u}_1^{n-1}(\delta,k)}\right|^{1/2} \tag{2.20}$$

we find by plugging (2.19) into (2.14) and (2.15) that

$$\boxed{\rho(k,\delta) = |\exp(-\lambda(k)\,\delta)|,} \tag{2.21}$$

where $\lambda(k)$ is given by (2.18), see Figure 2.3. By induction, the following result follows for even n:

$$|\hat{u}_1^n(0,k)| = |\rho(k,\delta)^n\,\hat{u}_1^0(0,k)|,\ |\hat{u}_2^n(0,k)| = |\rho(k,\delta)^n\,\hat{u}_2^0(0,k)|.$$

Remark 2.2.1. *We can distinguish a few features from the expression of the convergence factor; see Figure 2.3.*

- *The main difference with the elliptic case is that $\rho(k,\delta)$ is a complex valued function. The convergence will occur if the modulus of ρ is smaller than 1, $|\rho(k,\delta)| < 1$.*
- *For evanescent modes $|k| > \tilde{\omega}$, $\lambda(k)$ is negative and real, so these modes converge faster for larger overlaps. But we have $|\rho(k,\delta)| \equiv 1$ for propagative modes $|k| < \tilde{\omega}$ whatever the overlap is since there $\lambda(k) \in i\mathbb{R}$. This prevents the convergence of the Schwarz algorithm for the Helmholtz equation.*
- *A possible fix is the use of a relaxation parameter θ:*

$$u_i^{n+1} := \theta u_i^{n+1} + (1-\theta)u_i^n\,.$$

The choice of the optimal parameter θ is not easy and the overall convergence rate is not good in any case.

It is thus clear that the Schwarz method cannot be used safely or efficiently as a solver for the Helmholtz equation.

2.2.2 ▪ Després' algorithm for the Helmholtz equation

The algorithm of P. L. Lions can be applied to the Helmholtz equation with the choice $\alpha = i\tilde{\omega}$ ($i^2 = -1$). We give here a slight modification of the original algorithm given in [43]. Domain Ω is decomposed into N overlapping subdomains $(\Omega_j)_{1\le j\le N}$, and at each iteration a subdomain solve reads as

$$\begin{aligned} (-\tilde{\omega}^2 - \Delta)(u_j^{n+1}) &= f(x,y) && \text{in} \quad \Omega_j,\ 1\le j\le N, \\ \left(\frac{\partial}{\partial \mathbf{n}_j} + i\tilde{\omega}\right)(u_j^{n+1}) &= \left(\frac{\partial}{\partial \mathbf{n}_j} + i\tilde{\omega}\right)(u^n) && \text{on} \quad \partial\Omega_j \end{aligned} \tag{2.22}$$

and then, using a partition of unity $(\xi_j)_{1\le j\le N}$ related to $(\Omega_j)_{1\le j\le N}$,

$$u^{n+1} := \sum_{j=1}^{N} \xi_j\, u_j^{n+1}.$$

This algorithm fixes the two drawbacks of the Schwarz algorithm explained in section 2.2.1. First, note that the solution to the boundary value problem (2.22) is unique even if $\tilde{\omega}^2$ is an eigenvalue of the Laplace operator with Dirichlet boundary conditions. Indeed, suppose for some subdomain Ω_k ($1 \le k \le N$) we have a function $v : \Omega_k \to \mathbb{C}$ that satisfies

$$\begin{aligned} -\tilde{\omega}^2 v - \Delta v &= 0 && \text{in} \quad \Omega_k, \\ \left(\frac{\partial}{\partial \mathbf{n}_k} + i\tilde{\omega}\right)(v) &= 0 && \text{on} \quad \partial\Omega_k. \end{aligned}$$

Multiplying the equation by the conjugate of v and integrating by parts, we get

$$\int_{\Omega_k} -\tilde{\omega}^2|v|^2 + |\nabla v|^2 - \int_{\partial\Omega_k} \frac{\partial v}{\partial \mathbf{n}_k}\,\bar{v} = 0.$$

Taking into account the Robin condition on $\partial\Omega_k$, we have

$$\int_{\Omega_k} -\tilde{\omega}^2|v|^2 + |\nabla v|^2 + \int_{\partial\Omega_k} i\,\tilde{\omega}|v|^2 = 0.$$

Taking the imaginary part of the above equality, we get

$$\int_{\partial\Omega_k} \tilde{\omega}|v|^2 = 0,$$

so that v is zero on $\partial\Omega_k$. Using again the Robin condition, we have that $\partial v/\partial \mathbf{n}_k = 0$ as well on $\partial\Omega_k$. Together with the fact that $-\tilde{\omega}^2 v - \Delta v = 0$ in Ω_k, a unique continuation principle proves that v is identically zero in Ω. Thus the Després algorithm is always well defined.

The second point is that convergence of an algorithm was proved in [45] for a nonoverlapping decomposition of domain Ω. It is worth noting that it was the first time an iterative solver with proven convergence was proposed for the Helmholtz and subsequently

Maxwell equations [46]. For overlapping subdomains, no general convergence proof is available. But, in the case of the plane $\mathbb{R}^2$ decomposed into two overlapping half-planes, the convergence factor can be computed and a convergence result follows. The algorithm studied here can be written as

$$\begin{aligned} -\Delta u_1^{n+1} - \tilde{\omega}^2 u_1^{n+1} &= f(x,y), \quad x,y \in \Omega_1, \\ \left(\frac{\partial}{\partial \mathbf{n}_1} + i\tilde{\omega}\right)(u_1^{n+1})(\delta,y) &= \left(\frac{\partial}{\partial \mathbf{n}_1} + i\tilde{\omega}\right) u_2^n(\delta,y) \quad \forall y \in \mathbb{R}, \\ \lim_{r\to\infty} \sqrt{r}\left(\frac{\partial u_1^{n+1}}{\partial r} + i\tilde{\omega} u_1^{n+1}\right) &= 0 \end{aligned} \tag{2.23}$$

and

$$\begin{aligned} -\Delta u_2^{n+1} - \tilde{\omega}^2 u_2^{n+1} &= f(x,y), \quad x,y \in \Omega_2, \\ \left(\frac{\partial}{\partial \mathbf{n}_2} + i\tilde{\omega}\right) u_2^{n+1}(0,y) &= \left(\frac{\partial}{\partial \mathbf{n}_2} + i\tilde{\omega}\right) u_1^n(0,y) \quad \forall y \in \mathbb{R}, \\ \lim_{r\to\infty} \sqrt{r}\left(\frac{\partial u_2^{n+1}}{\partial r} + i\tilde{\omega} u_2^{n+1}\right) &= 0. \end{aligned} \tag{2.24}$$

Performing a convergence analysis as in section 2.1, the convergence factor is

$$\boxed{\rho(k,\delta) = \left|\frac{\lambda(k) - i\tilde{\omega}}{\lambda(k) + i\tilde{\omega}}\right| \left|\exp(-\lambda(k)\,\delta)\right|,} \tag{2.25}$$

where $\lambda(k)$ is given by formula (2.18). Note that we have

$$\rho(k,\delta) = \begin{cases} \left|\dfrac{\sqrt{\tilde{\omega}^2 - k^2} - \tilde{\omega}}{\sqrt{\tilde{\omega}^2 - k^2} + \tilde{\omega}}\right| & \text{for } |k| < |\tilde{\omega}|, \\ \exp(-\sqrt{k^2 - \tilde{\omega}^2}\,\delta) \ \text{for } |k| < |\tilde{\omega}|; \end{cases}$$

see Figure 2.4. Note that for propagative modes $|k| < |\tilde{\omega}|$, $|\exp(-\lambda(k)\delta)| = 1$ since $\lambda(k)$ is purely imaginary. Therefore the convergence comes from the Robin condition independently of the overlap. As for vanishing modes $|k| > |\tilde{\omega}|$, λ is real. The modulus of the rational fraction in (2.25) is 1, and convergence comes from the overlap. Thus for all Fourier modes except $|k| = \tilde{\omega}$, the convergence factor is smaller than 1. Thus, the method converges for almost every Fourier number k.

2.3 ▪ Implementation issues

A Robin boundary value problem is different from standard Dirichlet or Neumann problems but poses no difficulty as long as the right-hand side of the Robin condition is known. But as we shall see, the discretization of the right-hand side (2.26) on an interface is not a trivial task. Fortunately, it may be eased by tricks explained in sections 2.3.1 and 2.3.2.

Consider, for example, problem (2.1) in Lions' algorithm, and let g_2^n denote the right-hand side on the interface of subdomain Ω_1:

$$g_2^n := \left(\frac{\partial}{\partial \mathbf{n}_1} + \alpha\right)(u_2^n) \text{ on } \partial\Omega_1 \cap \bar{\Omega}_2. \tag{2.26}$$

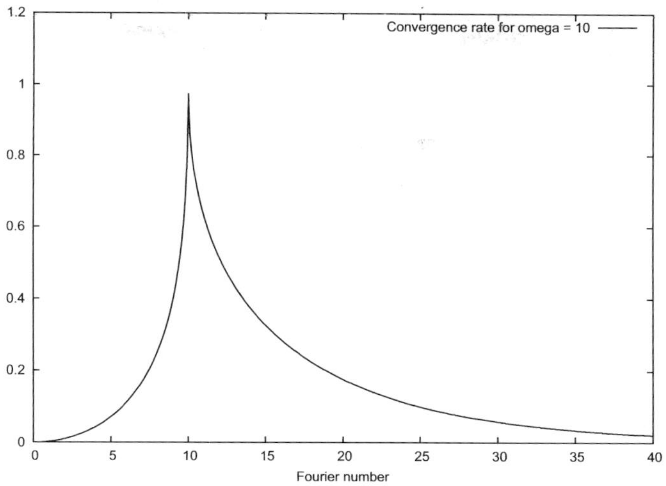

Figure 2.4. *Convergence factor* (2.25) *of Després' method for the Helmholtz equation vs. the Fourier number* ($\delta = 0.1$).

Then, the bilinear form associated to the variational formulation of (2.1) is

$$\begin{aligned} a_{Robin} : H^1(\Omega_1) \times H^1(\Omega_1) &\longrightarrow \mathbb{R}, \\ (u,v) &\longmapsto \int_{\Omega_1} (\eta u v + \nabla u \cdot \nabla v) + \int_{\partial\Omega_1 \cap \overline{\Omega_2}} \alpha u\, v\,. \end{aligned} \tag{2.27}$$

Note that the boundary integral term containing α is positive when $v = u$. Thus the bilinear form a_{Robin} is even more coercive than the one associated with the Neumann boundary value problem which contains only the subdomain integral term. The linear form related to (2.1) reads as

$$\begin{aligned} l_{Robin} : H^1(\Omega_1) &\longrightarrow \mathbb{R}, \\ v &\longmapsto \int_{\Omega_1} f\, v + \int_{\partial\Omega_1 \cap \overline{\Omega_2}} g_2^n\, v\,. \end{aligned} \tag{2.28}$$

The variational formulation of the Robin boundary value problem reads as follows:

$$\text{Find } u \in H^1(\Omega_1) \text{ such that}$$
$$\forall v \in H^1(\Omega_1),\ a_{Robin}(u,v) = l_{Robin}(v)\,.$$

The finite element discretization is simply obtained by replacing the infinite dimensional Sobolev space $H^1(\Omega_1)$ by a finite dimensional space, for example a classical $P1$ finite element space denoted by $V_h(\Omega_1)$:

$$\text{Find } u \in V_h(\Omega_1) \text{ such that}$$
$$\forall v_h \in V_h(\Omega_1),\ a_{Robin}(u_h,v_h) = l_{Robin}(v_h)\,.$$

From an algebraic point of view, we consider a finite element basis

$$\{\phi_k | k \in \mathcal{N}_1\} \subset V_h(\Omega_1),$$

where $\mathcal{N}_1$ is the set of degrees of freedom. The matrix form of $a_{Robin,h}$ is denoted by $A_{Robin,1}$:

$$[A_{Robin,1}]_{kl} := a_{Robin,h}(\phi_l,\phi_k) = \int_{\Omega_1} (\eta \phi_l \phi_k + \nabla\phi_l \cdot \nabla\phi_k) + \int_{\partial\Omega_1 \cap \overline{\Omega_2}} \alpha \phi_l\, \phi_k\,. \tag{2.29}$$

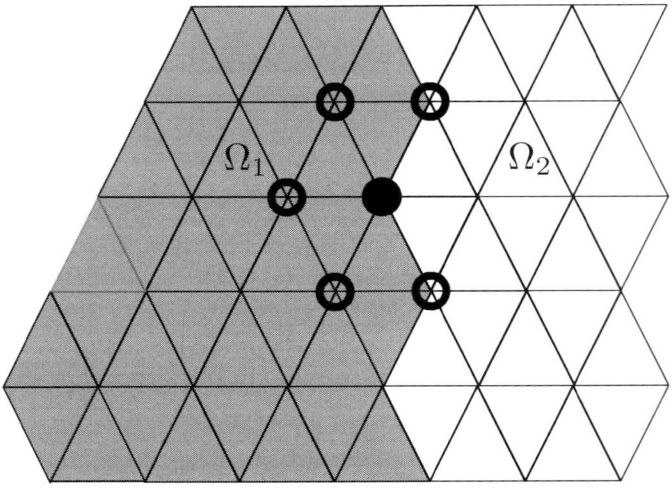

Figure 2.5. *Stencil for the outward normal derivative ($\partial u_1/\partial \mathbf{n}_1$) at the interface between two nonoverlapping subdomains Ω_1 and Ω_2.*

Matrix $A_{Robin,1}$ is actually the sum of the mass matrix, a stiffness matrix which corresponds to a Neumann problem and of a matrix that has zero entries for interior degrees of freedom ($\overset{\circ}{\text{Supp}}\, \phi_k \cap (\partial\Omega_1 \cap \overline{\Omega_2}) = \emptyset$); see also (2.39).

Note that the same technique can be applied in the case of the Després' algorithm for the Helmholtz equation with minimal adjustments due to the form of the problem. We simply replace η by $-\tilde{\omega}^2$ and multiply by the complex conjugate of the test function in the variational form.

The only difficulty lies now in the discretization of the right-hand side g_2^n on the interface. Suppose we have a $P1$ (piecewise linear) finite element so that we can identify a degree of freedom with a vertex for a triangular mesh in two dimensions. The variational form a_{Robin} implicitly defines a discretization scheme for the outward normal derivative $\partial u_1^{n+1}/\partial \mathbf{n}_1$. The corresponding stencil can only involve vertices that belong to domain $\bar{\Omega}_1$; see Figure 2.5. In order to ensure that the domain decomposition algorithm leads to a solution which is identical (up to the tolerance of the iterative solver) to that which would be obtained by the original discretization scheme on the whole domain, it seems necessary to discretize $g_2^n (= \partial u_2^n/\partial \mathbf{n}_1)$ using the same stencil. However, the function u_2^n is defined by the degrees of freedom that are in $\bar{\Omega}_2$ and thus cannot be those defining $\partial u_1^{n+1}/\partial \mathbf{n}_1$. In the overlapping case, a discretization of the right-hand side based on the same stencil points could be done but at the expense of identifying the stencil implicitly defined by the variational formulation. This is not so easy to implement in code.

In the next two sections, we give two tricks that ease the implementation so that the converged solution is equal to that which would be obtained by the original discretization scheme on the whole domain. One trick relates to a nonoverlapping decomposition and the other to the overlapping case only.

2.3.1 ▪ Two-domain nonoverlapping decomposition

This section is concerned with the finite element implementation of the Robin-type interface conditions for a nonoverlapping decomposition of the domain. We present the discretization scheme for a decomposition of a domain Ω into two nonoverlapping subdomains Ω_1 and Ω_2 with interface Γ_{12}. We first consider the optimized Schwarz algorithm at the continuous level,

$$\begin{aligned}
(\eta-\Delta)u_1^{n+1} &= f && \text{in } \Omega_1,\\
\frac{\partial u_1^{n+1}}{\partial \mathbf{n}_1}+\alpha u_1^{n+1} &= -\frac{\partial u_2^n}{\partial \mathbf{n}_2}+\alpha u_2^n && \text{on } \Gamma_{12},\\
(\eta-\Delta)u_2^{n+1} &= f && \text{in } \Omega_2,\\
\frac{\partial u_2^{n+1}}{\partial \mathbf{n}_2}+\alpha u_2^{n+1} &= -\frac{\partial u_1^n}{\partial \mathbf{n}_1}+\alpha u_1^n && \text{on } \Gamma_{12},
\end{aligned} \tag{2.30}$$

where we have used that on the interface between nonoverlapping subdomains, we have $\mathbf{n}_1 = -\mathbf{n}_2$; see Figure 2.1. A direct discretization would require the computation of the normal derivatives along the interfaces in order to evaluate the right-hand sides in the transmission conditions of (2.30). This can be avoided by renaming the problematic quantities:

$$\lambda_1^n = -\frac{\partial u_2^n}{\partial \mathbf{n}_2}+\alpha\, u_2^n \qquad \text{and} \qquad \lambda_2^n = -\frac{\partial u_1^n}{\partial \mathbf{n}_1}+\alpha\, u_1^n .$$

Then algorithm (2.30) becomes

$$\begin{aligned}
(\eta-\Delta)u_1^{n+1} &= f \quad \text{in } \Omega_1,\\
\frac{\partial u_1^{n+1}}{\partial \mathbf{n}_1}+\alpha\, u_1^{n+1} &= \lambda_1^n \quad \text{on } \Gamma_{12},\\
(\eta-\Delta)u_2^{n+1} &= f \quad \text{in } \Omega_2,\\
\frac{\partial u_2^{n+1}}{\partial \mathbf{n}_2}+\alpha\, u_2^{n+1} &= \lambda_2^n \quad \text{on } \Gamma_{12},\\
\lambda_1^{n+1} &= -\lambda_2^n+2\alpha\, u_2^{n+1},\\
\lambda_2^{n+1} &= -\lambda_1^n+2\alpha\, u_1^{n+1}.
\end{aligned} \tag{2.31}$$

The last two update relations of (2.31) follow from

$$\lambda_1^{n+1} := -\frac{\partial u_2^{n+1}}{\partial \mathbf{n}_2}+\alpha\, u_2^{n+1} = -\left(\frac{\partial u_2^{n+1}}{\partial \mathbf{n}_2}+\alpha\, u_2^{n+1}\right)+2\alpha\, u_2^{n+1} = -\lambda_2^n+2\alpha\, u_2^{n+1}, \tag{2.32}$$

and similarly

$$\lambda_2^{n+1} := -\frac{\partial u_1^{n+1}}{\partial \mathbf{n}_1}+\alpha\, u_1^{n+1} = -\left(\frac{\partial u_1^{n+1}}{\partial \mathbf{n}_1}+\alpha\, u_1^{n+1}\right)+2\alpha\, u_1^{n+1} = -\lambda_1^n+2\alpha\, u_1^{n+1}. \tag{2.33}$$

Equations (2.32) and (2.33) can be interpreted as a fixed-point algorithm in the new variables λ_j, $j = 1,2$, to solve the substructured problem

$$\begin{aligned}
\lambda_1 &= -\lambda_2+2\alpha\, u_2(\lambda_2, f),\\
\lambda_2 &= -\lambda_1+2\alpha\, u_1(\lambda_1, f),
\end{aligned} \tag{2.34}$$

where $u_j = u_j(\lambda_j, f)$, $j = 1,2$, are solutions of

$$(\eta - \Delta)u_j = f \text{ in } \Omega_j,$$
$$\frac{\partial u_j}{\partial \mathbf{n}_j} + \alpha u_j = \lambda_j \text{ on } \Gamma_{12}.$$

Instead of solving the substructured problem (2.34) by the fixed-point iteration (2.31), one usually uses a Krylov subspace method to solve the substructured problem. This corresponds to using the optimized Schwarz method as a preconditioner for the Krylov subspace method.

At this point, we can introduce the algebraic counterparts of the continuous quantities. A finite element discretization of the substructured problem (2.34) leads to the linear system

$$\begin{aligned} \lambda_1 &= -\lambda_2 + 2\alpha B_2 \boldsymbol{u}_2, \\ \lambda_2 &= -\lambda_1 + 2\alpha B_1 \boldsymbol{u}_1, \end{aligned} \tag{2.35}$$

where $\boldsymbol{u}_j = \boldsymbol{u}_j(\lambda_j, \boldsymbol{f}_j)$, $j = 1,2$, are solutions of

$$\begin{aligned} A_{Robin,1}\boldsymbol{u}_1 &= \boldsymbol{f}_1 + B_1^T \lambda_1, \\ A_{Robin,2}\boldsymbol{u}_2 &= \boldsymbol{f}_2 + B_2^T \lambda_2. \end{aligned} \tag{2.36}$$

We detail now the new matrices and vectors we have introduced:

- Vectors $\boldsymbol{u}_1$, $\boldsymbol{u}_2$ contain the degrees of freedom of the subdomain solutions.
- Vectors $\boldsymbol{f}_1$, $\boldsymbol{f}_2$ are the degrees of freedom related to f.
- Matrices B_1 and B_2 are the trace operators of the domains Ω_1 and Ω_2 on the interface Γ_{12}.

In order to define matrices $A_{Robin,j}$, we first split the two vectors $\mathbf{u}_1$ and $\mathbf{u}_2$ into interior and boundary degrees of freedom:

$$\boldsymbol{u}_j = \begin{bmatrix} \boldsymbol{u}_j^i \\ \boldsymbol{u}_j^b \end{bmatrix}, \quad j = 1,2, \tag{2.37}$$

where the indices i and b correspond to interior and interface degrees of freedom, respectively, for domain Ω_j. Then the discrete trace operators B_1 and B_2 are just the Boolean matrices corresponding to the decomposition (2.37) and they can be written as

$$B_j = \begin{bmatrix} 0 & I \end{bmatrix}, \quad j = 1,2, \tag{2.38}$$

where I denotes the identity matrix of appropriate size. For example,

$$B_1 \boldsymbol{u}_1 = \boldsymbol{u}_1^b \text{ and } B_2 \boldsymbol{u}_2 = \boldsymbol{u}_2^b.$$

Matrices $A_{Robin,1}$ and $A_{Robin,2}$ arise from the discretization of the local $\eta - \Delta$ operators along with the interface conditions $\partial_n + \alpha$,

$$A_{Robin,j} = K_j + B_j^T \alpha M_{\Gamma_{12}} B_j, \quad j = 1,2. \tag{2.39}$$

Here K_j are local matrices of the problem (as a combination of stiffness and mass matrices) and $M_{\Gamma_{12}}$ is the interface mass matrix

$$[M_{\Gamma_{12}}]_{nm} = \int_{\Gamma_{12}} \phi_n \phi_m \, d\xi \,. \tag{2.40}$$

The functions ϕ_n and ϕ_m are the basis functions associated with the degrees of freedom n and m on the interface Γ_{12}.

For given λ_1 and λ_2, the functions $\boldsymbol{u}_1$ and $\boldsymbol{u}_2$ can be computed by solving equations (2.36). By eliminating $\boldsymbol{u}_1$ and $\boldsymbol{u}_2$ in (2.35) using (2.36), we obtain the substructured linear system

$$F\lambda = \boldsymbol{d}, \tag{2.41}$$

where $\lambda = (\lambda_1, \lambda_2)$ and the matrix F and the right-hand side $\boldsymbol{d}$ are

$$F = \begin{pmatrix} I & I - 2\alpha B_2 A^{-1}_{Robin,2} B_2^T \\ I - 2\alpha B_1 A^{-1}_{Robin,1} B_1^T & I \end{pmatrix}, \qquad \boldsymbol{d} = \begin{pmatrix} 2\alpha B_1 A^{-1}_{Robin,1} f_1 \\ 2\alpha B_2 A^{-1}_{Robin,2} f_2 \end{pmatrix}. \tag{2.42}$$

The linear system (2.41) is solved by a Krylov subspace method; see Chapter 3. The matrix-vector product amounts to solving a subproblem in each subdomain and sending interface data between subdomains.

The general case of a decomposition into an arbitrary number of subdomains is treated in [91] for the case of the Helmholtz equations. It is also possible to discretize the interface conditions by using Lagrange multipliers. The method is then named FETI 2 LM (two-Lagrange multipliers); see [167].

2.3.2 ▪ Overlapping domain decomposition

The trick explained in the nonoverlapping case cannot be used in the overlapping case; see Figure 2.1. Indeed, even in the simple case of two subdomains, the normal derivatives are computed on two distinct interfaces, $\partial\Omega_1 \cap \bar{\Omega}_2$ and $\partial\Omega_2 \cap \bar{\Omega}_1$. Here also, it is possible to write an algorithm equivalent to P. L. Lions' algorithm (2.1)–(2.2), but where the iterate is a function $u^n : \Omega \to \mathbb{R}$, i.e., it is not a pair of functions (u_1^n, u_2^n).

Let an overlapping domain decomposition $\Omega = \Omega_1 \cup \Omega_2$, with partition of unity functions χ_1 and χ_2:

$$\chi_1 + \chi_2 = 1, \ \operatorname{supp}\chi_i \subset \Omega_i, \ i = 1,2.$$

Let $u^n : \Omega \to \mathbb{R}$ be an approximate solution to a Poisson equation. Then the update u^{n+1} is computed by Algorithm 2.1 below, which is the continuous version of the optimized restricted additive Schwarz (ORAS) algorithm; see [181]. A parallel can be drawn with section 1.3, where an equivalence between the original Schwarz algorithm and the RAS method was shown at the continuous level.

Lemma 2.3 (Equivalence between Lions' algorithm and ORAS). *The algorithm defined by* (2.43), (2.44), *and* (2.45) *is equivalent to the Lions' algorithm.*

Proof. The approach is very similar to that of Lemma 1.6, where we proved the equivalence between the Schwarz algorithm and the RAS method. Here, we have to prove the equality

$$u^n = E_1(\chi_1 u_1^n) + E_2(\chi_2 u_2^n),$$

Algorithm 2.1 ORAS algorithm at the continuous level

1. Compute the residual $r^n : \Omega \to \mathbb{R}$:
$$r^n := f - (\eta - \Delta) u^n. \tag{2.43}$$
2. For $i = 1,2$ solve for a local correction v_i^n:
$$\begin{aligned} (\eta - \Delta) v_i^n &= r^n && \text{in } \Omega_i, \\ v_i^n &= 0 && \text{on } \partial\Omega_i \cap \partial\Omega, \\ \left(\frac{\partial}{\partial \mathbf{n}_i} + \alpha\right)(v_i^n) &= 0 && \text{on } \partial\Omega_i \cap \bar{\Omega}_{3-i}. \end{aligned} \tag{2.44}$$
3. Compute an average of the local corrections and update u^n:
$$u^{n+1} = u^n + E_1(\chi_1 v_1^n) + E_2(\chi_2 v_2^n), \tag{2.45}$$
where $(\chi_i)_{i=1,2}$ and $(E_i)_{i=1,2}$ define a partition of unity as in defined in (1.4) of section 1.1.

---

where $u_{1,2}^n$ are given by (2.1)–(2.2) and u^n is given by (2.43)–(2.45). We assume that the property holds for the initial guesses
$$u^0 = E_1(\chi_1 u_1^0) + E_2(\chi_2 u_2^0)$$
and proceed by induction, assuming that the property holds at step n of the algorithm, i.e., $u^n = E_1(\chi_1 u_1^n) + E_2(\chi_2 u_2^n)$. From (2.45) we have
$$u^{n+1} = E_1(\chi_1(u^n + v_1^n)) + E_2(\chi_2(u^n + v_2^n)). \tag{2.46}$$
We prove now that $u^n_{|\Omega_1} + v_1^n = u_1^{n+1}$ by proving that $u^n_{|\Omega_1} + v_1^n$ satisfies (2.1) as u_1^{n+1} does. We first note that, using (2.44)–(2.43), we have
$$\begin{aligned} (\eta - \Delta)(u^n + v_1^n) &= (\eta - \Delta) u^n + r^n = (\eta - \Delta) u^n + f - (\eta - \Delta u^n) = f && \text{in } \Omega_1, \\ \left(\frac{\partial}{\partial \mathbf{n}_1} + \alpha\right)(u^n + v_1^n) &= \left(\frac{\partial}{\partial \mathbf{n}_1} + \alpha\right)(u^n) && \text{on } \partial\Omega_1 \cap \overline{\Omega_2}. \end{aligned} \tag{2.47}$$
It remains to prove that
$$\left(\frac{\partial}{\partial \mathbf{n}_1} + \alpha\right)(u^n) = \left(\frac{\partial}{\partial \mathbf{n}_1} + \alpha\right)(u_2^n) \text{ on } \partial\Omega_1 \cap \overline{\Omega_2}.$$
By the induction hypothesis we have $u^n = E_1(\chi_1 u_1^n) + E_2(\chi_2 u_2^n)$. On a neighborhood of $\partial\Omega_1 \cap \overline{\Omega_2}$ (zone (c) in Figure 2.6), we have $\chi_1 \equiv 0$ and thus $\chi_2 \equiv 1$. In this neighborhood, the derivatives are zero as well, so that on $\partial\Omega_1 \cap \overline{\Omega_2}$ we have
$$\left(\frac{\partial}{\partial \mathbf{n}_1} + \alpha\right)(u^n) = \left(\frac{\partial}{\partial \mathbf{n}_1} + \alpha\right)(\chi_1 u_1^n + \chi_2 u_2^n) = \left(\frac{\partial}{\partial \mathbf{n}_1} + \alpha\right)(u_2^n). \tag{2.48}$$

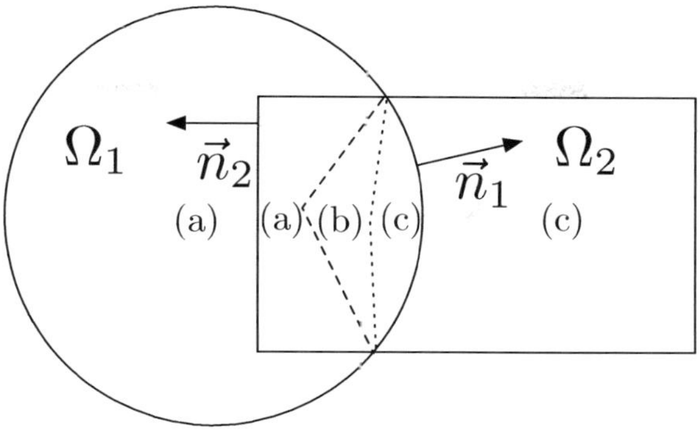

Figure 2.6. *Function of partition of unity* χ_1. *In zone* (a) $\chi_1 \equiv 1$, *in zone* (b) $0 < \chi_1 < 1$, *and in zone* (c) $\chi_1 \equiv 0$.

Finally, from (2.47) and (2.48) we can conclude that $u^n_{|\Omega_1} + v^n_1 = u^{n+1}_1$ satisfies problem (2.1) and is thus equal to u^{n+1}_1. The same holds for domain Ω_2, $u^n_{|\Omega_2} + v^n_2 = u^{n+1}_2$. Then equation (2.46) reads as

$$u^{n+1} = E_1(\chi_1 u_1^{n+1}) + E_2(\chi_2 u_2^{n+1}),$$

which ends the proof of the equivalence between P. L. Lions' algorithm and the continuous version of the ORAS algorithm. □

The advantage of the ORAS method over a direct implementation of P. L. Lions' algorithm is that the Robin boundary condition (2.44) has a right-hand side of zero that needs no discretization. The only drawback at the discretized level is that the actual overlap is reduced by a mesh size, compared to a full discretization of the original P. L. Lions algorithm. See [181] for more details.

We now give the algebraic definition of the ORAS method by giving the discrete counterparts of steps (2.43)–(2.45); see [181]. Let $\mathbf{U}^n \in \mathbb{R}^{\#\mathcal{N}}$ be an approximate solution to a linear system:

$$A\mathbf{U} = \mathbf{F}. \tag{2.49}$$

The set of degrees of freedom $\mathcal{N}$ is decomposed into two subsets $\mathcal{N}_1$ and $\mathcal{N}_2$. For $i = 1, 2$, let R_i denote the Boolean restriction matrix to $\mathcal{N}_i$, and let D_i define an algebraic partition of unity $\sum_{i=1}^2 R_i^T D_i R_i = \mathrm{Id}$, as in (1.25).

The above three-step algorithm can be written more compactly as follows.

Definition 2.4 (ORAS algorithm). *The iterative optimized restricted additive Schwarz (ORAS) algorithm is the preconditioned fixed-point iteration defined by*

$$\mathbf{U}^{n+1} = \mathbf{U}^n + M_{ORAS}^{-1}\mathbf{r}^n,\ \mathbf{r}^n := \mathbf{F} - A\mathbf{U}^n,$$

where the matrix

$$\boxed{M_{ORAS}^{-1} := \sum_{i=1}^{2} R_i^T D_i A_{i,Robin}^{-1} R_i} \tag{2.53}$$

is called the ORAS preconditioner.

The update $\mathbf{U}^{n+1}$ is computed in several steps by the following algorithm.

1. Compute the residual $\mathbf{r}^n \in \mathbb{R}^{\#\mathcal{N}}$:
$$\mathbf{r}^n := F - A\mathbf{U}^n. \tag{2.50}$$

2. For $i = 1,2$ solve for a local correction $\mathbf{V}_i^n$:
$$A_{i,Robin}\mathbf{V}_i^n = R_i\,\mathbf{r}^n, \tag{2.51}$$
where $A_{i,Robin}$ is the discretization matrix of a Robin problem as explained in (2.29) or in (2.39).

3. Compute an average of the local corrections and update $\mathbf{U}^n$ accordingly:
$$\mathbf{U}^{n+1} = \mathbf{U}^n + R_1^T D_1 \mathbf{V}_1^n + R_2^T D_2 \mathbf{V}_2^n. \tag{2.52}$$

In short, the only difference with the RAS method (1.29) consists in replacing the local Dirichlet matrices $R_i A R_i^T$ by matrices $A_{Robin,i}$ which correspond to local Robin subproblems.

As explained in Chapter 3, it is more profitable to use a Krylov method in place of the above iterative fixed-point algorithm. It consists in solving (2.49) by a Krylov method such as GMRES or BiCGSTAB preconditioned by operator M_{ORAS}^{-1}.

2.4 ▪ Optimal interface conditions

2.4.1 ▪ Optimal interface conditions and absorbing boundary conditions

Robin boundary conditions are not the most general interface conditions. Rather than give the general conditions in an a priori form, we shall derive them in this section so as to have the fastest convergence. We establish the existence of interface conditions which are optimal in terms of iteration counts. The corresponding interface conditions are pseudodifferential and are not practical. Nevertheless, this result is a guide for the choice of partial differential interface conditions. Moreover, this result establishes a link between the optimal interface conditions and artificial boundary conditions. This is also helpful when dealing with the design of interface conditions since it gives the possibility using the numerous papers and books published on the subject of artificial boundary conditions; see, e.g., [75, 99] or more generally [13, 30] on the truncation of infinite domains via the PML technique.

We consider a general linear second order elliptic partial differential operator $\mathcal{L}$ and the problem

$$\mathcal{L}(u) = f,\ \Omega,$$
$$u = 0,\ \partial\Omega.$$

The domain Ω is decomposed into two subdomains Ω_1 and Ω_2. We suppose that the problem is regular so that $u_i := u|_{\Omega_i}$, $i = 1,2$, is continuous and has continuous normal derivatives across the interface $\Gamma_i = \partial\Omega_i \cap \bar{\Omega}_j$, $i \neq j$. A modified Schwarz-type method is

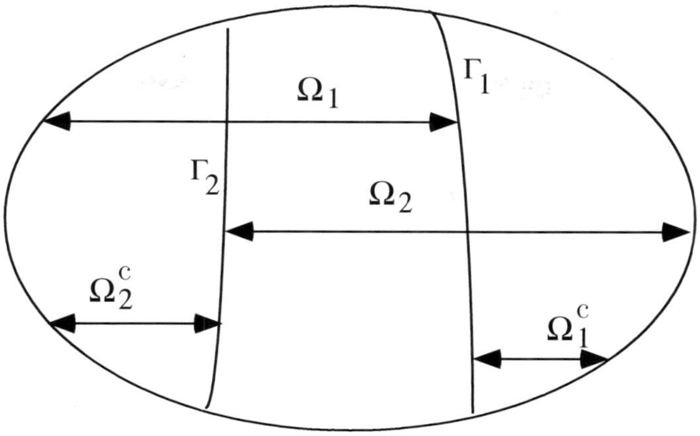

Figure 2.7. *Subdomains and their complements.*

considered,

$$\begin{aligned} \mathscr{L}(u_1^{n+1}) &= f \quad \text{in} \quad \Omega_1, \\ u_1^{n+1} &= 0 \quad \text{on} \quad \partial\Omega_1 \cap \partial\Omega, \\ \mu_1 \nabla u_1^{n+1}.\mathbf{n}_1 &+ \mathscr{B}_1(u_1^{n+1}) \\ = -\mu_1 \nabla u_2^n.\mathbf{n}_2 &+ \mathscr{B}_1(u_2^n) \quad \text{on} \quad \Gamma_1, \end{aligned} \qquad \begin{aligned} \mathscr{L}(u_2^{n+1}) &= f \quad \text{in} \quad \Omega_2, \\ u_2^{n+1} &= 0 \quad \text{on} \quad \partial\Omega_2 \cap \partial\Omega, \\ \mu_2 \nabla u_2^{n+1}.\mathbf{n}_2 &+ \mathscr{B}_2(u_2^{n+1}) \\ = -\mu_2 \nabla u_1^n.\mathbf{n}_1 &+ \mathscr{B}_2(u_1^n) \quad \text{on} \quad \Gamma_2, \end{aligned} \tag{2.54}$$

where μ_1 and μ_2 are real-valued functions and $\mathscr{B}_1$ and $\mathscr{B}_2$ are operators acting along the interfaces Γ_1 and Γ_2. For instance, $\mu_1 = \mu_2 = 0$ and $\mathscr{B}_1 = \mathscr{B}_2 = \text{Id}$ correspond to the original Schwarz algorithm (1.3); $\mu_1 = \mu_2 = 1$ and $\mathscr{B}_i = \alpha \in \mathbb{R}$, $i = 1,2$, were proposed in [130] by P. L. Lions.

The question we would like to answer is the following:

Are there other possibilities in order to have convergence in a minimal number of iterations?

In order to answer this question, we note that by linearity, the error e satisfies (we have supposed here with no loss of generality that $\mu_1 = \mu_2 = 1$)

$$\begin{aligned} \mathscr{L}(e_1^{n+1}) &= 0 \quad \text{in} \quad \Omega_1, \\ e_1^{n+1} &= 0 \quad \text{on} \quad \partial\Omega_1 \cap \partial\Omega, \\ \nabla e_1^{n+1}.\mathbf{n}_1 &+ \mathscr{B}_1(e_1^{n+1}) \\ = -\nabla e_2^n.\mathbf{n}_2 &+ \mathscr{B}_1(e_2^n) \quad \text{on} \quad \Gamma_1, \end{aligned} \qquad \begin{aligned} \mathscr{L}(e_2^{n+1}) &= 0 \quad \text{in} \quad \Omega_2, \\ e_2^{n+1} &= 0 \quad \text{on} \quad \partial\Omega_2 \cap \partial\Omega, \\ \nabla e_2^{n+1}.\mathbf{n}_2 &+ \mathscr{B}_2(e_2^{n+1}) \\ = -\nabla e_1^n.\mathbf{n}_1 &+ \mathscr{B}_2(e_1^n) \quad \text{on} \quad \Gamma_2. \end{aligned}$$

The initial guess e_i^0 is arbitrary so that it is impossible to have convergence at step 1 of the algorithm. Convergence needs at least two iterations. Having $e_1^2 \equiv 0$ requires

$$-\nabla e_2^1.\mathbf{n}_2 + \mathscr{B}_1(e_2^1) \equiv 0.$$

The only meaningful information on e_2^1 is that

$$\mathscr{L}(e_2^1) = 0 \quad \text{in} \quad \Omega_2.$$

In order to use this information, we introduce the Dirichlet-to-Neumann (DtN) map (a.k.a. Steklov–Poincaré):

$$\begin{aligned} u_0 &: \Gamma_1 \to \mathbb{R}, \\ \mathrm{DtN}_2(u_0) &:= \nabla v.\mathbf{n}_{2|\partial\Omega_1\cap\bar{\Omega}_2}, \end{aligned} \tag{2.55}$$

where $\mathbf{n}_2$ is the outward normal to $\Omega_2 \setminus \bar{\Omega}_1$, and v satisfies the following boundary value problem:

$$\begin{aligned} \mathscr{L}(v) &= 0 \quad \text{in} \quad \Omega_2 \setminus \bar{\Omega}_1, \\ v &= 0 \quad \text{on} \quad \partial\Omega_2 \cap \partial\Omega, \\ v &= u_0 \quad \text{on} \quad \partial\Omega_1 \cap \bar{\Omega}_2. \end{aligned}$$

If we take

$$\mathscr{B}_1 := \mathrm{DtN}_2,$$

we see that this choice is optimal since we have

$$-\nabla e_2^1.\mathbf{n}_2 + \mathscr{B}_1(e_2^1) \equiv 0.$$

Indeed, in $\Omega_2 \setminus \bar{\Omega}_1 \subset \Omega_2$, e_2^1 satisfies

$$\mathscr{L}(e_2^1) = 0.$$

Hence,

$$\begin{aligned} \nabla e_2^1.\mathbf{n}_2 &= \mathrm{DtN}_2(e_2^1), \\ \nabla e_2^1.\mathbf{n}_2 &= \mathscr{B}_1(e_2^1) \quad (\mathscr{B}_1 = \mathrm{DtN}_2). \end{aligned}$$

We have formally proved the following.

Result 2.4.1. *The use of $\mathscr{B}_i = \mathrm{DtN}_j$ $(i \neq j)$ as interface conditions in* (2.54) *is optimal: we have (exact) convergence in two iterations.*

The two-domain case for an operator with constant coefficients was first treated in [107]. The multidomain case for a variable coefficient operator with both positive results [146] and negative conjectures for arbitrary domain decompositions [152] has been considered as well.

Remark 2.4.1. *The main feature of this result is that it is very general since it does not depend on the exact form of the operator $\mathscr{L}$ and can also be extended to systems or coupled systems of equations with proper care as to the well-posedness of the algorithm.*

As an application, we take $\Omega = \mathbb{R}^2$ and $\Omega_1 =]-\infty, 0[\, \times \mathbb{R}$. Using the same Fourier technique that was used to study algorithm (2.5)–(2.6), it is possible to give the explicit form of the DtN operator for a constant coefficient operator.

- If $\mathscr{L} = \eta - \Delta$, the action of the DtN map can be computed as

$$\boxed{\mathrm{DtN}u_0 = \int_{\mathbb{R}} \sqrt{\eta + k^2}\, \hat{u}_0(k) e^{iky}\, dk.}$$

We say that DtN is a pseudodifferential operator whose symbol is

$$\boxed{\widehat{\mathrm{DtN}} := \sqrt{\eta + k^2}.}$$

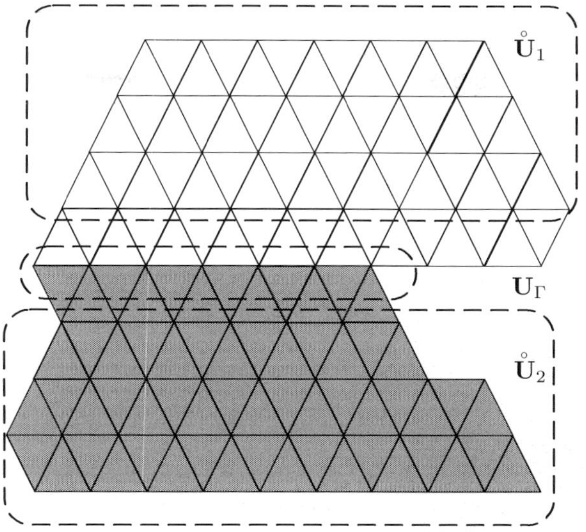

Figure 2.8. *Geometric partition into two subdomains and the corresponding partition of the degrees of freedom.*

- If $\mathscr{L}$ is a convection-diffusion operator $\mathscr{L} := \eta + \mathbf{a}\nabla - \nu\Delta$, the symbol of the DtN map is

$$\boxed{\widehat{\mathrm{DtN}} := \frac{-\mathbf{a}.\mathbf{n}_1 + \sqrt{(\mathbf{a}.\mathbf{n}_1)^2 + 4\nu(\eta + \mathbf{a}.\tau_1 k\nu + \nu^2 k^2)}}{2\nu},}$$

  where $\mathbf{n}_1$ is the outward normal to subdomain Ω_1 and τ_1 is the tangential derivative along the interface.

- If $\mathscr{L} = -\tilde{\omega}^2 - \Delta$ is the Helmholtz operator, the symbol of DtN is

$$\boxed{\widehat{\mathrm{DtN}} := \sqrt{k^2 - \tilde{\omega}^2}.}$$

These symbols are not polynomials in the Fourier variable k so that the corresponding operators and hence the optimal interface conditions are not partial differential operators. They correspond to exact absorbing conditions. These conditions are used on the artificial boundary resulting from the truncation of a computational domain. The solution on the truncated domain depends on the choice of this artificial condition. We say that it is *an exact absorbing boundary condition* if the solution computed on the truncated domain is the restriction of the solution of the original problem. Surprisingly enough, the notions of exact absorbing conditions for domain truncation and optimal interface conditions in domain decomposition methods coincide.

As the above examples show, the optimal interface transmission conditions are pseudodifferential. Therefore they are difficult to implement. Moreover, in the general case of a variable coefficient operator and/or a curved boundary, the exact form of these operators is not known, although they can be approximated by partial differential operators which are easier to implement. The approximation of the DtN has been addressed by many authors since the seminal paper [75] by Engquist and Majda on this question.

2.4.2 ▪ Optimal algebraic interface conditions

The issue of optimal interface conditions is considered here at the matrix level for a nonoverlapping domain decomposition. When a partial differential problem is discretized by a finite element method, for instance, it yields a linear system of the form $A\mathbf{U} = \mathbf{F}$, where $\mathbf{F}$ is a given right-hand side and $\mathbf{U}$ is the set of unknowns. Corresponding to the domain decomposition, the set of unknowns $\mathbf{U}$ is decomposed into interior nodes of the subdomains $\overset{\circ}{\mathbf{U}}_1$ and $\overset{\circ}{\mathbf{U}}_2$, and to the unknowns $\mathbf{U}_\Gamma$ associated to the interface Γ. This leads to a block decomposition of the linear system

$$\begin{pmatrix} A_{11} & A_{1\Gamma} & 0 \\ A_{\Gamma 1} & A_{\Gamma\Gamma} & A_{\Gamma 2} \\ 0 & A_{2\Gamma} & A_{22} \end{pmatrix} \begin{pmatrix} \overset{\circ}{\mathbf{U}}_1 \\ \mathbf{U}_\Gamma \\ \overset{\circ}{\mathbf{U}}_2 \end{pmatrix} = \begin{pmatrix} \overset{\circ}{\mathbf{F}}_1 \\ \mathbf{F}_\Gamma \\ \overset{\circ}{\mathbf{F}}_2 \end{pmatrix}. \tag{2.56}$$

In order to write an optimized Schwarz method, we have to introduce two square matrices T_1 and T_2 which act on vectors of the type $\mathbf{U}_\Gamma$. Then the OSM algorithm reads as

$$\begin{pmatrix} A_{11} & A_{1\Gamma} \\ A_{\Gamma 1} & A_{\Gamma\Gamma} + T_2 \end{pmatrix} \begin{pmatrix} \overset{\circ}{\mathbf{U}}{}_1^{n+1} \\ \mathbf{U}_{\Gamma,1}^{n+1} \end{pmatrix} = \begin{pmatrix} F_1 \\ F_\Gamma + T_2 \mathbf{U}_{\Gamma,2}^n - A_{\Gamma 2} \overset{\circ}{\mathbf{U}}{}_2^n \end{pmatrix}, \tag{2.57}$$

$$\begin{pmatrix} A_{22} & A_{2\Gamma} \\ A_{\Gamma 2} & A_{\Gamma\Gamma} + T_1 \end{pmatrix} \begin{pmatrix} \overset{\circ}{\mathbf{U}}{}_2^{n+1} \\ \mathbf{U}_{\Gamma,2}^{n+1} \end{pmatrix} = \begin{pmatrix} F_2 \\ F_\Gamma + T_1 \mathbf{U}_{\Gamma,1}^n - A_{\Gamma 1} \overset{\circ}{\mathbf{U}}{}_1^n \end{pmatrix}. \tag{2.58}$$

Lemma 2.5. *Assume $A_{\Gamma\Gamma} + T_1 + T_2$ is invertible and problem* (2.56) *is well-posed. Then if the algorithm* (2.57)–(2.58) *converges, it converges to the solution of* (2.56). *That is, if we denote by $\left(\overset{\circ}{\mathbf{U}}{}_1^\infty, \mathbf{U}_{\Gamma,1}^\infty, \overset{\circ}{\mathbf{U}}{}_2^\infty, \mathbf{U}_{\Gamma,2}^\infty\right)$ the limit as n goes to infinity of the sequence $\left(\overset{\circ}{\mathbf{U}}{}_1^n, \mathbf{U}_{\Gamma,1}^n, \overset{\circ}{\mathbf{U}}{}_2^n, \mathbf{U}_{\Gamma,2}^n\right)_{n\geq 0}$,*

$$\mathbf{U}_i^\infty = \mathbf{U}_i \ \text{ and } \ \mathbf{U}_{\Gamma,1}^\infty = \mathbf{U}_{\Gamma,2}^\infty = \mathbf{U}_\Gamma,\ i = 1,2.$$

Proof. Note first that we have a duplication of the interface unknowns $\mathbf{U}_\Gamma$ into $\mathbf{U}_{\Gamma,1}$ and $\mathbf{U}_{\Gamma,2}$.

We subtract the last line of (2.57) from the last line of (2.58), take the limit as n goes to infinity, and get

$$(A_{\Gamma\Gamma} + T_1 + T_2)(\mathbf{U}_{\Gamma,1}^\infty - \mathbf{U}_{\Gamma,2}^\infty) = 0,$$

which proves that $\mathbf{U}_{\Gamma,1}^\infty = \mathbf{U}_{\Gamma,2}^\infty$. Then, taking the limit of (2.57) and (2.58) as n goes to infinity shows that $\left(\overset{\circ}{\mathbf{U}}{}_1^\infty, \mathbf{U}_{\Gamma,1}^\infty = \mathbf{U}_{\Gamma,2}^\infty, \overset{\circ}{\mathbf{U}}{}_2^\infty\right)^T$ is a solution to (2.56) which is unique by assumption. ◻

As in section 2.4, we can use the optimal interface conditions.

Lemma 2.6. *Assume A_{ii} is invertible for $i = 1,2$. Then, in algorithm* (2.57)–(2.58), *taking*

$$T_1 := -A_{\Gamma 1} A_{11}^{-1} A_{1\Gamma} \ \text{ and } \ T_2 := -A_{\Gamma 2} A_{22}^{-1} A_{2\Gamma}$$

yields a convergence in two steps.

Proof. Note that in this case, the bottom-right blocks of the 2×2 block matrices in (2.57) and (2.58),

$$A_{\Gamma\Gamma} - A_{\Gamma 1} A_{11}^{-1} A_{1\Gamma} \text{ and } A_{\Gamma\Gamma} - A_{\Gamma 2} A_{22}^{-1} A_{2\Gamma},$$

are actually Schur complements. It is classical that the subproblems (2.57) and (2.58) are well-posed. By linearity, in order to prove convergence, it is sufficient to consider the convergence to zero in the case $\left(\mathring{\mathbf{F}}_1, \mathbf{F}_\Gamma, \mathring{\mathbf{F}}_2\right)^T = 0$. At step 1 of the algorithm, we have

$$A_{11}\,\mathring{\mathbf{U}}_1^1 + A_{1\Gamma}\mathbf{U}_{\Gamma,1}^1 = 0,$$

or equivalently, by applying $-A_{\Gamma 1}A_{11}^{-1}$,

$$-A_{\Gamma 1}\,\mathring{\mathbf{U}}_1^1 - A_{\Gamma 1}A_{11}^{-1}A_{1\Gamma}\mathbf{U}_{\Gamma,1}^1 = 0,$$

so the right-hand side of (2.58) is zero at step 2 (i.e., $n = 1$). We thus have convergence to zero in domain 2. The same holds for domain 1. □

Matrices T_1 and T_2 are in general dense matrices whose computation and use as interface conditions is very costly. The relationship between optimal algebraic interface conditions and the continuous ones is simply that the last lines of block matrices in (2.57) and (2.58) are the correct discretizations of the optimal interface introduced in section 2.4.1.

2.5 ▪ Optimized interface conditions

As we have already seen in the previous sections, three different reasons led to the development of the new transmission conditions. The first was to obtain Schwarz algorithms that are convergent without overlap; see [130]. The second motivation for changing the transmission conditions was to obtain a convergent Schwarz method for the Helmholtz equation, where the classical overlapping Schwarz algorithm is not convergent [42, 46]. The third motivation was that the convergence of the classical Schwarz method is rather slow and depends strongly on the size of the overlap. For an introduction to optimized Schwarz methods, see [86].

2.5.1 ▪ Optimized interface conditions for $\eta - \Delta$

In section 2.1.2, we have proved convergence of the domain decomposition method with interface conditions of the type

$$\partial_n + \alpha - \partial_\tau(\beta \partial_\tau) \tag{2.59}$$

for a general but nonoverlapping domain decomposition.

In section 2.4, we have exhibited interface conditions which are optimal in terms of iteration counts but are pseudodifferential operators which are difficult to use in practice. These results are not sufficient for the design of effective boundary conditions which for the sake of simplicity must have the form (2.59). From section 2.4, we know that the parameters α and β must somehow be such that (2.59) approximates the optimal interface conditions

$$\frac{\partial}{\partial \mathbf{n}_i} + \mathrm{DtN}.$$

At first sight, it seems that the approximations proposed in the field of artificial boundary conditions are also relevant in the context of domain decomposition methods. Actually, this is not the case, as was proved for the convection-diffusion equation; see [113, 114].

In order to clarify the situation, we need an estimate of the convergence factor as a function of the parameters α and β, the size of the overlap, and the coefficients of the partial differential operator. In particular, it will provide a means for choosing the interface conditions in an optimal way. This type of study is limited to a very simple situation: a constant coefficient operator and a whole space decomposed into two half-spaces. But let us insist on the fact that these limitations concern only this theoretical study. The optimized values of the parameters of the interface conditions can be used with success in complex applications; see section 2.5.2. The robustness of the approach comes from the general convergence result of section 2.1.2 and from the replacement of the fixed-point algorithm on the interface by a Krylov method as explained for (2.41) and (2.49). The efficiency comes from the study below which is made possible by the use of Fourier techniques similar to those used in artificial boundary conditions.

Optimization of the interface conditions

It is possible to optimize the choice of the parameter α in Robin interface conditions in order to minimize the maximum convergence factor in the physical space

$$\max_k \rho(k,\delta;\alpha).$$

When the subdomains overlap, we have seen that the convergence factor (2.7) is bounded from above by a positive constant so that it can be checked that the min-max problem

$$\boxed{\max_k \rho(k,\delta;\alpha_{\text{opt}}) = \min_\alpha \max_k \rho(k,\delta;\alpha)}$$

admits a unique solution.

When the subdomains do not overlap, then for any choice of α we have

$$\max_k \rho(k,0;\alpha) = 1$$

so that the above min-max problem is ill-posed.

In any case, the purpose of domain decomposition methods is not to solve partial differential equations directly. They are used to solve the corresponding linear systems arising from their discretizations. It is possible to study the convergence factor of the related domain decomposition methods at the discrete level based on the discretization scheme; see [143]. The Fourier transform is replaced by discrete Fourier series, i.e., the decomposition on the vectors

$$V_k = \left(e^{ij\Delta y\, k}\right)_{j\in\mathbb{Z}},\ k \in \pi/(\mathbb{Z}\Delta y),$$

with Δy the mesh size in the y-direction. This convergence factor depends as before not only on the parameters of the continuous problem but also on the discrete parameters: mesh size in x and y. The resulting formula is quite complex and would be very difficult to optimize.

Nevertheless, comparison with the continuous case and numerical experiments prove that a semicontinuous approach is sufficient for finding an optimal value for the parameter α. This is of course due to the fact that as the discretization parameters go to zero, the discrete convergence factor tends to its continuous counterpart.

Table 2.1. *Number of iterations for different values of the mesh size and two possible choices for* α.

$1/\Delta y$	10	20	40	80
α^{sc}_{opt}	6	7	10	16
$\alpha = 1$	27	51	104	231

For the sake of simplicity, we consider only the nonoverlapping case, $\delta = 0$. We keep the formula of the convergence factor in the continuous case:

$$\rho(k,\alpha) := \left| \frac{\lambda(k) - \alpha}{\lambda(k) + \alpha} \right| \tag{2.60}$$

with $\lambda(k) = \sqrt{\eta + k^2}$. But we observe that the mesh induces a truncation in the frequency domain. We have $|k| < \pi/\Delta y := k_{\max}$. For a parameter α, the convergence factor is approximated by

$$\rho_h(\alpha) = \max_{|k|<\pi/\Delta y} \rho(k;\alpha).$$

The optimization problem reads as

$$\begin{gathered} \text{Find } \alpha^{sc}_{opt} \text{ such that} \\ \rho_h(\alpha^{sc}_{opt}) = \min_{\alpha} \max_{k<\pi/\Delta y} \rho(k;\alpha). \end{gathered} \tag{2.61}$$

It can be checked that the optimum is given by the relation

$$\rho(0;\alpha^{sc}_{opt}) = \rho(k_{\max};\alpha^{sc}_{opt}).$$

Let $\lambda_m = \lambda(0)$ and $\lambda_M = \lambda(k_{\max})$; then we have

$$\alpha^{sc}_{opt} = \sqrt{\lambda_m \lambda_M}. \tag{2.62}$$

It can then easily be checked that in the limit of small Δy,

$$\rho_h(\alpha^{sc}_{opt}) \simeq 1 - 2\sqrt{\frac{\sqrt{\eta}\Delta y}{\pi}} \quad \text{and} \quad \alpha^{sc}_{opt} \simeq \eta^{1/4}\frac{\pi}{\Delta y},$$

whereas for α independent of Δy, we have

$$\rho_h(\alpha) \simeq 1 - 2\frac{\alpha \Delta y}{\pi}$$

for small Δy. Numerical tests on the model problem of a rectangle divided into two half-rectangles and a finite difference discretization shows good agreement with the above formulas. In Table 2.1, the iteration counts are given for two possible choices of the parameter α, $\alpha = 1$ or α^{sc}_{opt} given by formula (2.62) and for a reduction of the error by a factor of 10^{-6}. For refined mesh, the iteration count is reduced by a factor larger than 10. The optimized Schwarz method is thus quite sensitive to the choice of the interface condition. When the method is used as a preconditioner in Krylov methods, as explained in section 2.3.1, performance is less dependent on the optimal choice of the interface condition. Typically, the iteration count is reduced by a factor of 3; see Table 2.3. Since taking optimal interface conditions is beneficial in terms of iteration counts and has no extra cost, it is a good thing to do, especially for wave propagation phenomena that we consider in the following.

2.5.2 ▪ Optimized conditions for the Helmholtz equation

The complete study can be found in [91]. In order to have a self-consistent paragraph, we write down once again the considered Helmholtz equation

$$\mathscr{L}(u) := (-\tilde{\omega}^2 - \Delta)(u) = f(x,y), \quad x,y \in \Omega.$$

The difficulty comes from the negative term of order zero of the operator.

Although the following analysis could be carried out on rectangular domains as well, we prefer for simplicity to present the analysis in the domain $\Omega = \mathbb{R}^2$ with the Sommerfeld radiation condition at infinity,

$$\lim_{r\to\infty} \sqrt{r}\left(\frac{\partial u}{\partial r} + i\tilde{\omega} u\right) = 0,$$

where $r = \sqrt{x^2+y^2}$. We decompose the domain into two nonoverlapping subdomains $\Omega_1 = (-\infty, 0\,] \times \mathbb{R}$ and $\Omega_2 = [\,0, \infty) \times \mathbb{R}$ and consider the Schwarz algorithm

$$\begin{aligned} -\Delta u_1^{n+1} - \tilde{\omega}^2 u_1^{n+1} &= f(x,y), \quad x,y \in \Omega_1, \\ \mathscr{B}_1(u_1^{n+1})(0) &= \mathscr{B}_1(u_2^n)(0) \end{aligned} \tag{2.63}$$

and

$$\begin{aligned} -\Delta u_2^{n+1} - \tilde{\omega}^2 u_2^{n+1} &= f(x,y), \quad x,y \in \Omega_2, \\ \mathscr{B}_2(u_2^{n+1})(0) &= \mathscr{B}_2(u_1^n)(0), \end{aligned} \tag{2.64}$$

where $\mathscr{B}_j$, $j = 1,2$, are two linear operators. Note that for the classical Schwarz method $\mathscr{B}_j$ is the identity, $\mathscr{B}_j = I$, and without overlap the algorithm cannot converge. But even with overlap in the case of the Helmholtz equation, only the evanescent modes in the error are damped, while the propagating modes are unaffected by the Schwarz algorithm [91]. One possible remedy is to use a relatively fine coarse grid [22] or Robin transmission conditions; see, for example, [44, 19]. We consider here transmission conditions which lead to a convergent nonoverlapping version of the Schwarz method. We assume that the linear operators $\mathscr{B}_j$ are of the form

$$\mathscr{B}_j := \partial_x + \mathscr{T}_j, \quad j = 1,2,$$

for two linear operators $\mathscr{T}_1$ and $\mathscr{T}_2$ acting in the tangential direction on the interface. Our goal is to use these operators to optimize the convergence factor of the algorithm. For the analysis it suffices by linearity to consider the case $f(x,y) = 0$ and to analyze convergence to the zero solution. Taking a Fourier transform in the y-direction we obtain

$$\begin{aligned} -\frac{\partial^2 \hat{u}_1^{n+1}}{\partial x^2} - (\tilde{\omega}^2 - k^2)\hat{u}_1^{n+1} &= 0, \\ (\partial_x + \sigma_1(k))(\hat{u}_1^{n+1})(0) &= (\partial_x + \sigma_1(k))(\hat{u}_2^n)(0), \end{aligned} \quad x < 0,\ k \in \mathbb{R}, \tag{2.65}$$

and

$$\begin{aligned} -\frac{\partial^2 \hat{u}_2^{n+1}}{\partial x^2} - (\tilde{\omega}^2 - k^2)\hat{u}_2^{n+1} &= 0, \\ (\partial_x + \sigma_2(k))(\hat{u}_2^{n+1})(0) &= (\partial_x + \sigma_2(k))(\hat{u}_1^n)(0), \end{aligned} \quad x > 0,\ k \in \mathbb{R}, \tag{2.66}$$

where $\sigma_j(k)$ denotes the symbol of the operator $\mathscr{T}_j$, and k is the Fourier variable, which we also call the frequency. The general solutions of these ODEs are

$$\hat{u}_j^{n+1} = A_j e^{\lambda(k)x} + B_j e^{-\lambda(k)x}, \quad j = 1,2,$$

where $\lambda(k)$ denotes the root of the characteristic equation $\lambda^2 + (\tilde{\omega}^2 - k^2) = 0$ with positive real or imaginary part,

$$\begin{aligned} \lambda(k) &= \sqrt{k^2 - \tilde{\omega}^2} && \text{for} \quad |k| \geq \tilde{\omega}, \\ \lambda(k) &= i\sqrt{\tilde{\omega}^2 - k^2} && \text{for} \quad |k| < \tilde{\omega}. \end{aligned} \tag{2.67}$$

Since the Sommerfeld radiation condition excludes growing solutions as well as incoming modes at infinity, we obtain the solutions

$$\begin{aligned} \hat{u}_1^{n+1}(x,k) &= \hat{u}_1^{n+1}(0,k)e^{\lambda(k)x}, \\ \hat{u}_2^{n+1}(x,k) &= \hat{u}_2^{n+1}(0,k)e^{-\lambda(k)x}. \end{aligned}$$

Using the transmission conditions and the fact that

$$\begin{aligned} \frac{\partial \hat{u}_1^{n+1}}{\partial x} &= \lambda(k)\hat{u}_1^{n+1}, \\ \frac{\partial \hat{u}_2^{n+1}}{\partial x} &= -\lambda(k)\hat{u}_2^{n+1} \end{aligned}$$

we obtain over one step of the Schwarz iteration

$$\begin{aligned} \hat{u}_1^{n+1}(x,k) &= \frac{-\lambda(k)+\sigma_1(k)}{\lambda(k)+\sigma_1(k)} e^{\lambda(k)x} \hat{u}_2^n(0,k), \\ \hat{u}_2^{n+1}(x,k) &= \frac{\lambda(k)+\sigma_2(k)}{-\lambda(k)+\sigma_2(k)} e^{-\lambda(k)x} \hat{u}_1^n(0,k). \end{aligned}$$

Evaluating the second equation at $x = 0$ for iteration index n and inserting it into the first equation, we get after evaluating again at $x = 0$

$$\hat{u}_1^{n+1}(0,k) = \frac{-\lambda(k)+\sigma_1(k)}{\lambda(k)+\sigma_1(k)} \cdot \frac{\lambda(k)+\sigma_2(k)}{-\lambda(k)+\sigma_2(k)} \hat{u}_1^{n-1}(0,k).$$

Defining the convergence factor ρ by

$$\boxed{\rho(k) := \frac{-\lambda(k)+\sigma_1(k)}{\lambda(k)+\sigma_1(k)} \cdot \frac{\lambda(k)+\sigma_2(k)}{-\lambda(k)+\sigma_2(k)}} \tag{2.68}$$

we find by induction that

$$\hat{u}_1^{2n}(0,k) = \rho(k)^n \hat{u}_1^0(0,k),$$

and by a similar calculation on the second subdomain,

$$\hat{u}_2^{2n}(0,k) = \rho(k)^n \hat{u}_2^0(0,k).$$

Choosing in the Fourier transformed domain

$$\sigma_1(k) := \lambda(k), \qquad \sigma_2(k) := -\lambda(k)$$

corresponds to using exact absorbing boundary conditions as interface conditions. So we get $\rho(k) \equiv 0$ and the algorithm converges in two steps independently of the initial guess. Unfortunately this choice becomes difficult to use in the real domain where computations take place, since the optimal choice of the symbols $\sigma_j(k)$ leads to nonlocal operators $\mathcal{T}_j$ in the real domain, caused by the square root in the symbols. We have to construct local approximations for the optimal transmission conditions.

In Després' algorithm [43], the approximation consists of $\mathcal{T}_j \equiv i\tilde{\omega}$ $(i^2 = -1)$. In [75], the approximation valid for the truncation of an infinite computational domain is obtained via Taylor expansions of the symbol in the vicinity of $k = 0$,

$$\mathcal{T}_j^{\text{app}} = \pm i\left(\tilde{\omega} - \frac{1}{2\tilde{\omega}}\partial_{\tau\tau}\right),$$

which leads to the zeroth- or second-order Taylor transmission conditions, depending on whether one keeps only the constant term or the second-order term as well. But these transmission conditions are only effective for the low-frequency components of the error. This is sufficient for the truncation of a domain since there is an exponential decay of the high-frequency part (large k) of the solution away from the artificial boundary.

In domain decomposition, however, what is important is the convergence factor, which is given by the maximum over k of $\rho(k)$. Since there is no overlap between the subdomains, it is not possible to profit from any decay. We present now an approximation procedure suited to domain decomposition methods. To avoid an increase in the bandwidth of the local discretized subproblems, we take polynomials of degree at most 2, which leads to the transmission operators $\mathcal{T}_j^{\text{app}}$, which are at most second-order partial differential operators acting along the interface. By symmetry of the Helmholtz equation there is no interest in a first-order term. We therefore approximate the operators $\mathcal{T}_j$, $j = 1, 2$, in the form

$$\mathcal{T}_j^{\text{app}} = \pm(a + b\,\partial_{\tau\tau})$$

with $a, b \in \mathbf{C}$ and where τ denotes the tangent direction at the interface.

Optimized Robin interface conditions

We approximate the optimal operators $\mathcal{T}_j$, $j = 1, 2$, in the form

$$\mathcal{T}_j^{\text{app}} = \pm(p + qi), \quad p, q \in \mathbb{R}^+. \tag{2.69}$$

The non-negativity of p, q comes from the Shapiro–Lopatinski necessary condition for the well-posedness of the local subproblems (2.14)–(2.15). Inserting this approximation into the convergence factor (2.68) we find

$$\rho(p,q,k) = \begin{cases} \dfrac{p^2 + \left(q - \sqrt{\tilde{\omega}^2 - k^2}\right)^2}{p^2 + \left(q + \sqrt{\tilde{\omega}^2 - k^2}\right)^2}, & \tilde{\omega}^2 \geq k^2, \\[2ex] \dfrac{q^2 + \left(p - \sqrt{k^2 - \tilde{\omega}^2}\right)^2}{q^2 + \left(p + \sqrt{k^2 - \tilde{\omega}^2}\right)^2}, & \tilde{\omega}^2 < k^2. \end{cases} \tag{2.70}$$

First note that for $k^2 = \tilde{\omega}^2$ the convergence factor $\rho(p, q, \tilde{\omega}) = 1$, no matter what one chooses for the free parameters p and q. In the Helmholtz case, one cannot uniformly minimize the convergence factor over all relevant frequencies, as in the case of positive definite problems; see [112, 91, 114]. However, the point $k = \tilde{\omega}$ represents only one single mode in the spectrum, and a Krylov method will easily take care of this when the Schwarz method is used as a preconditioner, as our numerical experiments will show. We therefore consider the optimization problem

$$\min_{p,q\in\mathbb{R}^+}\left(\max_{k\in(k_{\min},\tilde{\omega}_-)\cup(\tilde{\omega}_+,k_{\max})}|\rho(p,q,k)|\right), \tag{2.71}$$

where $\tilde{\omega}_-$ and $\tilde{\omega}_+$ are parameters to be chosen, $k_{\min}$ denotes the smallest frequency relevant to the subdomain, and $k_{\max}$ denotes the largest frequency supported by the numerical grid. This largest frequency is of the order π/h. For example, if the domain Ω is a strip of height L with homogeneous Dirichlet conditions on the top and bottom, the solution can be expanded in a Fourier series with the harmonics $\sin(j\pi y/L)$, $j\in\mathbb{N}$. Hence the relevant frequencies are $k = j\pi/L$. They are equally distributed with a spacing π/L, and thus choosing $\tilde{\omega}_- = \tilde{\omega} - \pi/L$ and $\tilde{\omega}_+ = \tilde{\omega} + \pi/L$ leaves precisely one frequency $k = \tilde{\omega}$ for the Krylov method and treats all the others by the optimization. If $\tilde{\omega}$ falls in between the relevant frequencies, say $j\pi/L < \tilde{\omega} < (j+1)\pi/L$, then we can even get the iterative method to converge by choosing $\tilde{\omega}_- = j\pi/L$ and $\tilde{\omega}_+ = (j+1)\pi/L$, which will allow us to directly verify our asymptotic analysis numerically without the use of a Krylov method. How to choose the optimal parameters p and q is given in the following.

Theorem 2.7 (Optimized Robin conditions). *Under the three assumptions*

$$2\tilde{\omega}^2 \le \tilde{\omega}_-^2 + \tilde{\omega}_+^2, \quad \tilde{\omega}_- < \tilde{\omega}, \tag{2.72}$$

$$2\tilde{\omega}^2 > k_{\min}^2 + \tilde{\omega}_+^2, \tag{2.73}$$

$$2\tilde{\omega}^2 < k_{\min}^2 + k_{\max}^2, \tag{2.74}$$

the solution to the min-max problem (2.71) *is unique and the optimal parameters are given by*

$$p^* = q^* = \sqrt{\frac{\sqrt{\tilde{\omega}^2-\tilde{\omega}_-^2}\,\sqrt{k_{\max}^2-\tilde{\omega}^2}}{2}}. \tag{2.75}$$

The optimized convergence factor (2.71) *is then given by*

$$\max_{k\in(k_{\min},\tilde{\omega}_-)\cup(\tilde{\omega}_+,k_{\max})}\rho(p^*,q^*,k) = \frac{1-\sqrt{2}\left(\frac{\tilde{\omega}^2-\tilde{\omega}_-^2}{k_{\max}^2-\tilde{\omega}^2}\right)^{1/4} + \sqrt{\frac{\tilde{\omega}^2-\tilde{\omega}_-^2}{k_{\max}^2-\tilde{\omega}^2}}}{1+\sqrt{2}\left(\frac{\tilde{\omega}^2-\tilde{\omega}_-^2}{k_{\max}^2-\tilde{\omega}^2}\right)^{1/4} + \sqrt{\frac{\tilde{\omega}^2-\tilde{\omega}_-^2}{k_{\max}^2-\tilde{\omega}^2}}}. \tag{2.76}$$

For the proof, see [91].

Optimized second-order interface conditions

We approximate the operators $\mathcal{T}_j$, $j = 1,2$, in the form $\mathcal{T}_1^{app} = -\mathcal{T}_2^{app} = a + b\partial_{\tau\tau}$ with $a, b \in \mathbb{C}$ and τ denoting the tangent direction at the interface. The design of the optimized second-order transmission conditions is simplified by the following.

Lemma 2.8. *Let u_1 and u_2 be two functions which satisfy*

$$\mathcal{L}(u_j) \equiv (-\tilde{\omega}^2 - \Delta)(u_j) = f \quad in\ \Omega_j,\ j = 1,2,$$

and the transmission condition

$$\left(\frac{\partial}{\partial \mathbf{n}_1}+\alpha\right)\left(\frac{\partial}{\partial \mathbf{n}_1}+\beta\right)(u_1)=\left(-\frac{\partial}{\partial \mathbf{n}_2}+\alpha\right)\left(-\frac{\partial}{\partial \mathbf{n}_2}+\beta\right)(u_2) \tag{2.77}$$

with $\alpha,\beta\in\mathbb{C}$, $\alpha+\beta\neq 0$, *and* $\mathbf{n}_j$ *denoting the unit outward normal to domain* Ω_j. *Then the second-order transmission condition*

$$\left(\frac{\partial}{\partial \mathbf{n}_1}+\frac{\alpha\beta-\tilde{\omega}^2}{\alpha+\beta}-\frac{1}{\alpha+\beta}\frac{\partial^2}{\partial \tau_1^2}\right)(u_1)=\left(-\frac{\partial}{\partial \mathbf{n}_2}+\frac{\alpha\beta-\tilde{\omega}^2}{\alpha+\beta}-\frac{1}{\alpha+\beta}\frac{\partial^2}{\partial \tau_2^2}\right)(u_2) \tag{2.78}$$

is satisfied as well.

Proof. Expanding the transmission condition (2.77) yields

$$\left(\frac{\partial^2}{\partial \mathbf{n}_1^2}+(\alpha+\beta)\frac{\partial}{\partial \mathbf{n}_1}+\alpha\beta\right)(u_1)=\left(\frac{\partial^2}{\partial \mathbf{n}_2^2}-(\alpha+\beta)\frac{\partial}{\partial \mathbf{n}_2}+\alpha\beta\right)(u_2).$$

Now, using the equation $\mathscr{L}(u_1)=f$, we can substitute $-(\frac{\partial^2}{\partial \tau_1^2}+\tilde{\omega}^2)(u_1)-f$ for $\frac{\partial^2}{\partial \mathbf{n}_1^2}(u_1)$, and similarly we can substitute $-(\frac{\partial^2}{\partial \tau_2^2}+\tilde{\omega}^2)(u_2)-f$ for $\frac{\partial^2}{\partial \mathbf{n}_2^2}(u_2)$. Hence, we get

$$\begin{aligned}\left(-\frac{\partial^2}{\partial \tau_1^2}-\tilde{\omega}^2+(\alpha+\beta)\frac{\partial}{\partial \mathbf{n}_1}+\alpha\beta\right)(u_1)-f&\\ =\left(-\frac{\partial^2}{\partial \tau_2^2}-\tilde{\omega}^2-(\alpha+\beta)\frac{\partial}{\partial \mathbf{n}_2}+\alpha\beta\right)(u_2)-f.&\end{aligned}$$

Now the terms f on both sides cancel and a division by $\alpha+\beta$ yields (2.78). □

Note that Higdon has already proposed approximations to absorbing boundary conditions in factored form in [110]. In our case, this special choice of approximating $\sigma_j(k)$ by

$$\sigma_1^{app}(k)=-\sigma_2^{app}(k)=\frac{\alpha\beta-\tilde{\omega}^2}{\alpha+\beta}+\frac{1}{\alpha+\beta}k^2 \tag{2.79}$$

leads to a particularly elegant formula for the convergence factor. Inserting $\sigma_j^{app}(k)$ into the convergence factor (2.68) and simplifying, we obtain

$$\begin{aligned}\rho(k;\alpha,\beta)&:=\left(\frac{\lambda(k)-\sigma_1}{\lambda(k)+\sigma_1}\right)^2=\left(\frac{-(\alpha+\beta)\lambda(k)+\alpha\beta+k^2-\tilde{\omega}^2}{(\alpha+\beta)\lambda(k)+\alpha\beta+k^2-\tilde{\omega}^2}\right)^2\\ &=\left(\frac{\lambda(k)^2-(\alpha+\beta)\lambda(k)+\alpha\beta}{\lambda(k)^2+(\alpha+\beta)\lambda(k)+\alpha\beta}\right)^2=\left(\frac{\lambda(k)-\alpha}{\lambda(k)+\alpha}\right)^2\left(\frac{\lambda(k)-\beta}{\lambda(k)+\beta}\right)^2,\end{aligned} \tag{2.80}$$

where $\lambda(k)$ is defined in (2.67) and the two parameters $\alpha,\beta\in\mathbb{C}$ can be used to optimize the performance. By the symmetry of $\lambda(k)$ with respect to k, it suffices to consider only positive k to optimize performance. We thus need to solve the min-max problem

$$\min_{\alpha,\beta\in\mathbb{C}}\left(\max_{k\in(k_{\min},\tilde{\omega}_-)\cup(\tilde{\omega}_+,k_{\max})}|\rho(k;\alpha,\beta)|\right), \tag{2.81}$$

where $\tilde{\omega}_-$ and $\tilde{\omega}_+$ are again the parameters to exclude the frequency $k = \tilde{\omega}$ where the convergence factor equals 1, as in the zeroth-order optimization problem. The convergence factor $\rho(k;\alpha,\beta)$ consists of two factors, and λ is real for vanishing modes and imaginary for propagative modes. If we choose $\alpha \in i\mathbb{R}$ and $\beta \in \mathbb{R}$, then for λ real the first factor is of modulus one and the second one can be optimized using β. If λ is imaginary, then the second factor is of modulus one and the first one can be optimized independently using α. Hence for this choice of α and β the min-max problem decouples. We therefore consider here the simpler min-max problem

$$\min_{\alpha\in i\mathbb{R},\beta\in\mathbb{R}} \left(\max_{k\in(k_{\min},\tilde{\omega}_-)\cup(\tilde{\omega}_+,k_{\max})} |\rho(k;\alpha,\beta)| \right) \tag{2.82}$$

which has an elegant analytical solution. Note, however, that the original minimization problem (2.81) might have a solution with a better convergence factor, an issue investigated in [87].

Theorem 2.9 (Optimized second-order conditions). *The solution of the min-max problem* (2.82) *is unique, and the optimal parameters are given by*

$$\alpha^* = i\left((\tilde{\omega}^2 - k_{\min}^2)(\tilde{\omega}^2 - \tilde{\omega}_-^2)\right)^{1/4} \in i\mathbb{R} \tag{2.83}$$

and

$$\beta^* = \left((k_{\max}^2 - \tilde{\omega}^2)(\tilde{\omega}_+^2 - \tilde{\omega}^2)\right)^{1/4} \in \mathbb{R}. \tag{2.84}$$

The convergence factor (2.82) *is then for the propagating modes given by*

$$\max_{k\in(k_{\min},\tilde{\omega}_-)} |\rho(k,\alpha^*,\beta^*)| = \left(\frac{(\tilde{\omega}^2-\tilde{\omega}_-^2)^{1/4} - (\tilde{\omega}^2 - k_{\min}^2)^{1/4}}{(\tilde{\omega}^2-\tilde{\omega}_-^2)^{1/4} + (\tilde{\omega}^2 - k_{\min}^2)^{1/4}} \right)^2, \tag{2.85}$$

and for the evanescent modes it is

$$\max_{k\in(\tilde{\omega}_+,k_{\max})} \rho(k,\alpha^*,\beta^*) = \left(\frac{(k_{\max}^2 - \tilde{\omega}^2)^{1/4} - (\tilde{\omega}_+^2 - \tilde{\omega}^2)^{1/4}}{(k_{\max}^2 - \tilde{\omega}^2)^{1/4} + (\tilde{\omega}_+^2 - \tilde{\omega}^2)^{1/4}} \right)^2. \tag{2.86}$$

Proof. For $k \in (k_{\min}, \tilde{\omega}_-)$ we have $\left|\frac{i\sqrt{\tilde{\omega}^2-k^2}-\beta}{i\sqrt{\tilde{\omega}^2-k^2}+\beta}\right| = 1$ since $\beta \in \mathbb{R}$ and thus $|\rho(k;\alpha,\beta)| = \left|\frac{i\sqrt{\tilde{\omega}^2-k^2}-\alpha}{i\sqrt{\tilde{\omega}^2-k^2}+\alpha}\right|^2$ depends only on α. Similarly, for $k \in (\tilde{\omega}_+, k_{\max})$ we have $\left|\frac{\sqrt{k^2-\tilde{\omega}^2}-\alpha}{\sqrt{k^2-\tilde{\omega}^2}+\alpha}\right| = 1$ since $\alpha \in i\mathbb{R}$ and therefore $|\rho(k;\alpha,\beta)| = \left|\frac{\sqrt{k^2-\tilde{\omega}^2}-\beta}{\sqrt{k^2-\tilde{\omega}^2}+\beta}\right|^2$ depends only on β. The solution (α, β) of the minimization problem (2.82) is thus given by the solution of the two independent minimization problems

$$\min_{\alpha\in i\mathbb{R},} \left(\max_{k\in(k_{\min},\tilde{\omega}_-)} \left| \frac{i\sqrt{\tilde{\omega}^2-k^2}-\alpha}{i\sqrt{\tilde{\omega}^2-k^2}+\alpha} \right| \right) \tag{2.87}$$

and

$$\min_{\beta\in\mathbb{R}} \left(\max_{k\in(\tilde{\omega}_+,k_{\max})} \left| \frac{\sqrt{k^2-\tilde{\omega}^2}-\beta}{\sqrt{k^2-\tilde{\omega}^2}+\beta} \right| \right). \tag{2.88}$$

We show the solution for the second problem (2.88) only; the solution for the first problem (2.87) is similar. First note that the maximum of $|\rho_\beta| := \left|\frac{\sqrt{k^2-\tilde{\omega}^2}-\beta}{\sqrt{k^2-\tilde{\omega}^2}+\beta}\right|$ is attained on the boundary of the interval $[\tilde{\omega}_+, k_{\max}]$, because the function ρ_β (but not $|\rho_\beta|$) is monotonically increasing with $k \in [\tilde{\omega}_+, k_{\max}]$. On the other hand, as a function of β, $|\rho_\beta(\tilde{\omega}_+)|$ grows monotonically with β while $|\rho_\beta(k_{\max})|$ decreases monotonically with β. The optimum is therefore reached when we balance the two values on the boundary, $\rho_\beta(\tilde{\omega}_+) = -\rho_\beta(k_{\max})$, which implies that the optimal β satisfies the equation

$$\frac{\sqrt{k_{\max}^2 - \tilde{\omega}^2} - \beta}{\sqrt{k_{\max}^2 - \tilde{\omega}^2} + \beta} = -\frac{\sqrt{\tilde{\omega}_+^2 - \tilde{\omega}^2} - \beta}{\sqrt{\tilde{\omega}_+^2 - \tilde{\omega}^2} + \beta} \tag{2.89}$$

whose solution is given in (2.84). □

The optimization problem (2.88) arises also for symmetric positive definite problems when an optimized Schwarz algorithm without overlap and Robin transmission conditions are used. The present solution can be found in [188].

Note that the optimization of the interface conditions was performed for the convergence factor of a fixed-point method and not for a particular Krylov method applied to the substructured problem. In the positive definite case, one can show that minimizing the convergence factor is equivalent to minimizing the condition number of the substructured problem [113]. Numerical experiments in the next section indicate that for the Helmholtz equation our optimization also leads to parameters close to the best ones for the preconditioned Krylov method.

Numerical results

We present two sets of numerical experiments. The first set corresponds to the model problem analyzed above, and the results obtained illustrate the analysis and confirm the asymptotic convergence results. The second numerical experiment comes from industry and consists of analyzing the noise levels in the interior of a VOLVO S90.

We study a two-dimensional cavity on the unit square Ω with homogeneous Dirichlet conditions on the top and bottom and Robin-type radiation conditions on the left and right. We thus have the Helmholtz problem

$$\begin{aligned}
-\Delta u - \tilde{\omega}^2 u &= f, && 0 < x, y < 1,\\
u &= 0, && 0 < x < 1,\ y = 0, 1,\\
\frac{\partial u}{\partial x} - i\tilde{\omega} u &= 0, && x = 0,\ 0 < y < 1,\\
-\frac{\partial u}{\partial x} - i\tilde{\omega} u &= 0, && x = 1,\ 0 < y < 1.
\end{aligned} \tag{2.90}$$

We decompose the unit square into two subdomains of equal size, and we use a uniform rectangular mesh for the discretization. We perform all our experiments directly on the error equations, $f = 0$, and choose the initial guess of the Schwarz iteration so that all the frequencies are present in the error.

We show two sets of experiments. The first one is with $\tilde{\omega} = 9.5\pi$, thus excluding $\tilde{\omega}$ from the k frequencies relevant in this setting, $k = n\pi$, $n = 1, 2, \ldots$. This allows us to test the iterative Schwarz method directly, since with optimization parameters $\tilde{\omega}_- = 9\pi$ and

Table 2.2. *Number of iterations for different transmission conditions and different mesh parameters for the model problem.*

	Order Zero				Order Two			
	Iterative		Krylov		Iterative		Krylov	
h	Taylor	Optimized	Taylor	Optimized	Taylor	Optimized	Taylor	Optimized
1/50	-	457	26	16	-	22	28	9
1/100	-	126	34	21	-	26	33	10
1/200	-	153	44	26	-	36	40	13
1/400	-	215	57	34	-	50	50	15
1/800	-	308	72	43	-	71	61	19

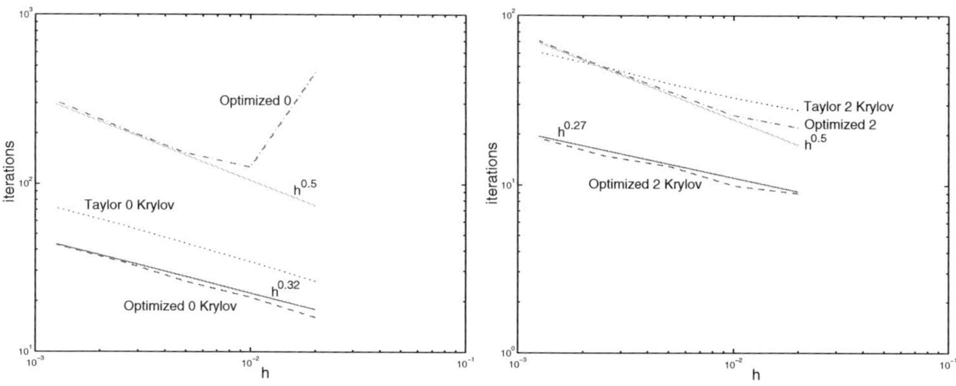

Figure 2.9. *Asymptotic behavior for the zeroth-order transmission conditions (left) and for the second-order transmission conditions (right).*

$\tilde{\omega}_+ = 10\pi$ we obtain a convergence factor which is uniformly less than 1 for all k. Table 2.2 shows the number of iterations needed for different values of the mesh parameter h for both the zeroth- and second-order transmission conditions. The Taylor transmission conditions do not lead to a convergent iterative algorithm, because for all frequencies $k > \tilde{\omega}$, the convergence factor equals 1. However, with Krylov acceleration, GMRES in this case, the methods converge. Note, however, that the second-order Taylor condition is only a little better than the zeroth-order Taylor conditions. The optimized transmission conditions already lead, in the case where $\tilde{\omega}$ lies between two frequencies, to a convergent iterative algorithm. The iterative version even beats the Krylov accelerated Taylor conditions in the second-order case. It is no wonder that the optimized conditions lead to by far the best algorithms when they are accelerated by a Krylov method: the second-order optimized Schwarz method is faster than any Taylor method by more than a factor of three. Note that the only difference in cost of the various transmission conditions consists of different entries in the interface matrices, without enlarging the bandwidth of the matrices. Figure 2.9 shows the asymptotic behavior of the methods considered, on the left for zeroth-order conditions and on the right for second-order conditions. Note that the scale on the right for the second-order transmission conditions is different by an order of magnitude. In both cases the asymptotic analysis is confirmed for the iterative version of the optimized methods. In addition, one can see that the Krylov method improves the asymptotic rate by almost an additional square root, as expected from the analysis in ideal situations. Note the outlier of the zeroth-order optimized transmission condition for $h = 1/50$. It is due to the discrepancy between the spectrum of the continuous and the

Table 2.3. *Number of iterations for different transmission conditions and different mesh parameters for the model problem when $\tilde{\omega}$ lies precisely on a frequency of the problem and thus Krylov acceleration is mandatory.*

	Order Zero		Order Two	
h	Taylor	Optimized	Taylor	Optimized
1/50	24	15	27	9
1/100	35	21	35	11
1/200	44	26	41	13
1/400	56	33	52	16
1/800	73	43	65	20

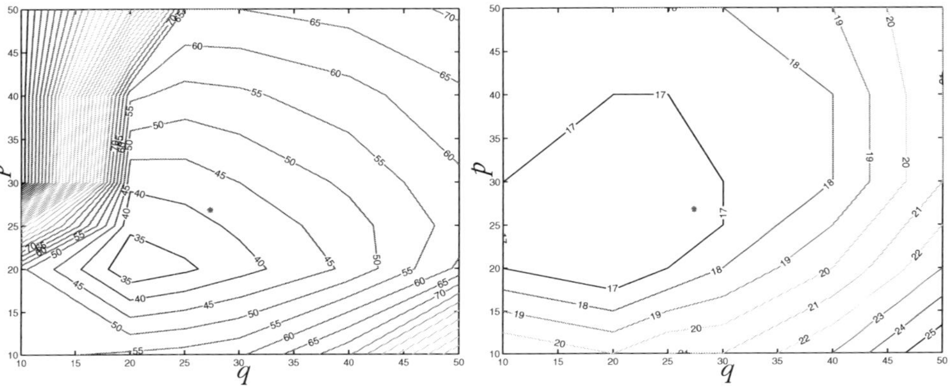

Figure 2.10. *Number of iterations needed to achieve a certain precision as a function of the optimization parameters p and q in the zeroth-order transmission conditions for the iterative algorithm (left) and for the Krylov accelerated algorithm (right). The star denotes the optimized parameters p^* and q^* found by our Fourier analysis.*

discrete operator: $\tilde{\omega} = 9.5\pi$ lies precisely in between two frequencies 9π and 10π at the continuous level, but for the discrete Laplacian with $h = 1/50$ this spectrum is shifted to 8.88π and 9.84π, and thus the frequency 9.84π falls into the range $[9\pi, 10\pi]$ neglected by the optimization. Note, however, that this is of no importance when Krylov acceleration is used, so it is not worthwhile to consider this issue further.

In the second experiment we put $\tilde{\omega}$ directly onto a frequency of the model problem, $\tilde{\omega} = 10\pi$, so that the iterative methods can no longer be considered, since for that frequency the convergence factor equals 1. The Krylov accelerated versions, however, are not affected by this, as one can see in Table 2.3. The number of iterations does not differ from the case where $\tilde{\omega}$ was chosen to lie between two frequencies, which shows that with Krylov acceleration the method is robust for any values of $\tilde{\omega}$. We finally tested for the smallest resolution of the model problem how well Fourier analysis predicts the optimal parameters. Since we want to test both the iterative and the Krylov versions, we again need to put the frequency $\tilde{\omega}$ in between two problem frequencies, and in this case it is important to be precise. We therefore choose $\tilde{\omega}$ to be exactly between two frequencies of the discrete problem, $\tilde{\omega} = 9.3596\pi$, and optimized using $\tilde{\omega}_- = 8.8806\pi$ and $\tilde{\omega}_+ = 9.8363\pi$. Figure 2.10 shows the number of iterations the algorithm needs to achieve a residual of $10e-6$ as a function of the optimization parameters p and q of the zeroth-order transmission conditions, on the left in the iterative version and on the right for the Krylov accelerated version. The Fourier analysis shows well where the optimal parameters lie,

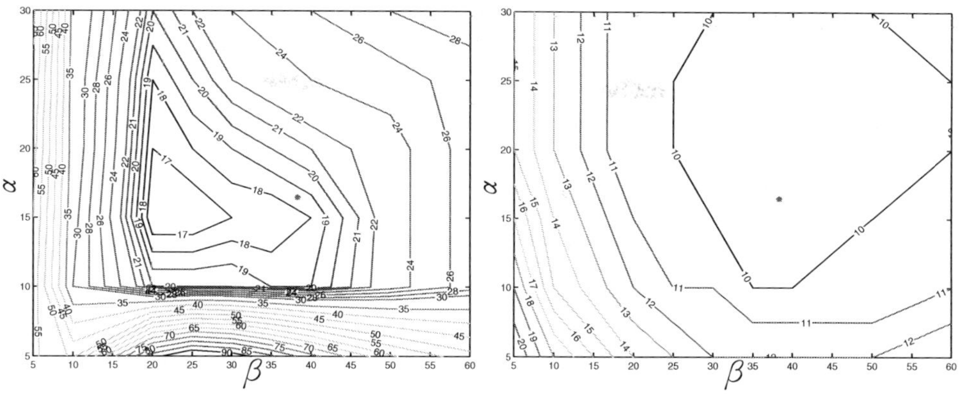

Figure 2.11. *Number of iterations needed to achieve a certain precision as a function of the optimization parameters α and β in the second-order transmission conditions for the iterative algorithm (left) and for the Krylov accelerated algorithm (right). The star denotes the optimized parameters α^* and β^* found by our Fourier analysis.*

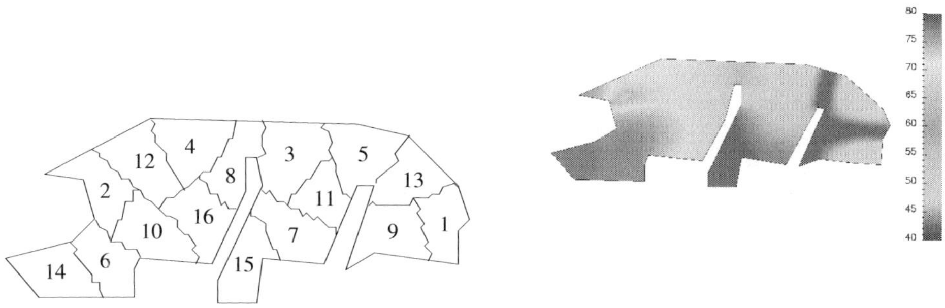

Figure 2.12. *Decomposition into* 16 *subdomains (left) and acoustic field in the passenger compartment (right) of the VOLVO S90.*

and when a Krylov method is used, the optimized Schwarz method is very robust with respect to the choice of the optimization parameter. The same holds also for the second-order transmission conditions, as Figure 2.11 shows.

Noise levels in a VOLVO S90

We analyze the noise level distribution in the passenger cabin of a VOLVO S90. The vibrations stem from the part of the car called the firewall. This example is representative of a large class of industrial problems where one tries to determine the acoustic response in the interior of a cavity caused by vibrating parts. We perform a two-dimensional simulation on a vertical cross section of the car. Figure 2.12 shows the decomposition of the car into 16 subdomains. The problem is characterized by $\tilde{\omega} a = 18.46$, which corresponds to a frequency of 1000 Hz in the car of length a. To solve the problem, the optimized Schwarz method was used as a preconditioner for the Krylov method ORTHODIR, and for the convergence criterion we used

$$||\tilde{K}u - f||_{L_2} \leq 10^{-6}||f||_{L_2}. \tag{2.91}$$

When using zeroth-order Taylor conditions and a decomposition into 16 subdomains, the method needed 105 iterations to converge, whereas when using second-order optimized transmission conditions, the method converged in 34 iterations, confirming that also in real applications the optimized Schwarz method is faster by about a factor of three, as we found for the model problem earlier. Figure 2.12 shows the acoustic field obtained in the passenger compartment of the VOLVO S90.

2.5.3 ▪ Optimized conditions for other equations

Over the last few years, a lot of results have been obtained for different classes of equations, and optimized algorithms, based on carefully chosen parameters in the transmission conditions, have been derived. For steady-state symmetric problems, we can cite [41, 86]. For advection–diffusion-type problems, see [115, 113, 114, 131, 143, 144]. As far as the Helmholtz equations are concerned, optimized transmission conditions were developed in [28, 29, 34, 91, 89]. There are also works for problems involving discontinuous coefficients in [67, 82, 96]. Other particular aspects have been discussed, such as the influence of geometry [85], coarse grid corrections [68], cross points [90], or circular domains [93]. In fluid dynamics, optimized transmission conditions have been studied in [55, 53, 54]. See also [188] for porous flow media. Like the Helmholtz equations, high-frequency time-harmonic Maxwell equations are difficult to solve by classical iterative methods since they are indefinite in nature. Optimized methods have been developed for both the first-order formulation of Maxwell equations in [56, 57, 50] or [72, 73] and also for the second-order formulation in [28, section 4.7], [34, 4, 163, 162, 166].

All the previous methods were based on optimized polynomial approximations of the symbol of transparent boundary conditions. An alternative method is the use of rational approximations of Padé type; see [6]. Perfectly matched layers (PML) [13, 30] are also used in domain decomposition methods [171, 6]. Note as well that approximate factorization methods for the Helmholtz operator can be designed using absorbing boundary conditions [92] or PML [164]. Applications to real-life problems using the discontinuous Galerkin method can be found in [56, 57, 74]. For finite element based nonoverlapping and nonconforming domain decomposition methods for the computation of multiscale electromagnetic radiation and scattering problems we refer the reader to [126, 163, 162, 166]. In a recent work the first-order and second-order formulations were presented in a unified framework in [52, 51].

Note that the method of deriving optimized transmission conditions is quite general and can in principle be applied to other types of equations.

2.6 ▪ FreeFem++ implementation of ORAS

As a numerical illustration, we solve the Helmholtz equation in a square Ω of size 6 wavelengths with the first-order absorbing boundary condition

$$-k^2 u - \Delta u = f \text{ in } \Omega, \quad \frac{\partial u}{\partial \mathbf{n}} + iku = 0 \text{ on } \partial\Omega.$$

The right-hand side f is a Gaussian function. The real part of the solution is plotted in Figure 2.13. After discretization, the resulting linear system (2.49) is solved by a domain decomposition method. As explained in Chapter 3, it is more profitable to use a Krylov method preconditioned by the ORAS operator M_{ORAS}^{-1} (2.53) in place of the iterative fixed-

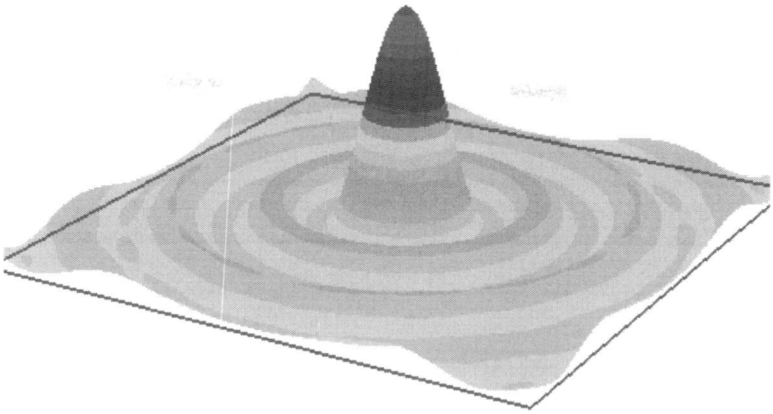

Figure 2.13. *Real part of a solution of the Helmholtz equation.*

point algorithm (2.22). We solve (2.49) by a preconditioned GMRES method. We need to provide the FreeFem++ implementation for the ORAS preconditioner. As in the case of the Schwarz algorithm, we need to create a specific data file `dataHelmholtz.edp`, a partition of unity, and the restriction and extension operators.

```
load "metis"
load "medit"
int nn=4,mm=4;      // number of the domains in each direction
int npart= nn*mm;     // total number of domains
int nloc = 40; // local no of dofs per domain in one direction
bool withmetis = 0; // =1 (Metis decomp) =0 (uniform decomp)
int sizeovr = 2; // size of the overlap
real allong = real(nn)/real(mm); // aspect ratio of the global domain
// Mesh of a rectangular domain
mesh Th=square(nn*nloc,mm*nloc,[x*allong,y]);
fespace Vh(Th,P1);
fespace Ph(Th,P0);
Ph part; // piecewise constant function
int[int] lpart(Ph.ndof); // giving the decomposition
// Domain decomposition data structures
mesh[int] aTh(npart); // sequence of ovr. meshes
matrix[int] Rihreal(npart); // local restriction operators
matrix[int] Dihreal(npart); // partition of unity operators
matrix<complex>[int] Rih(npart); // local restriction operators
matrix<complex>[int] Dih(npart); // partition of unity operators
int[int] Ndeg(npart); // number of dofs for each mesh
real[int] AreaThi(npart); // area of each subdomain
matrix<complex>[int] aA(npart); // local Dirichlet matrices
matrix<complex>[int] aR(npart); // local Robin matrices
// Definition of the problem to solve
// -k^2*u - Delta (u) = f,
// u = g, Dirichlet boundary (top and bottom)
// dn(u) + i k u = 0, Robin boundary (left and right)
int Dirichlet=1, Robin=2;
int[int] chlab=[1, Robin, 2, Robin, 3, Robin, 4, Robin];
Th=change(Th,refe=chlab);
macro Grad(u) [dx(u),dy(u)] // EOM
real k=12.*pi;
func f = exp(-((x-.5)^2+(y-.5)^2)*120.);//1; // right-hand side
```

```
func g = 0; // Dirichlet data
varf vaglobal(u,v) = int2d(Th)(-k^2*u*v+Grad(u)'*Grad(v))
                       + int1d(Th,Robin)(1i*k*u*v) - int2d(Th)(f*v)+ on(Dirichlet,u=g) ;
matrix<complex> Aglobal;
Vh<complex> rhsglobal,uglob; // rhs and solution of the global problem
complex alpha = 1i*k; // Despres algorithm
// Optimal alpha:
// h12: the mesh size for a rectangular domain of dimension 2x1,
// that is obtained with nnxmm = 2x1,4x2,8x4
real h21 = 1./(mm*nloc);
real kplus = k+pi/1; // the rectangle height is 1
real kminus = k-pi/1;
real Cw = min(kplus^2-k^2,k^2-kminus^2);
real alphaOptR = pow(Cw/h21,1/3.)/2;
complex alphaOpt = alphaOptR + 1i*alphaOptR;
//alpha = alphaOpt;
//alpha=1.0e30;
// Iterative solver
real tol=1e-4; // tolerance for the iterative method
int maxit=1000; // maximum number of iterations
```

Listing 2.1. `./FreefemCommon/dataHelmholtz.edp`

In the following script file, note that the GMRES routine has to be adapted to complex numbers.

```
/*# debutPartition #*/
include "../../FreefemCommon/dataHelmholtz.edp"
include "../../FreefemCommon/decomp.idp"
include "../../FreefemCommon/createPartition.idp"
SubdomainsPartitionUnity(Th,part[],sizeovr,aTh,Rihreal,Dihreal,Ndeg,AreaThi);
for (int i=0; i<npart; i++) {
  Rih[i] = Rihreal[i];
  Dih[i] = Dihreal[i];
}
/*# endPartition #*/
/*# debutGlobalData #*/
Aglobal = vaglobal(Vh,Vh,solver = UMFPACK); // global matrix
rhsglobal[] = vaglobal(0,Vh); // global rhs
uglob[] = Aglobal^-1*rhsglobal[];
Vh realuglob = real(uglob);
plot(realuglob,dim=3,wait=1,cmm="Exact solution", value =1,fill=1);
/*# finGlobalData #*/
/*# debutLocalData #*/
for(int i = 0;i<npart;++i)
  {
    cout << " Domain :" << i << "/" << npart << endl;
    mesh Thi = aTh[i];
    fespace Vhi(Thi,P1);
    varf RobinInt(u,v) = int2d(Thi)(-k^2*u*v+Grad(u)'*Grad(v))
                          + int1d(Thi,Robin)(1i*k*u*v) + on(Dirichlet, u=g)
                          + int1d(Thi,10)(alpha*u*v);
    aR[i] = RobinInt(Vhi,Vhi);
    set(aR[i],solver = UMFPACK); // direct solvers using UMFPACK
  }
/*# finLocalData #*/
/*# debutGMRESsolve #*/
include "GMRES.idp"
Vh<complex> un = 0, sol, er; // initial guess, final solution and error
sol[] = GMRES(un[], tol, maxit);
plot(sol,dim=3,wait=1,cmm="Final solution",value =1,fill=1);
```

```
er[] = sol[]−uglob[];
cout << "Final scaled error = " << er[].linfty/uglob[].linfty << endl;
/*# finGMRESsolve #*/
```

Listing 2.2. `./CIO/FreefemProgram/ORAS-GMRES.edp`

The domain is decomposed into 6×6 regular subdomains. Convergence curves are displayed in Figure 2.14(left) using either Dirichlet interface conditions (240 iterations) or first-order absorbing boundary conditions (100 iterations). These results are obtained by Després' one-level algorithm, that is, without a coarse space correction. We can improve them by adding a specially designed coarse space for Helmholtz equations [36]. The effect can be seen in Figure 2.14(right). We plot the convergence curves for the one-level Després algorithm and its acceleration using either 5 or 10 vectors per subdomain in the coarse space.

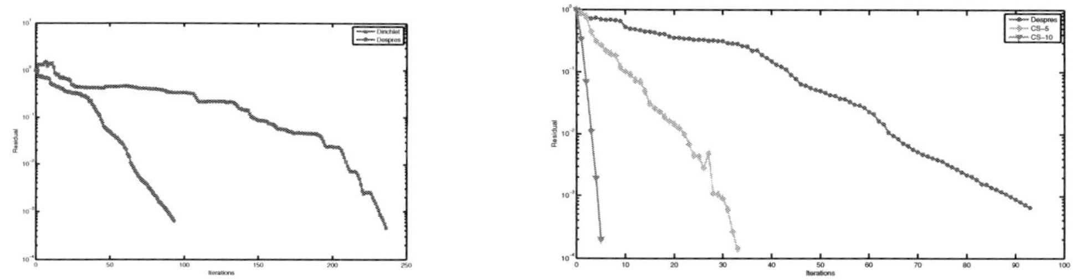

Figure 2.14. *Left: Comparison between Schwarz and Després algorithms. Right: Comparison between the one-level Després algorithm and its acceleration with a coarse space* [36].

Chapter 3

Krylov methods

In Chapter 1, we introduced Schwarz methods as iterative solvers. The aim of this chapter is to explain the benefit of using domain decomposition methods as preconditioners for Krylov methods such as CG (conjugate gradient), GMRES, or BiCGSTAB. In section 3.1, we present fixed-point methods. In section 3.2, we introduce the fundamental notion of the Krylov subspace. This yields two celebrated algorithms: the conjugate gradient (CG) method for symmetric positive definite problems in section 3.3 and the Generalized Minimal RESidual (GMRES) method for nonsymmetric linear systems in section 3.4. At the end of section 3.4, we explain why Krylov methods are always preferable to fixed-point iterations. In section 3.5, the case where the system is not invertible is dealt with. Finally, in section 3.6, we give FreeFem++ implementations and the use of Krylov methods in domain decomposition methods. It should be noted that the detailed and sometimes technical derivation of these methods is not necessary to be able to use them with success.

Recall first that the iterative versions of the Schwarz methods introduced in Chapter 1 can be written as preconditioned fixed-point iterations,

$$\mathbf{U}^{n+1} = \mathbf{U}^n + M^{-1}\mathbf{r}^n, \quad \mathbf{r}^n := \mathbf{F} - A\,\mathbf{U}^n,$$

where M^{-1} depends on the method used (RAS or ASM). The key point in what follows is to prove that fixed-point methods (3.3) are intrinsically slower than Krylov methods; for more details see [40]. Since this result is valid for any fixed-point method, we will start by placing ourselves in an abstract linear algebra framework and then apply the results to the Schwarz methods.

3.1 ▪ Fixed-point iterations

Consider the following well-posed but difficult to solve linear system

$$A\,\mathbf{x} = \mathbf{b},\ \mathbf{x} \in \mathbb{R}^N, \tag{3.1}$$

and M, an "easy to invert" matrix of the same size as A. Actually, by easy to invert we mean that the matrix-vector of M^{-1} by a residual vector $\mathbf{r} = \mathbf{b} - A\,\mathbf{x}$ is cheap. Then, (3.1) can be solved by an iterative method:

$$M\,\mathbf{x}^{n+1} = M\,\mathbf{x}^n + (\mathbf{b} - A\,\mathbf{x}^n) \Leftrightarrow M\,\mathbf{x}^{n+1} = M\,\mathbf{x}^n + \mathbf{r}^n\,. \tag{3.2}$$

This suggests the following definition.

Definition 3.1 (Fixed-point method). *The following algorithm, equivalent to* (3.2),

$$\boxed{\mathbf{x}^{n+1} = \mathbf{x}^n + M^{-1}(\mathbf{b} - A\mathbf{x}^n) \Leftrightarrow \mathbf{x}^{n+1} = \mathbf{x}^n + M^{-1}\mathbf{r}^n} \tag{3.3}$$

is called a fixed-point algorithm and the solution $\mathbf{x}$ *of* (3.1) *is a fixed point of the operator:*

$$\mathbf{x} \longmapsto \mathbf{x} + M^{-1}(\mathbf{b} - A\mathbf{x}).$$

The difference between matrices A and M is denoted by $P := M - A$. When convergent, the iteration (3.3) will converge to the solution of the preconditioned system

$$M^{-1}A\mathbf{x} = M^{-1}\mathbf{b}.$$

The above system, which has the same solution as the original system, is called a preconditioned system; here M^{-1} is called the preconditioner. In other words, a splitting method is equivalent to a fixed-point iteration on a preconditioned system. We see that the fixed-point iterations (3.3) can be written as

$$\begin{aligned}
\mathbf{x}^{n+1} &= \mathbf{x}^n + M^{-1}(\mathbf{b} - A\mathbf{x}^n) \\
&= (I - M^{-1}(M - P))\mathbf{x}^n + M^{-1}\mathbf{b} \\
&= M^{-1}P\mathbf{x}^n + M^{-1}\mathbf{b} \\
&= M^{-1}P\mathbf{x}^n + M^{-1}A\mathbf{x} = M^{-1}P\mathbf{x}^n + M^{-1}(M - P)\mathbf{x} \\
&= \mathbf{x} + M^{-1}P(\mathbf{x}^n - \mathbf{x}).
\end{aligned} \tag{3.4}$$

From (3.4) it is easy to see that the error vector $\mathbf{e}^n := \mathbf{x}^n - \mathbf{x}$ verifies

$$\mathbf{e}^{n+1} = M^{-1}P\mathbf{e}^n$$

so that

$$\mathbf{e}^{n+1} = (M^{-1}P)^{n+1}\mathbf{e}^0. \tag{3.5}$$

For this reason $M^{-1}P$ is called the *iteration matrix* associated with (3.4) and since the expression of the iteration does not change w.r.t. n, we call (3.4) a *stationary iteration*.

We recall below the well-known convergence results in the case of stationary iterations.

Lemma 3.2 (Convergence of the fixed-point iterations). *The fixed-point iteration* (3.4) *converges for arbitrary initial error* $\mathbf{e}^0$ *(that is,* $\mathbf{e}^n \to 0$ *as* $n \to \infty$*) if and only if the spectral radius of the iteration matrix is inferior to* 1, *that is,* $\rho(M^{-1}P) < 1$, *where*

$$\rho(B) = \max\{|\lambda|, \lambda \textit{ eigenvalue of } B\}.$$

For a detailed proof see, for example, [103].

In general, it is not easy to ensure that the spectral radius of the iteration matrix $M^{-1}P$ is smaller than 1, except in the case of M-matrices, for which any regular splitting $A = M-P$ (M nonsingular, M^{-1} and P non-negative) leads to a convergent fixed-point iteration (see Chapter 4 in [169] for details).

These drawbacks of fixed-point methods motivate the use of Krylov methods, which are much more efficient in terms of robustness and speed.

3.2 ▪ Krylov spaces

In this section we will show that the solution of a fixed-point method belongs to an affine space called the Krylov space. We start by showing that the solution of a fixed-point iteration can be written as a series.

Lemma 3.3 (Fixed-point solution as a series). *Let*

$$\mathbf{z}^n := M^{-1}\mathbf{r}^n = M^{-1}(\mathbf{b} - A\mathbf{x}^n)$$

be the residual preconditioned by M at the iteration n for the fixed-point iteration

$$\mathbf{x}^{n+1} = \mathbf{x}^n + \mathbf{z}^n = \mathbf{x}^n + M^{-1}(\mathbf{b} - A\mathbf{x}^n). \tag{3.6}$$

Then, the fixed-point iteration is equivalent to

$$\boxed{\mathbf{x}^{n+1} = \mathbf{x}^0 + \sum_{i=0}^{n} (M^{-1}P)^i\, \mathbf{z}^0 .} \tag{3.7}$$

Proof. Note that we have that

$$\mathbf{x}^{n+1} = \mathbf{x}^n + M^{-1}\mathbf{r}^n = \mathbf{x}^n + \mathbf{z}^n \Rightarrow \mathbf{x}^{n+1} = \mathbf{x}^0 + \mathbf{z}^0 + \mathbf{z}^1 + \cdots + \mathbf{z}^n. \tag{3.8}$$

From (3.5) we see that the residual vector $\mathbf{r}^n = \mathbf{b} - A\mathbf{x}^n = -A(\mathbf{x}^n - \mathbf{x})$ verifies

$$\begin{aligned} \mathbf{r}^n &= -A\mathbf{e}^n = (P - M)(M^{-1}P)^n\mathbf{e}^0 = (PM^{-1})^n P\mathbf{e}^0 - (PM^{-1})^{n-1}P\mathbf{e}^0 \\ &= (PM^{-1})^n(P\mathbf{e}^0 - M\mathbf{e}^0) = (PM^{-1})^n\mathbf{r}^0. \end{aligned} \tag{3.9}$$

From (3.9) we have that

$$\mathbf{z}^n = M^{-1}\mathbf{r}^n = M^{-1}(PM^{-1})^n\mathbf{r}^0 = M^{-1}(PM^{-1})^n M\mathbf{z}^0 = (M^{-1}P)^n\mathbf{z}^0. \tag{3.10}$$

From (3.8) and (3.10) we obtain

$$\mathbf{x}^{n+1} = \mathbf{x}^0 + \mathbf{z}^0 + (M^{-1}P)\mathbf{z}^0 + \cdots + (M^{-1}P)^n\mathbf{z}^0 = \mathbf{x}^0 + \sum_{i=0}^{n}(M^{-1}P)^i\,\mathbf{z}^0, \tag{3.11}$$

which leads to the conclusion. Thus the error $\mathbf{x}^{n+1} - \mathbf{x}^0$ is a geometric series of common ratio $M^{-1}P$. Note that (3.11) can also be written in terms of the residual vector.

$$\begin{aligned} \mathbf{x}^{n+1} &= \mathbf{x}^0 + M^{-1}\mathbf{r}^0 + (M^{-1}P)M^{-1}\mathbf{r}^0 + \cdots + (M^{-1}P)^n M^{-1}\mathbf{r}^0 \\ &= \mathbf{x}^0 + \sum_{i=0}^{n}(M^{-1}P)^i M^{-1}\,\mathbf{r}^0. \qquad \square \end{aligned} \tag{3.12}$$

The previous result leads to the following remark.

Remark 3.2.1. *The correction $\mathbf{x}^{n+1} - \mathbf{x}^0$ is given by*

$$\mathbf{x}^{n+1} - \mathbf{x}^0 = S_n(M^{-1}P)M^{-1}\,\mathbf{r}^0 = S_n(M^{-1}P)\,\mathbf{z}^0, \tag{3.13}$$

where S_n is the polynomial given by $S_n(t) = 1 + t + \cdots + t^n$. Moreover, from (3.12) *we see that*

$$\begin{aligned} \mathbf{x}^{n+1} - \mathbf{x}^0 &\in \mathrm{Span}\{M^{-1}\mathbf{r}^0, (M^{-1}P)M^{-1}\mathbf{r}^0, \ldots, (M^{-1}P)^n M^{-1}\mathbf{r}^0\} \\ &= \mathrm{Span}\{\mathbf{z}^0, (M^{-1}P)\mathbf{z}^0, \ldots, (M^{-1}P)^n \mathbf{z}^0\}. \end{aligned} \tag{3.14}$$

In conclusion, the solution of a fixed-point iteration is generated in a space spanned by powers of the iteration matrix $M^{-1}P = I - M^{-1}A$ applied to a given vector. The main computational cost is thus given by multiplication by the matrix A and by the application of M^{-1}. At nearly the same cost, we could generate better approximations in the same space.

Definition 3.4 (Krylov spaces). *For a given matrix B and a vector $\mathbf{y}$ we define the Krylov subspaces of dimension n associated to B and $\mathbf{y}$ by*

$$\boxed{\mathcal{K}^n(B, \mathbf{y}) := \mathrm{Span}\{\mathbf{y}, B\mathbf{y}, \ldots, B^{n-1}\mathbf{y}\}.}$$

Therefore, a Krylov space is a space of polynomials of a matrix times a vector.

According to the previous definition, we see from (3.14) that

$$\mathbf{x}^n - \mathbf{x}^0 \in \mathcal{K}^n(M^{-1}P, M^{-1}\mathbf{r}^0) = \mathcal{K}^n(M^{-1}P, \mathbf{z}^0).$$

By the same kind of reasoning, we also have that

$$\mathbf{e}^n = (M^{-1}P)^n \mathbf{e}^0 \Rightarrow \mathbf{e}^n \in \mathcal{K}^{n+1}(M^{-1}P, \mathbf{e}^0),$$

$$\mathbf{z}^n = (M^{-1}P)^n \mathbf{z}^0 \Rightarrow \mathbf{z}^n \in \mathcal{K}^{n+1}(M^{-1}P, \mathbf{z}^0).$$

But since $M^{-1}P = I - M^{-1}A$ we see that

$$\mathcal{K}^n(M^{-1}P, \mathbf{z}^0) = \mathcal{K}^n(M^{-1}A, \mathbf{z}^0).$$

A reasonable way to compute a solution iteratively would be to look for an optimal element in the above-mentioned space. This element will necessarily be better and will approach the solution more quickly than a fixed-point method. Krylov methods differ by the way an "optimality" condition is imposed in order to uniquely define $\mathbf{x}^n$ and by the actual algorithm that implements it.

Remark 3.2.2. *Note also that the difference between two successive iterates is given by*

$$\begin{aligned} \mathbf{x}^{n+1} - \mathbf{x}^n &= (M^{-1}P)\mathbf{z}^n = (M^{-1}P)^n M^{-1}\mathbf{r}^0 \\ &= (I - M^{-1}A)^n M^{-1}\mathbf{r}^0 \in \mathcal{K}^{n+1}(M^{-1}A, M^{-1}\mathbf{r}^0). \end{aligned}$$

This sequence can be further improved by searching for a better solution in this Krylov space.

Example 3.5. *From the family of fixed-point methods we can mention the Richardson iterations with a relaxation parameter:*

$$\mathbf{x}^{n+1} = \mathbf{x}^n + \alpha \mathbf{r}^n = \mathbf{x}^n + \alpha(\mathbf{b} - A\mathbf{x}^n). \tag{3.15}$$

The parameter α can be chosen in such a way as to have the best convergence factor over the iterations. In the case of a symmetric positive definite matrix, the value of α which minimizes the convergence rate of the algorithm is

$$\alpha_{opt} = \frac{2}{\lambda_{\min}(A) + \lambda_{\max}(A)},$$

and in this case the error behaves as

$$||\mathbf{e}^n||_A \leq \left(\frac{\kappa_2(A)-1}{\kappa_2(A)+1}\right)^n ||\mathbf{e}^0||_A,$$

where

$$\kappa_2(A) = \frac{\lambda_{\max}(A)}{\lambda_{\min}(A)}$$

is the condition number of A, $\lambda_{\max}(A)$ and $\lambda_{\min}(A)$ being the extreme eigenvalues of A. In practice, it is very difficult to accurately estimate $\lambda_{\min}$ or $\lambda_{\max}$ and thus α_{opt}. This motivates the introduction of the gradient method in the next section.

3.2.1 ▪ Gradient methods

This suggests the definition of a new class of methods for symmetric positive matrices.

Definition 3.6 (Gradient methods). *A descent method is an iteration of the form*

$$\mathbf{x}^{n+1} = \mathbf{x}^n + \alpha_n \mathbf{p}^n,\ \alpha_n \geq 0, \tag{3.16}$$

where α_n is chosen such that it minimizes $||\mathbf{e}^{n+1}||_A^2 = ||\mathbf{x}^{n+1} - \mathbf{x}||_A^2$ (the square of the A-norm for the error norm at each iteration, $||\mathbf{e}^{n+1}||_A^2 := (A\,\mathbf{e}^{n+1}, \mathbf{e}^{n+1})_2$) and vector $\mathbf{p}^n$ is called the descent direction. If $\mathbf{p}^n = \mathbf{r}^n$, the resulting method is called the optimal gradient method.

Consider function f,

$$\alpha \longmapsto f(\alpha) := ||\mathbf{x}^n + \alpha\,\mathbf{p}^n - \mathbf{x}||_A^2,$$

and the minimization problem

$$\begin{aligned}
f(\alpha_n) &:= ||\underbrace{\mathbf{x}^n + \alpha_n \mathbf{p}^n}_{\mathbf{x}^{n+1}} - \mathbf{x}||_A^2 \\
&= \min_{\alpha} ||\mathbf{x}^n + \alpha \mathbf{p}^n - \mathbf{x}||_A^2 = \min_{\alpha} (A(\mathbf{x}^n + \alpha \mathbf{p}^n - \mathbf{x}), \mathbf{x}^n + \alpha \mathbf{p}^n - \mathbf{x}) \\
&= \min_{\alpha} \left[\alpha^2 (A\mathbf{p}^n, \mathbf{p}^n) + 2\alpha (A\mathbf{p}^n, \mathbf{x}^n - \mathbf{x}) + (A(\mathbf{x}^n - \mathbf{x}), \mathbf{x}^n - \mathbf{x})\right] \\
&= \min_{\alpha} [\alpha^2 (A\mathbf{p}^n, \mathbf{p}^n) - 2\alpha (\mathbf{p}^n, \underbrace{A(\mathbf{x} - \mathbf{x}^n)}_{\mathbf{r}^n})] + (A(\mathbf{x}^n - \mathbf{x}), \mathbf{x}^n - \mathbf{x}).
\end{aligned}$$

We have that the optimal parameter α_n is characterized by

$$\frac{\partial f}{\partial \alpha}(\alpha_n) = 0 \Rightarrow \alpha_n = \frac{(\mathbf{p}^n, \mathbf{r}^n)}{(A\mathbf{p}^n, \mathbf{p}^n)}. \tag{3.17}$$

Note that in the case of the gradient method ($\mathbf{p}^n = \mathbf{r}^n$), α_n becomes

$$\alpha_n = \frac{||\mathbf{r}^n||_2^2}{||\mathbf{r}^n||_A^2} = \frac{(\mathbf{e}^n, \mathbf{r}^n)_A}{||\mathbf{r}^n||_A^2}. \tag{3.18}$$

We have the following convergence result.

Lemma 3.7 (Optimal gradient convergence). *The optimal step gradient method* $\mathbf{x}^{n+1} = \mathbf{x}^n + \alpha_n \mathbf{r}^n$*, with* α_n *chosen in* (3.18)*, converges and the error estimate verifies the estimate*

$$\|\mathbf{e}^n\|_A \le \left(1 - \frac{1}{\kappa_2(A)}\right)^{\frac{n}{2}} \|\mathbf{e}^0\|_A,$$

where $\kappa_2(A) = \frac{\lambda_{\max}(A)}{\lambda_{\min}(A)}$ *is the condition number of* A.

Proof. From (3.16) we see that

$$\begin{aligned}
\|\mathbf{e}^{n+1}\|_A^2 &= \|\mathbf{e}^n + \alpha_n \mathbf{r}^n\|_A^2 = \|\mathbf{e}^n\|_A^2 + 2\alpha_n \underbrace{(\mathbf{e}^n, \mathbf{r}^n)_A}_{-\alpha_n \|\mathbf{r}^n\|_A^2} + \alpha_n^2 \|\mathbf{r}^n\|_A^2 \\
&= \|\mathbf{e}^n\|_A^2 - \alpha_n^2 \|\mathbf{r}^n\|_A^2 = \|\mathbf{e}^n\|_A^2 - \alpha_n^2 (A\mathbf{r}^n, \mathbf{r}^n) = \|\mathbf{e}^n\|_A^2 - \frac{(\mathbf{r}^n, \mathbf{r}^n)^2}{(A\mathbf{r}^n, \mathbf{r}^n)} \\
&= \|\mathbf{e}^n\|_A^2 \left(1 - \frac{(\mathbf{r}^n, \mathbf{r}^n)}{(A^{-1}\mathbf{r}^n, \mathbf{r}^n)} \cdot \frac{(\mathbf{r}^n, \mathbf{r}^n)}{(A\mathbf{r}^n, \mathbf{r}^n)}\right),
\end{aligned}$$

where we have used the identity

$$\|\mathbf{e}^n\|_A^2 = (\mathbf{e}^n, A\mathbf{e}^n) = (A^{-1}\mathbf{r}^n, \mathbf{r}^n).$$

Matrix A being symmetric positive definite we have

$$\lambda_{\min}(A) \le \frac{(\mathbf{r}^n, \mathbf{r}^n)}{(A^{-1}\mathbf{r}^n, \mathbf{r}^n)} \le \lambda_{\max}(A) \text{ and } \frac{1}{\lambda_{\max}(A)} \le \frac{(\mathbf{r}^n, \mathbf{r}^n)}{(A\mathbf{r}^n, \mathbf{r}^n)} \le \frac{1}{\lambda_{\min}(A)},$$

which leads to the estimate

$$\|\mathbf{e}^{n+1}\|_A^2 \le \left(1 - \frac{\lambda_{\min}(A)}{\lambda_{\max}(A)}\right) \|\mathbf{e}^n\|_A^2 \Rightarrow \|\mathbf{e}^n\|_A \le \left(1 - \frac{1}{\kappa_2(A)}\right)^{\frac{n}{2}} \|\mathbf{e}^0\|_A,$$

and this ends the proof. □

3.3 ▪ The conjugate gradient method

We present here a Krylov method that applies to symmetric positive definite (SPD) matrices. Suppose that we want to solve $A\mathbf{x} = \mathbf{b}$ with A an SPD matrix. The idea is that the solution at each iteration solves a minimization problem in the A-norm over a Krylov space:

$$\text{Find } \mathbf{y}^n \in \mathcal{K}^n(A, \mathbf{r}^0) \text{ such that}$$

$$\|\mathbf{e}^n\|_A = \|\mathbf{x} - \underbrace{(\mathbf{x}_0 + \mathbf{y}^n)}_{\mathbf{x}^n}\|_A = \min_{\mathbf{w} \in \mathcal{K}^n(A, \mathbf{r}^0)} \|\mathbf{x} - (\mathbf{x}_0 + \mathbf{w})\|_A. \tag{3.19}$$

In the following we will define a new method that extends the idea of gradient methods and will prove afterwards that it is exactly the Krylov method whose iterates are given by (3.19).

Definition 3.8 (Conjugate gradient). *Starting from an initial guess* $\mathbf{x}^0$ *and an initial descent direction* $\mathbf{p}^0 = \mathbf{r}^0 = \mathbf{b} - A\mathbf{x}^0$, *a conjugate gradient method is an iteration of the form*

$$\begin{aligned} \mathbf{x}^{n+1} &= \mathbf{x}^n + \alpha_n \mathbf{p}^n, \\ \mathbf{r}^{n+1} &= \mathbf{r}^n - \alpha_n A\mathbf{p}^n, \\ \mathbf{p}^{n+1} &= \mathbf{r}^{n+1} + \beta_{n+1}\mathbf{p}^n, \end{aligned} \tag{3.20}$$

where α_n *and* β_n *are chosen such that they minimize the norm of the error* $\|\mathbf{e}^{n+1}\|_A^2 = \|\mathbf{x}^{n+1} - \mathbf{x}\|_A^2$ *at each iteration.*

Coefficients α_n and β_{n+1} are easy to find. First, from (3.18) we see that the coefficient α_n which minimizes $\|\mathbf{e}^{n+1}\|_A^2$ is necessarily given by

$$\alpha_n = \frac{(\mathbf{p}^n, \mathbf{r}^n)}{(A\mathbf{p}^n, \mathbf{p}^n)}. \tag{3.21}$$

By taking the dot product of the second relation of (3.20) by $\mathbf{p}^n$ and by using (3.21), we see that

$$(\mathbf{r}^{n+1}, \mathbf{p}^n) = (\mathbf{r}^n, \mathbf{p}^n) - \alpha_n (A\mathbf{p}^n, \mathbf{p}^n) = 0.$$

By taking the dot product of the third relation of (3.20) by $\mathbf{r}^{n+1}$ and by using the previous orthogonality relation we obtain

$$(\mathbf{p}^{n+1}, \mathbf{r}^{n+1}) = (\mathbf{r}^{n+1}, \mathbf{r}^{n+1}) + \beta_{n+1}(\mathbf{r}^{n+1}, \mathbf{p}^n) = (\mathbf{r}^{n+1}, \mathbf{r}^{n+1}) \quad \forall n \geq 0.$$

By substituting the last relation taken in n into (3.21) we get

$$\boxed{\alpha_n = \frac{(\mathbf{r}^n, \mathbf{r}^n)}{(A\mathbf{p}^n, \mathbf{p}^n)} = \frac{\|\mathbf{r}^n\|_2^2}{\|\mathbf{p}^n\|_A^2}.} \tag{3.22}$$

In a similar way we can find β_{n+1}. If we substitute (3.22) taken in $n+1$ into $\|\mathbf{e}^{n+2}\|_A^2$ and use the third relation of (3.20) to replace $\mathbf{p}^{n+1}$ we get

$$\begin{aligned} \|\mathbf{e}^{n+2}\|_A^2 &= \|\mathbf{x}^{n+2} - \mathbf{x}\|_A^2 = \|\mathbf{x}^{n+1} - \mathbf{x} + \alpha_{n+1}\mathbf{p}^{n+1}\|_A^2 \\ &= \|\mathbf{e}^{n+1}\|_A^2 + 2(\underbrace{A\mathbf{e}^{n+1}}_{-\mathbf{r}^{n+1}}, \alpha_{n+1}\mathbf{p}^{n+1}) + \alpha_{n+1}^2 \|\mathbf{p}^{n+1}\|_A^2 \\ &= \|\mathbf{e}^{n+1}\|_A^2 - 2\alpha_{n+1}(\mathbf{r}^{n+1}, \mathbf{r}^{n+1}) + \alpha_{n+1}^2 \|\mathbf{p}^{n+1}\|_A^2 \\ &= \|\mathbf{e}^{n+1}\|_A^2 - \frac{(\mathbf{r}^{n+1}, \mathbf{r}^{n+1})^2}{(A\mathbf{p}^{n+1}, \mathbf{p}^{n+1})} \\ &= \|\mathbf{e}^{n+1}\|_A^2 - \frac{(\mathbf{r}^{n+1}, \mathbf{r}^{n+1})^2}{(A(\mathbf{r}^{n+1} + \beta_{n+1}\mathbf{p}^n), (\mathbf{r}^{n+1} + \beta_{n+1}\mathbf{p}^n))}. \end{aligned}$$

Therefore, minimizing $\|\mathbf{e}^{n+2}\|_A^2$ with respect to β_{n+1} is equivalent to minimizing the quadratic form $(A(\mathbf{r}^{n+1} + \beta_{n+1}\mathbf{p}^n), (\mathbf{r}^{n+1} + \beta_{n+1}\mathbf{p}^n))$ with respect to β_{n+1}, which gives

$$\beta_{n+1} = -\frac{(A\mathbf{r}^{n+1}, \mathbf{p}^n)}{(A\mathbf{p}^n, \mathbf{p}^n)}. \tag{3.23}$$

By taking the A-dot product of the third relation of (3.20) by $\mathbf{p}^n$ and by introducing (3.23) into it, we see that

$$(A\mathbf{p}^{n+1},\mathbf{p}^n) = (A\mathbf{r}^{n+1},\mathbf{p}^n) + \beta_{n+1}(A\mathbf{p}^n,\mathbf{p}^n) = 0.$$

Using this identity and taking the A-dot product of the third relation of (3.20) by $\mathbf{p}^{n+1}$ we get

$$\begin{aligned}(A\mathbf{p}^{n+1},\mathbf{p}^{n+1}) &= (A\mathbf{r}^{n+1},\mathbf{p}^{n+1}) + \beta_n(A\mathbf{p}^{n+1},\mathbf{p}^n)\\ &= (A\mathbf{r}^{n+1},\mathbf{p}^{n+1}) = (A\mathbf{p}^{n+1},\mathbf{r}^{n+1}).\end{aligned} \tag{3.24}$$

By introducing (3.24) into the dot product of the second relation of (3.20) by $\mathbf{r}^n$, we see that

$$(\mathbf{r}^{n+1},\mathbf{r}^n) = (\mathbf{r}^n,\mathbf{r}^n) - \alpha_n(A\mathbf{p}^n,\mathbf{r}^n) = (\mathbf{r}^n,\mathbf{r}^n) - \alpha_n(A\mathbf{p}^n,\mathbf{p}^n) = 0. \tag{3.25}$$

Finally, by taking the dot product of the second relation of (3.20) by $\mathbf{r}^{n+1}$ and by using (3.25), we see that

$$(\mathbf{r}^{n+1},A\mathbf{p}^n) = -\frac{\|\mathbf{r}^{n+1}\|_2^2}{\alpha_n}. \tag{3.26}$$

By plugging (3.26) into (3.23) we conclude by using the expression of (3.22) that

$$\boxed{\beta_{n+1} = \frac{\|\mathbf{r}^{n+1}\|_2^2}{\|\mathbf{r}^n\|_2^2}.} \tag{3.27}$$

The resulting algorithm is given in Algorithm 3.1; see [9].

Algorithm 3.1 CG algorithm

Compute $\mathbf{r}_0 := b - A\mathbf{x}_0$, $\mathbf{p}_0 = \mathbf{r}_0$
for $i = 0, 1, \ldots$ **do**
  $\rho_i = (\mathbf{r}_i, \mathbf{r}_i)_2$
  $\mathbf{q}_i = A\mathbf{p}_i$
  $\alpha_i = \dfrac{\rho_i}{(\mathbf{p}_i,\mathbf{q}_i)_2}$
  $\mathbf{x}_{i+1} = \mathbf{x}_i + \alpha_i \mathbf{p}_i$
  $\mathbf{r}_{i+1} = \mathbf{r}_i - \alpha_i \mathbf{q}_i$
  $\rho_{i+1} = (\mathbf{r}_{i+1}, \mathbf{r}_{i+1})_2$
  $\beta_{i+1} = \dfrac{\rho_{i+1}}{\rho_i}$
  $\mathbf{p}_{i+1} = \mathbf{r}_{i+1} + \beta_{i+1}\mathbf{p}_i$
  check convergence; continue if necessary
end for

We can see that the conjugate gradient algorithm has a few remarkable properties that can be stated in the following lemma.

Lemma 3.9 (Conjugate gradient as a Krylov method). *The descent directions* $\mathbf{p}^n$ *are* A*-orthogonal or conjugate*

$$(A\mathbf{p}^k,\mathbf{p}^l) = 0 \quad \forall k,l,\ k \neq l,$$

and the residual vectors are orthogonal between them and to the descent directions

$$(\mathbf{r}^k,\mathbf{r}^l) = 0 \quad \forall k,l,\ k\neq l,\ \textit{and}\ (\mathbf{p}^k,\mathbf{r}^l) = 0 \quad \forall k,l,\ k<l.$$

Moreover, both the descent directions and the residuals span the Krylov space,

$$\mathcal{K}^{n+1}(A,\mathbf{r}^0)=\mathrm{Span}\{\mathbf{r}^0,\mathbf{r}^1,\dots,\mathbf{r}^n\}=\mathrm{Span}\{\mathbf{p}^0,\mathbf{p}^1,\dots,\mathbf{p}^n\}, \tag{3.28}$$

and vector $\mathbf{x}^n$ *from* (3.20) *minimizes the error norm* $||\mathbf{e}^n||_A^2$ *on the affine space* $\mathbf{x}^0+\mathcal{K}^n(A,\mathbf{r}^0)$. *The algorithm will converge in at most* N *iterations,* N *being the size of matrix* A.

Proof. The proof is a direct consequence of the algebraic manipulations performed previously to find the coefficients α_n and β_{n+1}. The property (3.28) can be obtained by recurrence on the relations (3.20) and in the same way by the fact that $\mathbf{x}^n\in\mathbf{x}^0+\mathcal{K}^n(A,\mathbf{r}^0)$. The optimality of x^n is proved in [169] or [100], e.g.. Moreover, the matrix A being symmetric positive definite, there exists an A-orthogonal basis of size N, which explains why we cannot have more than N conjugate descent vectors. This leads to a convergence in at most N iterations. □

Lemma 3.10 (Convergence of the conjugate gradient method). *The conjugate gradient method defined in* (3.20) *converges to the solution of the linear system* $A\mathbf{x}=\mathbf{b}$, *and we have the error estimate*

$$||\mathbf{e}^n||_A\le 2\left(\frac{\sqrt{\kappa_2(A)}-1}{\sqrt{\kappa_2(A)}+1}\right)^n ||\mathbf{e}^0||_A, \tag{3.29}$$

where $\kappa_2(A)$ *is the condition number of the matrix* A.

Proof. As we have seen previously, the conjugate gradient method is a Krylov method. Since

$$\mathbf{e}^n\in\mathcal{K}^{n+1}(A,\mathbf{e}^0),$$

this means that building a sequence of iterates $\mathbf{x}^n$ is equivalent to building a sequence of polynomials $P_n\in\mathcal{P}_n$ with real coefficients such that

- $\mathbf{e}^n=P_n(A)\mathbf{e}^0$,
- $P_n(0)=1$, and
- P_n minimizes $J(Q)=||Q(A)\mathbf{e}^0||_A^2$ over all the polynomials $Q\in\mathcal{P}_n$ such that $Q(0)=1$.

Matrix A can be diagonalized using orthonormal eigenvectors $A=T\Lambda T^t$. For any given polynomial Q,

$$Q(A)=TQ(\Lambda)T^t.$$

Therefore we have for any monic (coefficient of the leading term is equal to 1) polynomial Q of degree n

$$||\mathbf{e}^n||_A=||P_n(A)\mathbf{e}^0||_A\le||Q(A)\mathbf{e}^0||_A=||TQ(\Lambda)T^t\mathbf{e}^0||_A\le\max_{\lambda\in\sigma(A)}|Q(\lambda)|||\mathbf{e}^0||_A$$

and the estimate yields

$$||\mathbf{e}^n||_A\le\min_{Q\in\mathcal{P}_n,Q(0)=1}\ \max_{\lambda\in[\lambda_1,\lambda_2]}|Q(\lambda)|||\mathbf{e}^0||_A, \tag{3.30}$$

where $\lambda_1=\lambda_{\min}(A)$ and $\lambda_2=\lambda_{\max}(A)$. We know from [26] that

$$\max_{x\in[\lambda_1,\lambda_2]}Q^*(x)=\min_{Q\in\mathcal{P}_n,Q(0)=1}\ \max_{x\in[\lambda_1,\lambda_2]}|Q(x)|,$$

where

$$Q^*(x) = \frac{T_n\left(\frac{2x-\lambda_1-\lambda_2}{\lambda_2-\lambda_1}\right)}{T_n\left(\frac{-\lambda_1-\lambda_2}{\lambda_2-\lambda_1}\right)},$$

with T_n being the Chebyshev polynomial of the first kind

$$T_n(x) = \begin{cases} \dfrac{(x+\sqrt{x^2-1})^n + (x-\sqrt{x^2-1})^n}{2}, & |x| > 1, \\ \cos(n \arccos(x)), & |x| \le 1. \end{cases}$$

We see that

$$\begin{aligned} \max_{x \in [\lambda_1, \lambda_2]} |Q^*(x)| &= \frac{1}{\left|T_n\left(\frac{-\lambda_1-\lambda_2}{\lambda_2-\lambda_1}\right)\right|} = \frac{1}{T_n\left(\frac{\kappa_2(A)+1}{\kappa_2(A)-1}\right)} \\ &= \frac{2}{\left(\frac{\sqrt{\kappa_2(A)}-1}{\sqrt{\kappa_2(A)}+1}\right)^{-n} + \left(\frac{\sqrt{\kappa_2(A)}-1}{\sqrt{\kappa_2(A)}+1}\right)^{n}} \le 2\left(\frac{\sqrt{\kappa_2(A)}-1}{\sqrt{\kappa_2(A)}+1}\right)^n. \end{aligned} \tag{3.31}$$

From (3.31) and (3.30) the conclusion follows. □

Remark 3.3.1. *We estimate the iteration counts to reach a reduction factor of the error by a factor $\epsilon \ll 1$ for a poorly conditioned system, $\kappa_2(A) \gg 1$. From Lemma* 3.7, *the optimal gradient method will converge in n_{OG} iterations with*

$$\epsilon = \left(1 - \frac{1}{\kappa_2(A)}\right)^{\frac{n_{OG}}{2}}.$$

Thus, we have

$$\log \epsilon = \frac{n_{OG}}{2} \log\left(1 - \frac{1}{\kappa_2(A)}\right),$$

that is,

$$n_{OG} \simeq 2\kappa_2(A)(-\log \epsilon).$$

From Lemma 3.10, *the conjugate gradient method will converge in n_{CG} iterations with*

$$n_{CG} \simeq \frac{\sqrt{\kappa_2(A)}}{2}(-\log \epsilon).$$

These estimates are clearly in favor of the CG algorithm over the optimal gradient method.

3.3.1 ▪ The preconditioned conjugate gradient method

In order to have a faster convergence, an SPD preconditioner, denoted here by M, is often used. In order to preserve the SPD properties, we use the fact that M^{-1} admits an SPD square root, denoted by $M^{-1/2}$, such that

$$M^{-1/2} \cdot M^{-1/2} = M^{-1}.$$

The CG method is applied to the symmetrized preconditioned system

$$(M^{-1/2} A M^{-1/2}) M^{1/2} \mathbf{x} = M^{-1/2} \mathbf{b}$$

instead of simply $M^{-1}Ax = M^{-1}b$ since the latter is generally nonsymmetric, but its spectrum is the same as that of matrix $M^{-1/2}AM^{-1/2}$. A naive and costly implementation would require the quite expensive computation of the square root of the inverse of operator M. Once again, a detailed analysis reveals that it can be bypassed.

Lemma 3.11 (Preconditioned conjugate gradient). *The conjugate gradient method applied to the system*

$$\tilde{A}\tilde{\mathbf{x}} = \tilde{\mathbf{b}}, \tag{3.32}$$

with

$$\tilde{A} = M^{-1/2}AM^{-1/2},\ \tilde{\mathbf{x}} = M^{1/2}\mathbf{x},\ \tilde{\mathbf{b}} = M^{-1/2}\mathbf{b},$$

can be rewritten as the following iterative method: starting from an initial guess $\mathbf{x}^0$ *and an initial descent direction*

$$\mathbf{p}^0 = M^{-1}\mathbf{r}^0 = M^{-1}(\mathbf{b} - A\mathbf{x}^0),$$

compute the sequence of approximations as

$$\begin{aligned}
\mathbf{x}^{n+1} &= \mathbf{x}^n + \alpha_n \mathbf{p}^n,\\
\mathbf{r}^{n+1} &= \mathbf{r}^n - \alpha_n A\mathbf{p}^n,\\
\mathbf{p}^{n+1} &= M^{-1}\mathbf{r}^{n+1} + \beta_{n+1}\mathbf{p}^n,
\end{aligned} \tag{3.33}$$

where α_n *and* β_{n+1} *are given by*

$$\boxed{\alpha_n = \frac{(M^{-1}\mathbf{r}^n, \mathbf{r}^n)}{\|\mathbf{p}^n\|_A^2},\ \beta_{n+1} = \frac{(M^{-1}\mathbf{r}^{n+1}, \mathbf{r}^{n+1})}{(M^{-1}\mathbf{r}^n, \mathbf{r}^n)}.} \tag{3.34}$$

Proof. Suppose we now apply the conjugate gradient method (3.20) to the system (3.32). This gives

$$\begin{aligned}
\tilde{\mathbf{x}}^{n+1} &= \tilde{\mathbf{x}}^n + \alpha_n \tilde{\mathbf{p}}^n,\\
\tilde{\mathbf{r}}^{n+1} &= \tilde{\mathbf{r}}^n - \alpha_n \tilde{A}\tilde{\mathbf{p}}^n,\\
\tilde{\mathbf{p}}^{n+1} &= \tilde{\mathbf{r}}^{n+1} + \beta_{n+1}\tilde{\mathbf{p}}^n,
\end{aligned} \tag{3.35}$$

with

$$\tilde{\mathbf{x}}^n = M^{1/2}\mathbf{x}^n,\ \tilde{\mathbf{p}}^n = M^{1/2}\mathbf{p}^n,\ \tilde{\mathbf{r}}^n = M^{-1/2}\mathbf{r}^n,\ \tilde{A} = M^{-1/2}AM^{-1/2} \tag{3.36}$$

and the coefficients

$$\begin{aligned}
\alpha_n &= \frac{(\tilde{\mathbf{r}}^n, \tilde{\mathbf{r}}^n)}{\|\tilde{\mathbf{p}}^n\|_{\tilde{A}}^2} = \frac{(M^{-1/2}\mathbf{r}^n, M^{-1/2}\mathbf{r}^n)}{((M^{-1/2}AM^{-1/2})M^{1/2}\mathbf{p}^n, M^{1/2}\mathbf{p}^n)} = \frac{(M^{-1}\mathbf{r}^n, \mathbf{r}^n)}{\|\mathbf{p}^n\|_A^2},\\
\beta_n &= \frac{(\tilde{\mathbf{r}}^{n+1}, \tilde{\mathbf{r}}^{n+1})}{(\tilde{\mathbf{r}}^n, \tilde{\mathbf{r}}^n)} = \frac{(M^{-1/2}\mathbf{r}^{n+1}, M^{-1/2}\mathbf{r}^{n+1})}{(M^{-1/2}\mathbf{r}^n, M^{-1/2}\mathbf{r}^n)} = \frac{(M^{-1}\mathbf{r}^{n+1}, \mathbf{r}^{n+1})}{(M^{-1}\mathbf{r}^n, \mathbf{r}^n)},
\end{aligned} \tag{3.37}$$

which are exactly the coefficients from (3.34).

By now substituting (3.36) into the three relations of (3.35), we obtain the iterations (3.33). The initial descent direction $\mathbf{p}^0$ is derived from

$$\tilde{\mathbf{p}}^0 = \tilde{\mathbf{r}}^0 \Leftrightarrow M^{1/2}\mathbf{p}^0 = M^{-1/2}\mathbf{r}^0 \Leftrightarrow \mathbf{p}^0 = M^{-1}\mathbf{r}^0,$$

which ends the proof. □

We have just proved that the preconditioned conjugate gradient (PCG) requires only the application of the preconditioner M^{-1} and that the computation of $M^{-1/2}$ is thus not required. Moreover, since the spectrums of $M^{-1/2}AM^{-1/2}$ and $M^{-1}A$ are the same, in the convergence estimate (3.29) we simply have to replace $\kappa_2(A)$ by $\kappa_2(M^{-1}A)$.

The resulting iterative procedure is given in Algorithm 3.2.

Algorithm 3.2 PCG algorithm

Compute $\mathbf{r}_0 := b - A\mathbf{x}_0$, $\mathbf{z}_0 = M^{-1}\mathbf{r}_0$, $\mathbf{p}_0 = \mathbf{z}_0$.
for $i = 0, 1, \ldots$ **do**
  $\rho_i = (\mathbf{r}_i, \mathbf{z}_i)_2$
  $\mathbf{q}_i = A\mathbf{p}_i$
  $\alpha_i = \dfrac{\rho_i}{(\mathbf{p}_i, \mathbf{q}_i)_2}$
  $\mathbf{x}_{i+1} = \mathbf{x}_i + \alpha_i \mathbf{p}_i$
  $\mathbf{r}_{i+1} = \mathbf{r}_i - \alpha_i \mathbf{q}_i$
  $\mathbf{z}_{i+1} = M^{-1}\mathbf{r}_{i+1}$
  $\rho_{i+1} = (\mathbf{r}_{i+1}, \mathbf{z}_{i+1})_2$
  $\beta_{i+1} = \dfrac{\rho_{i+1}}{\rho_i}$
  $\mathbf{p}_{i+1} = \mathbf{z}_{i+1} + \beta_{i+1}\mathbf{p}_i$
  check convergence; continue if necessary
end for

Remark 3.3.2. *There is another way of deriving the preconditioned form of the conjugate gradient method by noting that even if $M^{-1}A$ is not symmetric, it is self-adjoint for the M-inner product (recall that M is symmetric positive definite), that is,*

$$(M^{-1}Ax, y)_M = (Ax, y) = (x, Ay) = (x, M(M^{-1}A)y) = (x, M^{-1}Ay)_M,$$

where $(x,y)_M := (Mx, y)$. Therefore, an alternative is to replace the usual Euclidean inner product in the conjugate gradient algorithm by the M-inner product. This idea is considered by Saad (see Chapter 9 of [169] for details).

3.4 ▪ The GMRES method for nonsymmetric problems

If either operator A or preconditioner M is not symmetric definite positive, the PCG algorithm cannot be used. The bad news is that in the nonsymmetric case there is no method with a fixed cost per iteration that leads to an optimal choice for $\mathbf{x}^n$. In this case, two popular methods are the GMRES [170] and BiCGSTAB [186]. With left preconditioning, they are applied to the preconditioned system

$$M^{-1}A\mathbf{x} = M^{-1}\mathbf{b}.$$

These methods differ by the way they pick an approximation $\mathbf{x}^n$ to the solution $\mathbf{x}$ in the Krylov subspace $\mathcal{K}^n((M^{-1}A), \mathbf{z}^0)$ with $\mathbf{z}^0 = M^{-1}\mathbf{r}^0 = M^{-1}(\mathbf{b} - A\mathbf{x}^0)$ being the initial preconditioned residual.

In order to fix the ideas we will concentrate on the solution of the unpreconditioned system $A\mathbf{x} = \mathbf{b}$. In the preconditioned case A needs to be replaced by $M^{-1}A$ and $\mathbf{b}$ by $M^{-1}\mathbf{b}$.

Note that it makes sense to minimize the residual, because the error is in general unknown, and we can only directly minimize special norms of the error. Moreover, the norm of the error is bounded by a constant times the norm of the residual

$$||\mathbf{e}^n||_2 = ||A^{-1}\mathbf{r}^n||_2 \leq ||A^{-1}||||\mathbf{r}^n||_2.$$

Finally we can note that

$$||\mathbf{r}^n||_2 = (A\mathbf{e}^n, A\mathbf{e}^n) = (A^*A\mathbf{e}^n, \mathbf{e}^n) = ||\mathbf{e}^n||^2_{A^*A}.$$

In what follows we will concentrate on the GMRES method by explaining in detail the principles on which it is based and some ideas about its convergence properties. Let us introduce first the general framework of a minimal residual method.

Definition 3.12 (Minimal residual method). *Given an initial iterate* $\mathbf{x}^0$ *and the initial residual* $\mathbf{r}^0 = \mathbf{b} - A\mathbf{x}^0$, *a minimal residual method applied to a system* $A\mathbf{x} = \mathbf{b}$ *will build the iterates* $\mathbf{x}^n = \mathbf{x}^0 + \mathbf{y}^n$, *where*

$$\boxed{\begin{aligned} &\mathbf{y}^n \in \mathcal{K}^n(A, \mathbf{r}^0) \text{ such that} \\ &||\mathbf{r}^n||_2 = ||\mathbf{r}^0 - A\mathbf{y}^n||_2 = \min_{\mathbf{w} \in \mathcal{K}^n(A,\mathbf{r}^0)} ||A\mathbf{w} - \mathbf{r}^0||_2 . \end{aligned}} \tag{3.38}$$

We say that they minimize the Euclidean norm of the residual.

We see from the previous definition that we are looking for the vector $\mathbf{x}^n = \mathbf{x}^0 + \mathbf{y}^n$ such that $\mathbf{y}^n$ achieves the minimum of $||\mathbf{r}^0 - A\mathbf{y}||$ over $\mathcal{K}^n(A, \mathbf{r}^0)$. If the dimension of $\mathcal{K}^n(A, \mathbf{r}^0)$ is n and $\{\mathbf{v}_1, \mathbf{v}_2, \ldots, \mathbf{v}_n\}$ is a basis of it, then we can look for $\mathbf{y}^n$ under the form

$$\mathbf{y}^n = \sum_{j=1}^{n} y_j \mathbf{v}_j.$$

If we denote the function to minimize by

$$f(y_1, y_2, \ldots, y_n) = \left\| \mathbf{r}^0 - A\sum_{i=1}^{n} y_i \mathbf{v}_i \right\|_2^2 = \left\| \mathbf{r}^0 - \sum_{i=1}^{n} y_i A\mathbf{v}_i \right\|_2^2,$$

then the optimality conditions $\frac{\partial f}{\partial y_i} = 0$ translate into

$$\sum_{j=1}^{n} (A\mathbf{v}_j, A\mathbf{v}_i) y_j = (\mathbf{r}^0, A\mathbf{v}_i). \tag{3.39}$$

If V denotes the matrix containing the column vectors $\mathbf{v}_j$, then (3.39) is equivalent to the system

$$(V^*A^*AV)\mathbf{Y}^n = \mathbf{F}, \text{ where } \mathbf{Y}^n = (y_1, y_2, \ldots, y_n)^T. \tag{3.40}$$

Note that the size of system (3.40) increases with the number of iterations, which means that the algorithmic complexity of computing the solution of (3.40) increases as the number of iterations becomes larger. On the other hand, if the basis of $\mathcal{K}^n(A, \mathbf{r}^0)$ has no particular structure, then system (3.40) can be too ill-conditioned. The GMRES method proposes a cheap solution by building a special basis of the Krylov space which makes the system easy to solve.

3.4.1 ▪ The GMRES method

The idea of the GMRES method is to build an orthonormal basis of the Krylov space by using a Gram–Schmidt algorithm. That is,

$$\text{Starting from } \mathbf{v}_1 = \frac{\mathbf{r}^0}{\|\mathbf{r}^0\|_2},$$
$$\text{Compute } \mathbf{v}_{n+1} = \frac{A\mathbf{v}_n - \sum_{i=1}^{n}(A\mathbf{v}_n, \mathbf{v}_i)\mathbf{v}_i}{\left\| A\mathbf{v}_n - \sum_{i=1}^{n}(A\mathbf{v}_n, \mathbf{v}_i)\mathbf{v}_i \right\|_2}, \; n \geq 1. \tag{3.41}$$

Computing the Gram–Schmidt orthogonalization of the basis of a Krylov space is called the *Arnoldi method*.

Supposing that the recurrence from (3.41) is well-defined (that is, the denominator is non-null), then it is equivalent to

$$A\mathbf{v}_n = \sum_{i=1}^{n}(A\mathbf{v}_n, \mathbf{v}_i)\mathbf{v}_i + \left\| A\mathbf{v}_n - \sum_{i=1}^{n}(A\mathbf{v}_n, \mathbf{v}_i)\mathbf{v}_i \right\|_2 \mathbf{v}_{n+1}. \tag{3.42}$$

If we denote by V_n the unitary matrix ($V_n^* V_n = I_n$), having as its columns the first n already computed orthonormal vectors $\mathbf{v}_1, \ldots, \mathbf{v}_n$, then V_{n+1} is obtained from V_n by adding the column vector $\mathbf{v}_{n+1}$. Thus (3.42) can be rewritten as

$$AV_n = V_{n+1}H_n, \tag{3.43}$$

where the entries of the rectangular matrix $H_n \in \mathcal{M}_{n+1,n}(\mathbb{C})$ are

$$(H_n)_{i,n} = (A\mathbf{v}_n, \mathbf{v}_i), \; i \leq n,$$
$$(H_n)_{n+1,n} = \left\| A\mathbf{v}_n - \sum_{i=1}^{n}(H_n)_{i,n}\mathbf{v}_i \right\|_2. \tag{3.44}$$

Moreover, since $(H_n)_{l,m} = 0$, $l > m+1$, we say that H_n *is an upper Hessenberg matrix.* Note also that the matrix H_{n+1} is obtained from H_n by adding a column given by the elements $(H_{n+1})_{i,n+1}$ from (3.44) and a line whose only nonzero element is $(H_{n+1})_{n+2,n+1}$.

With these notations and by using (3.43) we see that the problem (3.40) is equivalent to finding $\mathbf{Y}^n$ such that

$$V_n^* A^* A V_n \mathbf{Y}^n = \|\mathbf{r}^0\| [(A^*\mathbf{v}_1, \mathbf{v}_1), (A^*\mathbf{v}_1, \mathbf{v}_2), \ldots, (A^*\mathbf{v}_1, \mathbf{v}_n)]^T$$
$$\Leftrightarrow H_n^* \underbrace{V_{n+1}^* V_{n+1}}_{I_n} H_n \mathbf{Y}^n = \|\mathbf{r}^0\| \left[(H_n^*)_{1,1}, (H_n^*)_{2,1} \ldots, (H_n^*)_{n,1}\right]^T \tag{3.45}$$

and consequently

$$\boxed{H_n^* H_n \mathbf{Y}^n = \beta H_n^* \eta_{1,n+1}, \text{ where } \beta = \|\mathbf{r}^0\|_2,} \tag{3.46}$$

where $\eta_{1,n+1}$ is the first column vector of the canonical basis of $\mathbb{R}^{n+1}$. Solving such a system is straightforward if one knows the QR decomposition of H_n,

$$H_n = Q_n R_n,$$

with Q_n being a unitary matrix of order $n+1$ and R_n an upper triangular matrix of size $(n+1) \times n$ where the last line is null. This factorization is quite easy to compute in the

Algorithm 3.3 GMRES algorithm

Compute $\mathbf{r}_0 := b - A\mathbf{x}_0$, $\beta = \|\mathbf{r}_0\|_2$, $\mathbf{v}_1 = \mathbf{r}_0/\beta$, $\xi_1 = \beta$.
for $n = 1, 2, \ldots$ **do**
  $\mathbf{w}_{n+1} = A\mathbf{v}_n$
  for $i = 1, 2, \ldots, n$ **do**
    $(H_n)_{i,n} = (\mathbf{w}_{n+1}, \mathbf{v}_i)$.
    $\mathbf{w}_{n+1} = \mathbf{w}_{n+1} - (H_n)_{i,n}\mathbf{v}_i$
  end for
  $(H_n)_{n+1,n} = \|\mathbf{w}_{n+1}\|_2$.
  if $(H_n)_{n+1,n} \neq 0$ **then**
    $\mathbf{v}_{n+1} = \mathbf{w}_{n+1}/(H_n)_{n+1,n}$.
  end if
  for $i = 1, 2, \ldots, n-1$ **do**

$$\begin{pmatrix} (H_n)_{i,n} \\ (H_n)_{i+1,n} \end{pmatrix} = \begin{pmatrix} \bar{c}_i & s_i \\ -s_i & c_i \end{pmatrix} \begin{pmatrix} (H_n)_{i,n} \\ (H_n)_{i+1,n} \end{pmatrix}$$

  end for
  Compute Givens rotation

$$c_n = (H_n)_{n,n} / \sqrt{(H_n)_{n,n}^2 + (H_n)_{n+1,n}^2},\ s_n = (H_n)_{n+1,n} / \sqrt{(H_n)_{n,n}^2 + (H_n)_{n+1,n}^2}.$$

  Update $(H_n)_{n,n} = \bar{c}_n (H_n)_{n,n} + s_n (H_n)_{n+1,n}$, $(H_n)_{n+1,n} = 0$.
  Update $(\xi_n, \xi_{n+1}) = (\bar{c}_n \xi_n, -s_n \xi_n)$.
  Solve the triangular system $\tilde{H}\mathbf{y} = (\xi_1, \xi_2, \ldots, \xi_n)^T$.
  $\mathbf{x}_n = \mathbf{x}_0 + [\mathbf{v}_1 \mathbf{v}_2 \ldots \mathbf{v}_n]\mathbf{y}$
  check convergence on residual norm ξ_{n+1}; continue if necessary
end for

case of an upper Hessenberg matrix (since it is "almost" upper triangular). Supposing that this factorization is available, (3.46) reduces to the system

$$R_n^* \underbrace{(Q_n^* Q_n)}_{I_n} R_n \mathbf{Y}^n = \beta R_n^* Q_n^* \eta_{1,n+1} \Leftrightarrow R_n^* R_n \mathbf{Y}^n = \beta R_n^* Q_n^* \eta_{1,n+1}. \tag{3.47}$$

If we denote $\tilde{R}_n$ as the main block of size $n \times n$ of R_n obtained by deleting the last row, we have that

$$R_n^* R_n = \tilde{R}_n^* \tilde{R}_n \tag{3.48}$$

(the last row of R_n is zero). In the same way, $R_n^* Q_n^*$ is obtained from $\tilde{R}_n^* \tilde{Q}_n^*$ by appending a column vector ($\tilde{Q}_n$ is the main block of size $n \times n$ of Q_n obtained by deleting the last row and column). Thus

$$\beta R_n^* Q_n^* \eta_{1,n+1} = \beta \tilde{R}_n^* \tilde{Q}_n^* \eta_{1,n}, \tag{3.49}$$

where $\eta_{1,n}$ is the first column vector of the canonical basis of $\mathbb{R}^n$. We can conclude from (3.48) and (3.49) that solving (3.47) reduces to solving an upper triangular system of order n:

$$\boxed{\tilde{R}_n \mathbf{Y}^n = \tilde{\mathbf{g}}_n,\ \tilde{\mathbf{g}}_n = \beta \tilde{Q}_n^* \eta_{1,n}.} \tag{3.50}$$

The complexity of solving (3.50) is known to be of order n^2 operations.

To summarize, the iterate n obtained from the optimization problem (3.38) is given by

$$\mathbf{x}^n = \mathbf{x}^0 + \sum_{i=1}^{n} y_i \mathbf{v}_i = \mathbf{x}^0 + V_n \mathbf{y}_n = \mathbf{x}^0 + \tilde{R}_n^{-1} \tilde{\mathbf{g}}_n. \tag{3.51}$$

By using (3.43) and the fact that $\mathbf{Y}^n$ is a solution of (3.50), we see that the residual norm at iteration n verifies

$$\begin{aligned}
\|\mathbf{r}^n\|_2 &= \|\mathbf{b} - A\mathbf{x}^n\|_2 = \|\mathbf{b} - A(\mathbf{x}^0 + V_n \mathbf{Y}^n)\|_2 = \|\mathbf{r}^0 - AV_n \mathbf{Y}^n\|_2 \\
&= \|V_{n+1}(\beta \boldsymbol{\eta}_{1,n+1} - AV_n \mathbf{Y}^n\|_2 = \|\beta V_{n+1} \boldsymbol{\eta}_{1,n+1} - V_{n+1} H_n \mathbf{Y}^n\|_2 \\
&= \|\beta V_{n+1} \underbrace{(Q_n Q_n^*)}_{I_n} \boldsymbol{\eta}_{1,n+1} - V_{n+1} \underbrace{(Q_n Q_n^*)}_{I_n} \underbrace{Q_n R_n}_{H_n} \mathbf{Y}^n\|_2 \\
&= \|V_{n+1} Q_n \left[\beta Q_n^* \boldsymbol{\eta}_{1,n+1} - R_n \mathbf{Y}^n\right]\|_2 = \|V_{n+1} Q_n([\tilde{\mathbf{g}}_n, \gamma_{n+1}]^T - [\tilde{R}_n \mathbf{Y}^n, 0]^T)\|_2 \\
&= \|V_{n+1} Q_n [0, 0, \ldots, \gamma_{n+1}]^T\|_2 = |\gamma_{n+1}|,
\end{aligned} \tag{3.52}$$

where γ_{n+1} is the last element of $\beta Q_n^* \boldsymbol{\eta}_{1,n+1}$ (we used the fact that matrices V_{n+1} and Q_n are unitary). See [169] for more details of the proof.

The resulting iterative procedure is given in Algorithm 3.3.

Remark 3.4.1. *The GMRES method has the following properties.*

- *As in the case of the conjugate gradient, the GMRES method converges in at most N iterations, where N is the dimension of the linear system to be solved.*
- *The method requires storage in memory at the iteration n, the matrix V_n, and the QR factorization of H_n: the cost of such a method is quite high (of order nN for n iterations). When n is small w.r.t. N, the cost is marginal. When the iteration counts get large, its cost can be a problem.*
- *We can write a restarted version of the algorithm after each j iteration by using the current approximation of the solution as a starting point. In this case, however, the convergence property in a finite number of iterations is lost.*

3.4.2 ▪ Convergence of the GMRES algorithm

In the case where A is diagonalizable we can derive an estimate of the convergence based on the eigenvalues of A. Note that the GMRES method builds a sequence of polynomials $P_n \in \mathscr{P}_n$ which minimize at each iteration the quantity $\|Q(A)\mathbf{r}^0\|_2$, where $Q \in \mathscr{P}_n$ and $Q(0) = 1$.

In the case where A is diagonalizable we have

$$A = W \Lambda W^{-1} \Rightarrow Q(A) = W Q(\Lambda) W^{-1}$$

and we have the estimate on the residual

$$\begin{aligned}
\|\mathbf{r}^n\|_2 &= \min_{Q \in \mathscr{P}_n, Q(0)=1} \|W Q(\Lambda) W^{-1} \mathbf{r}^0\|_2 \\
&\le \|W\|_2 \|W^{-1}\|_2 \|\mathbf{r}^0\|_2 \min_{Q \in \mathscr{P}_n, Q(0)=1} \|Q(\Lambda)\|_2 \\
&= \kappa_2(W) \|\mathbf{r}^0\|_2 \min_{Q \in \mathscr{P}_n, Q(0)=1} \max_{\lambda \in \sigma(A)} |Q(\Lambda)|.
\end{aligned} \tag{3.53}$$

Estimate (3.53) may be very inaccurate since the polynomial minimizing $||WQ(\Lambda)W^{-1}||_2$ may be very different from the one minimizing $||Q(\Lambda)||_2$. We can distinguish the following cases:

- A is a normal matrix; that is, A commutes with its transpose. In this case there exists a complex unitary matrix W such that $A = W\Lambda W^{-1}$ (see, e.g., [100] or [111]), which means that

$$||WQ(\Lambda)W^{-1}||_2 = ||Q(\Lambda)||_2 \text{ and } \kappa_2(W) = 1.$$

  In this case the estimate (3.53) is very precise:

$$||\mathbf{r}^n||_2 \leq ||\mathbf{r}^0||_2 \min_{Q \in \mathscr{P}_n, Q(0)=1} \max_{\lambda \in \sigma(A)} |Q(\Lambda)|.$$

  Estimating the convergence rate is equivalent to finding a polynomial with $Q(0) = 1$ which approximates 0 on the spectrum of A. If A has a large number of eigenvalues close to 0, then the convergence will be very slow.

- If A is not a normal matrix but is diagonalizable such that the matrix W is well conditioned, then the estimate (3.53) remains reasonable.

- In the general case, there is no relation between the convergence rate of the GMRES algorithm and the eigenspectrum alone; see [104]. Some supplementary information is needed.

The BiConjuguate Gradient Stabilized (BiCGSTAB) method [186] (see Algorithm 3.4) takes a different approach with a fixed cost per iteration but at the expense of losing optimal properties of $\mathbf{x}^n$. Two mutually orthogonal sequences are computed, and $\mathbf{x}^n$ is such that the residual is orthogonal to one of the sequences.

Other facts are the following:

- As in the case of the conjugate gradient, in order to have a faster convergence, a preconditioner, denoted here by M, is often used. Since one does not need to preserve any symmetry, the preconditioned iteration is simply derived from the unpreconditioned one by replacing A by $M^{-1}A$ in algorithm (3.3).

- Krylov methods do not need the explicit form of the operators A or of the preconditioner M. They are based on applying these operators to vectors (so-called matrix-vector product operations) as is the case for the fixed-point method (3.3). These are matrix-free methods.

- There exist MATLAB functions for the main Krylov methods: `pcg` for the conjugate gradient and `gmres` for the GMRES method.

In conclusion, fixed-point iterations (3.3) generate an approximate solution $\mathbf{x}^n$ that belongs to the same Krylov space as the CG or GMRES iterate $\mathbf{x}^n_K$ but with "frozen" coefficients. Thus we have that $\mathbf{x}^n$ is less (actually much much less) optimal than $\mathbf{x}^n_K$. As a result, it is convenient to replace the fixed-point method by a Krylov-type method.

For more details on these Krylov methods, see [169] for a complete overview including implementation issues, [9] for their templates, [103] for a more mathematical introduction, and [141] for finite precision effects.

Algorithm 3.4 Preconditioned BiCGSTAB algorithm

Compute $\mathbf{r}_0 := \mathbf{b} - A\mathbf{x}_0$
Choose $\tilde{\mathbf{r}}$ (for example, $\tilde{\mathbf{r}} = \mathbf{r}_0$)
for $i = 1, 2, \ldots$ **do**
  $\rho_{i-1} = (\tilde{\mathbf{r}}, \mathbf{r}_{i-1})_2$
  if $\rho_{i-1} = 0$ **then**
    method fails
  end if
  if $i = 1$ **then**
    $\mathbf{p}_1 = \mathbf{r}_0$
  else
    $\beta_{i-1} = (\rho_{i-1}/\rho_{i-2})(\alpha_{i-1}/\omega_{i-1})$
    $\mathbf{p}_i = \mathbf{r}_{i-1} + \beta_{i-1}(\mathbf{p}_{i-1} - \omega_{i-1}\mathbf{v}_{i-1})$
  end if
  solve $M\hat{\mathbf{p}} = \mathbf{p}_i$
  $\mathbf{v}_i = A\hat{\mathbf{p}}$
  $\alpha_i = \dfrac{\rho_{i-1}}{(\tilde{\mathbf{r}}, \mathbf{v}_i)_2}$
  $\mathbf{s} = \mathbf{r}_{i-1} - \alpha_i \mathbf{v}_i$
  if norm of s is small enough **then**
    $\mathbf{x}_i = \mathbf{x}_{i-1} + \alpha_i \hat{\mathbf{p}}$ and stop
  end if
  solve $M\hat{\mathbf{s}} = \mathbf{s}$
  $\mathbf{t} = A\hat{\mathbf{s}}$
  $\omega_i = (\mathbf{t}, \mathbf{s})_2 / (\mathbf{t}, \mathbf{t})_2$
  $\mathbf{x}_i = \mathbf{x}_{i-1} + \alpha_i \hat{\mathbf{p}} + \omega_i \hat{\mathbf{s}}$
  $\mathbf{r}_i = \mathbf{s} - \omega_i \mathbf{t}$
  check convergence; continue if necessary
end for

3.5 ▪ Krylov methods for ill-posed problems

Since it will be needed in some substructuring methods in section 6.5, we consider here the case where the linear system

$$A\mathbf{x} = \mathbf{b} \tag{3.54}$$

is square but ill-posed. In order to have existence of a solution, we assume that $\mathbf{b} \in \text{range}(A)$. A solution is unique up to any element of $\ker(A)$. We shall see that if A is symmetric, a Krylov method will converge to a solution. When A is nonsymmetric, it may not converge to a solution.

We start with the symmetric case. It is known in this case that

$$\mathbb{R}^N = \ker(A) \oplus^{\perp} \text{range}(A)$$

and that there is a unique solution $\mathbf{x} \in \text{range}(A)$ to (3.54) denoted by $A^{\dagger}\mathbf{b}$ in the following. Thus we have that $A^{\dagger}$ is an isomorphism from range(A) onto itself.

The idea at the origin of Krylov methods can be found in the following lemma, which is true for an invertible matrix.

Lemma 3.13. *Let C be an invertible matrix of size $N \times N$. Then, there exists a polynomial $\mathcal{P}$ of degree $p < N$ such that*

$$C^{-1} = \mathcal{P}(C).$$

Proof. Let

$$\mathcal{M}(X) := \sum_{i=0}^{d} a_i X^i$$

be a minimal polynomial of C of degree $d \leq N$. We have that

$$\sum_{i=0}^{d} a_i C^i = 0$$

and there is no nonzero polynomial of lower degree that annihilates C. Thus, a_0 cannot be zero. Otherwise, we would have

$$C \sum_{i=1}^{d} a_i C^{i-1} = 0,$$

and since C is invertible,

$$\sum_{i=1}^{d} a_i C^{i-1} = 0.$$

Then, $\sum_{i=0}^{d-1} a_{i+1} X^i$ would be an annihilating polynomial of C of degree lower than d. By definition of a minimal polynomial, this is impossible. This proves that $a_0 \neq 0$, and we can divide by a_0:

$$I_d + C \sum_{i=1}^{d} \frac{a_i}{a_0} C^{i-1} = 0,$$

that is,

$$C^{-1} = \mathcal{P}(C), \mathcal{P}(C) := -\sum_{i=1}^{d} \frac{a_i}{a_0} C^{i-1}. \quad \square \tag{3.55}$$

Our first step is to extend Lemma 3.13 to the singular symmetric case.

Lemma 3.14. *Let A be a symmetric singular matrix of size $N \times N$. Then there exist coefficients $(a_i)_{2 \leq i \leq p}$ with $p \leq N$ such that*

$$A = \sum_{i=2}^{p} a_i A^i. \tag{3.56}$$

Proof. Let Λ denote the set of the eigenvalues of A without taking into account their multiplicities. Since matrix A is singular, $0 \in \Lambda$. Moreover, since it is symmetric, it is diagonalizable and it is easy to check that A cancels its characteristic polynomial, that is,

$$\prod_{\lambda \in \Lambda} (\lambda I_d - A) = 0. \tag{3.57}$$

The zeroth-order term of the above polynomial is zero, and the next term is nonzero since it takes the following value:

$$-\left(\prod_{\lambda \in \Lambda \setminus \{0\}} \lambda \right) A.$$

Then, it suffices to expand the polynomial from the left-hand side of (3.57), divide it by $\prod_{\lambda\in\Lambda\setminus\{0\}}\lambda$, and rearrange terms to end the proof. □

The consequence of this is the following lemma.

Lemma 3.15. *The unique solution to* $A\mathbf{y}=\mathbf{r}$ *that belongs to* range(A) *for a given* $\mathbf{r}\in$ range(A) *can be written as*

$$\mathbf{y}=\sum_{i=2}^{p} a_i A^{i-2}\mathbf{r}, \tag{3.58}$$

where the coefficients a_i *are given in* (3.56).

Proof. Let $\mathbf{r}=A\mathbf{t}\in$ range(A) for some $\mathbf{t}\in\mathbb{R}^N$. Note that since $\mathbf{t}$ may have a nonzero component in ker(A), we may have $\mathbf{y}\neq\mathbf{t}$. We apply Lemma 3.14 and right multiply (3.56) by vector $\mathbf{t}$:

$$\underbrace{A\left(\sum_{i=2}^{p} a_i A^{i-2}\right)}_{\mathbf{y}:=}\mathbf{r}=\mathbf{r}.$$

Thus, we conclude that $\mathbf{y}$ given by (3.58) is the unique solution to $A\mathbf{y}=\mathbf{r}$ that belongs to range(A). □

We now apply these results to the solving of $A\mathbf{x}=\mathbf{b}\in$ range(A) by a Krylov method with $\mathbf{x}^0$ as an initial guess. This result does not depend on the actual implementation of the Krylov method.

Lemma 3.16 (Krylov method for singular symmetric systems). *Let* $\mathbf{x}^0$ *be the initial guess decomposed into its component in* ker$A$ *and* range(A)*:*

$$\mathbf{x}^0=\mathbf{x}^0_{\text{ker}}+\mathbf{x}^0_{\text{range}}\,.$$

Then, the solution $\mathbf{x}$ *to* (3.54) *provided by the Krylov method is*

$$\boxed{\mathbf{x}=\mathbf{x}^0_{\text{ker}}+A^{\dagger}\mathbf{b}.}$$

In other words, the solution to (3.54) *"selected" by a Krylov method is the one whose component in the kernel of* A *is* $\mathbf{x}^0_{\text{ker}}$.

Proof. At step n, a Krylov method will seek an approximate solution to

$$A\mathbf{y}=\mathbf{b}-A\mathbf{x}^0:=\mathbf{r}^0\in\text{range}(A)$$

in $\mathcal{K}^n(A,\mathbf{r}^0)$. From (3.58), the method will converge in at most $p-1$ iterations and then the final solution is

$$\mathbf{x}=\mathbf{x}^0+\mathbf{y}=\mathbf{x}^0_{\text{ker}}+\underbrace{\mathbf{x}^0_{\text{range}}+\mathbf{y}}_{\tilde{\mathbf{y}}}$$

with $\tilde{\mathbf{y}}\in$ range(A). It is then obvious to see that

$$A\tilde{\mathbf{y}}=A(\mathbf{x}^0_{\text{range}}+\mathbf{y})=A\mathbf{x}^0_{\text{range}}+\mathbf{r}^0=\mathbf{b}$$

so that $\tilde{\mathbf{y}}=A^{\dagger}\mathbf{b}$. □

Thus, we have proved that a Krylov method applied to a singular symmetric linear system makes sense since for n large enough a solution to $A\mathbf{x} = \mathbf{b} \in \text{range}(A)$ can be found in $\{\mathbf{x}^0\} + \mathscr{K}^n(A, \mathbf{r}^0)$.

Remark 3.5.1. *This is not generally true in the nonsymmetric case. In [16], it is proved that if* $\ker(A) \cap \ker(A^T) = \{0\}$ *GMRES method will converge to a least square solution. We give now a counter-example. Consider the system*

$$\begin{pmatrix} 0 & 0 & 0 \\ 1 & 0 & 0 \\ 0 & 1 & 0 \end{pmatrix} \mathbf{x} = \mathbf{b} = \begin{pmatrix} 0 \\ 1 \\ 1 \end{pmatrix} \tag{3.59}$$

whose solutions are

$$\mathbf{x} = \begin{pmatrix} 1 \\ 1 \\ 0 \end{pmatrix} + \mathbb{R} \begin{pmatrix} 0 \\ 0 \\ 1 \end{pmatrix}.$$

In particular, the first component of any solution is equal to 1. *For the sake of simplicity, assume* $\mathbf{x}^0 = 0$. *Then, the first residual* $\mathbf{r}^0 = b$ *has its first component equal to zero. The same holds for* $A^k\,\mathbf{r}^0$ *for all* $k \geq 1$. *Thus for all* n, *any vector in the space* $\{\mathbf{x}^0\} + \mathscr{K}^n(A, \mathbf{r}^0)$ *has its first component equal to zero and cannot be a solution of system* (3.59).

3.6 ▪ Schwarz preconditioners using FreeFem++

In order to use Schwarz algorithms as preconditioners in Krylov methods, in addition to the usual ingredients (partition of the mesh, local matrices, and restriction operators), we need only the associated matrix-vector product to the global problem as well as the preconditioning method used.

In the following program Aglobal corresponds to the finite element discretization of the variational formulation of the problem to be solved. We then need the matrix-vector product of the operator Aglobal with a vector l.

```
func real[int] A(real[int] &x)
{
  // Matrix vector product with the global matrix
  Vh Ax;
  Ax[]= Aglobal*x;
  return Ax[];
}
```

Listing 3.1. `./FreefemCommon/matvecAS.idp`

The preconditioning method can be the additive Schwarz method (ASM), which can be found in the following routine:

```
// and the application of the preconditioner
func real[int] AS(real[int] &l)
{
  // Application of the ASM preconditioner
  // M^{-1}*y = \sum Ri^T*Ai^{-1}*Ri*y
  // Ri restriction operators, Ai =Ri*A*Ri^T local matrices
  Vh s = 0;
  for(int i=0;i<npart;++i)
    {
      real[int] bi = Rih[i]*l; // restricts rhs
```

```
      real[int] ui = aA[i]^-1 * bi; // local solves
      s[] += Rih[i]'*ui; // prolongation
    }
  return s[];
}
```

Listing 3.2. `./FreefemCommon/matvecAS.idp`

```
// Preconditioned Conjugate Gradient Applied to the system
// A(un[]) = rhsglobal[]
// preconditioned by the linear operator
// AS: r[] --> AS(r[])
func real[int] myPCG(real[int] xi,real eps, int nitermax)
{
  ofstream filei("Convprec.m");
  ofstream fileignu("Convprec.gnu");
    Vh r, un, p, zr, rn, w, er;
    un[] = xi;
    r[] = A(un[]);
    r[] -= rhsglobal[];
    r[] *= -1.0;
    zr[] = AS(r[]);
    real resinit=sqrt(zr[]'*zr[]);
    p = zr;
    for(int it=0;it<nitermax;++it)
      {
        plot(un,value=1,wait=1,fill=1,dim=3,cmm="Approximate solution at iteration "+it);
        real relres = sqrt(zr[]'*zr[])/resinit;
        cout << "It: "<< it << " Relative residual = " << relres << endl;
        int j = it+1;
        filei << "relres("+j+")=" << relres << ";" << endl;
        fileignu << relres << endl;
        if(relres < eps)
          {
            cout << "CG has converged in " + it + " iterations " << endl;
            cout << "The relative residual is " + relres << endl;
            break;
          }
        w[] = A(p[]);
        real alpha = r[]'*zr[];
        real aux2 = alpha;
        real aux3 = w[]'*p[];
        alpha /= aux3; // alpha = (rj,zj)/(Apj,pj);
        un[] += alpha*p[]; // xj+1 = xj + alpha*p;
        r[] -= alpha*w[]; // rj+1 = rj - alpha*Apj;
        zr[] = AS(r[]); // zj+1 = M^-1*rj+1;
        real beta = r[]'*zr[];
        beta /= aux2; // beta = (rj+1,zj+1)/(rj,zj);
        p[] *= beta;
        p[] += zr[];
      }
    return un[];
}
```

Listing 3.3. `./KRYLOV/FreefemProgram/PCG.idp`

The Krylov method applied in this case is the conjugate gradient method (we are in the symmetric case since both the original problem and the preconditioner are symmetric positive definite). The whole program in which these routines are called is `AS-PCG.edp`.

```
/*# debutPartition #*/
include "../../FreefemCommon/dataGENEO.edp"
include "../../FreefemCommon/decomp.idp"
include "../../FreefemCommon/createPartition.idp"
SubdomainsPartitionUnity(Th,part[],sizeovr,aTh,Rih,Dih,Ndeg,AreaThi);
/*# endPartition #*/
/*# debutGlobalData #*/
Aglobal = vaglobal(Vh,Vh,solver = UMFPACK); // global matrix
rhsglobal[] = vaglobal(0,Vh); // global rhs
uglob[] = Aglobal^-1*rhsglobal[];
/*# finGlobalData #*/
/*# debutLocalData #*/
for(int i = 0;i<npart;++i)
  {
    cout << " Domain :" << i << "/" << npart << endl;
    matrix aT = Aglobal*Rih[i]';
    aA[i] = Rih[i]*aT;
    set(aA[i],solver = UMFPACK); // direct solvers using UMFPACK
  }
/*# finLocalData #*/
/*# debutPCGSolve #*/
include "../../FreefemCommon/matvecAS.idp"
include "PCG.idp"
Vh un = 0, sol; // initial guess un and final solution
cout << "Schwarz Dirichlet algorithm" << endl;
sol[] = myPCG(un[], tol, maxit); // PCG with initial guess un
plot(sol,cmm=" Final solution", wait=1,dim=3,fill=1,value=1);
Vh er = sol-uglob;
cout << " Final error: " << er[].linfty << endl;
/*# finPCGSolve #*/
```

Listing 3.4. `./KRYLOV/FreefemProgram/AS-PCG.edp`

In the three-dimensional case the only thing that changes with respect to the main script is the preparation of the data.

```
include "../../FreefemCommon/data3d.edp"
include "../../FreefemCommon/decomp3d.idp"
include "../../FreefemCommon/createPartition3d.idp"
SubdomainsPartitionUnity3(Th,part[],sizeovr,aTh,Rih,Dih,Ndeg,VolumeThi);
```

Listing 3.5. `./KRYLOV/FreefemProgram/AS-PCG3d.edp`

We can also use RAS as a preconditioner in GMRES or BiCGSTAB, by slightly modifying the program as described in the following routines:

```
func real[int] RAS(real[int] &l)
{
   // Application of the RAS preconditioner
   // M^{-1}*y = \sum Ri^T*Di*Ai^{-1}*Ri*y
   // Ri restriction operators, Ai =Ri*A*Ri^T local matrices
   Vh s = 0;
   for(int i=0;i<npart;++i) {
        real[int] bi = Rih[i]*l; // restricts rhs
        real[int] ui = aA[i] ^-1 * bi; // local solves
        bi = Dih[i]*ui; // partition of unity
        s[] += Rih[i]'*bi; // prolongation
```

```
    }
  return s[];
}
```

Listing 3.6. `./FreefemCommon/matvecAS.idp`

```
func real[int] GMRES(real[int] x0, real eps, int nbiter)
{
  ofstream filei("Convprec.m");
  Vh r, z, v,w,er, un;
  Vh[int] V(nbiter); // orthonormal basis
  real[int,int] Hn(nbiter+2,nbiter+1); // Hessenberg matrix
  real[int,int] rot(2,nbiter+2);
  real[int] g(nbiter+1),g1(nbiter+1);
  r[]=A(x0);
  r[] -= rhsglobal[];
  r[] *= -1.0;
  z[] = RAS(r[]); // z= M^{-1}(b-A*x0)
  g[0]=sqrt(z[]'*z[]); // initial residual norm
  filei << "relres("+1+")=" << g[0] << ";" << endl;
  V[0][]=1/g[0]*z[]; // first basis vector
  for(int it =0;it<nbiter;it++) {
        v[]=A(V[it][]);
        w[]=RAS(v[]); // w = M^{-1}A*V_it
        for(int i =0;i<it+1;i++) {
          Hn(i,it)=w[]'*V[i][];
          w[] -= Hn(i,it)*V[i][]; }
        Hn(it+1,it)= sqrt(w[]'*w[]);
        real aux=Hn(it+1,it);
        for(int i =0;i<it;i++){ // QR decomposition of Hn
          real aa=rot(0,i)*Hn(i,it)+rot(1,i)*Hn(i+1,it);
          real bb=-rot(1,i)*Hn(i,it)+rot(0,i)*Hn(i+1,it);
          Hn(i,it)=aa;
          Hn(i+1,it)=bb; }
        real sq = sqrt(Hn(it+1,it)^2+Hn(it,it)^2);
        rot(0,it)=Hn(it,it)/sq;
        rot(1,it)=Hn(it+1,it)/sq;
        Hn(it,it)=sq;
        Hn(it+1,it)=0;
        g[it+1] = -rot(1,it)*g[it];
        g[it] = rot(0,it)*g[it];
        real[int] y(it+1); // Reconstruct the solution
        for(int i=it;i>=0;i--) {
          g1[i]=g[i];
          for(int j=i+1;j<it+1;j++)
               g1[i]=g1[i]-Hn(i,j)*y[j];
          y[i]=g1[i]/Hn(i,i); }
        un[]=x0;
        for(int i=0;i<it+1;i++)
               un[]= un[]+ y[i]*V[i][];
          er[]=un[]-uglob[];
          real relres = abs(g[it+1]);
          cout << "It: "<< it << " Residual = " << relres << " Error = "<< sqrt(er[]'*er[]) << endl;
          int j = it+2;
          filei << "relres("+j+")=" << relres << ";" << endl;
          if(relres < eps) {
               cout << "GMRES has converged in " + it + " iterations " << endl;
               cout << "The relative residual is " + relres << endl;
               break; }
          V[it+1][]=1/aux*w[]; }
```

```
    return un[];
}
```

Listing 3.7. `../KRYLOV/FreefemProgram/GMRES.idp`

```
func real[int] BiCGstab(real[int] xi, real eps, int nbiter){
  ofstream filei("Convprec.m");
  Vh r, rb, rs, er, un, p, yn, zn, z, v, t, tn;
  un[]=xi;
  r[]=A(un[]);
  r[] -= rhsglobal[];
  r[] *= -1.0;
  rb[]=r[];
  real rho=1.;
  real alpha=1.;
  real omega=1.;
  z[] = RAS(r[]);
  real resinit=sqrt(z[]'*z[]);
  p[]=0; v[]=0;
  real rhonew, beta;
  for(int it =1;it<nbiter;it++){
        real relres = sqrt(z[]'*z[])/resinit;
        cout << "It: "<< it << " Relative residual = " << relres << endl;
        int j = it+1;
        filei << "relres("+j+")=" << relres << ";" << endl;
        if(relres < eps){
            cout << "BiCGstab has converged in " + it + " iterations " << endl;
            cout << "The relative residual is " + relres << endl;
            break;
        }
        rhonew=rb[]'*r[]; // rhoi = (rb,rim1);
        beta=(rhonew/rho)*(alpha/omega); // beta = (rhoi/rhoim1)*(alpha/omega);
        p[] -=omega*v[]; // pi = rim1 + beta*(pim1 - omega*vim1);
        p[] *= beta;
        p[] += r[];
        yn[] = RAS(p[]); // y = Mm1*pi; vi = A*y;
        v[] = A(yn[]);
        alpha = rhonew/(rb[]'*v[]); // alpha = rhoi/(rb'*vi);
        rs[] = r[]; // s = rim1 - alpha*vi;
        rs[] -= alpha*v[];
        zn[] = RAS(rs[]); // z = Mm1*s; t = A*z;
        t[] = A(zn[]);
        tn[] = RAS(t[]); // tn = Mm1*t;
        omega = (tn[]'*zn[])/(tn[]'*tn[]); // omega = (Mm1*t,Mm1*s)/(Mm1*t,Mm1*t);
        un[] += alpha*yn[]; // xi = xim1 + alpha*y + omega*z;
        un[] += omega*zn[];
        r[] = rs[];
        r[] -= omega*t[];
        z[] = RAS(r[]);
        rho = rhonew;
    }
    return un[];
}
```

Listing 3.8. `../KRYLOV/FreefemProgram/BiCGstab.idp`

The whole program in which the GMRES is called is similar to `AS-PCG.edp` where the last part is modified as `RAS-GMRES.edp`.

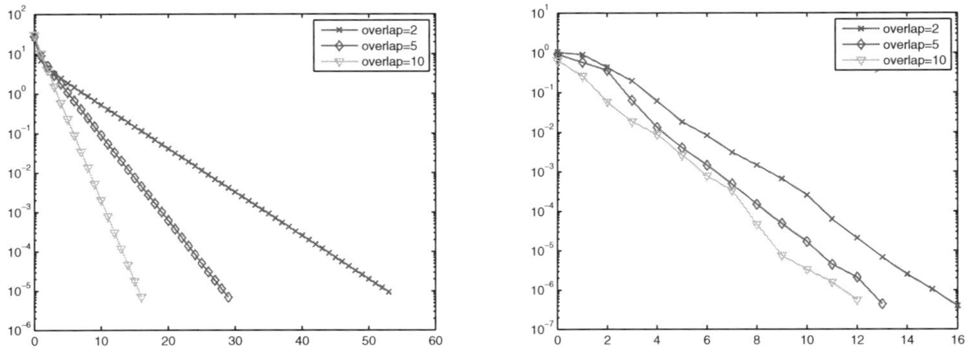

Figure 3.1. *Schwarz convergence as a solver (left) and as a preconditioner (right) for different overlaps.*

```
include "../../FreefemCommon/matvecAS.idp"
include "GMRES.idp"
Vh un = 0, sol, er; // initial guess, final solution and error
sol[] = GMRES(un[], tol, maxit);
plot(sol,dim=3,wait=1,cmm="Final solution",value =1,fill=1);
er[] = sol[]−uglob[];
cout << "Final error = " << er[].linfty/uglob[].linfty << endl;
```

Listing 3.9. `../KRYLOV/FreefemProgram/RAS-GMRES.edp`

In the case of the use of BiCGSTAB, only the last part has to be modified.

```
include "../../FreefemCommon/matvecAS.idp"
include "BiCGstab.idp"
Vh un = 0, sol, er; // initial guess, final solution and error
sol[] = BiCGstab(un[], tol, maxit);
plot(sol,dim=3,wait=1,cmm="Final solution",value =1,fill=1);
er[] = sol[]−uglob[];
cout << "Final error = " << er[].linfty/uglob[].linfty << endl;
```

Listing 3.10. `../KRYLOV/FreefemProgram/RAS-BiCGstab.edp`

The performance of Krylov methods is now much less sensitive to the overlap size compared to the iterative version as shown in Figure 3.1. Note also that in the case of the use of BiCGSTAB, an iteration involves two matrix-vector products, and thus the cost of a BiCGSTAB iteration is twice the cost of a CG iteration. In all previous scripts the convergence residuals can be stored in a file in MATLAB/Octave/Scilab form, and then the convergence histories can be plotted in one of these numeric languages.

Chapter 4
Coarse spaces

Theoretical and numerical results show that domain decomposition methods based solely on local subdomain solves are not scalable with respect to the number of subdomains. This can be fixed by using a two-level method in which the preconditioner is enriched by the solve of a coarse problem whose size is of the order of the number of subdomains. In section 4.1, we see that one-level Schwarz methods are not scalable since the iteration count grows with the number of subdomains in one direction. We present a fix introduced by Nicolaides [151]. A numerical implementation is given in section 4.2.1.

4.1 ▪ Need for a two-level method

The results presented so far concern only a small number of subdomains, two for the theoretical results of section 1.5 and four for the numerical results of section 3.6. When the number of subdomains is large, plateaus appear in the convergence of Schwarz domain decomposition methods. This is the case even for a simple model such as the Poisson problem:

$$\begin{cases} -\Delta u = f \text{ in } \Omega, \\ \quad u = 0 \text{ on } \partial\Omega. \end{cases}$$

The problem of the one-level method comes from the fact that in the Schwarz methods of section 1 there is a lack of a global exchange of information. Data are exchanged only from one subdomain to its direct neighbors. But the solution in each subdomain depends on the right-hand side in all subdomains. Let us denote by N_d the number of subdomains in one direction. Then, for instance, the leftmost domain of Figure 4.1 needs at least N_d iterations before being made aware of the value of the right-hand side f in the rightmost subdomain. The length of the plateau is thus typically related to the number of subdomains in one direction and therefore to the notion of *scalability* met in the context of high-performance computing.

To be more precise, there are two common notions of scalability.

Definition 4.1 (Strong scalability). *Strong scalability is defined as how the solution time varies with the number of cores for a fixed total problem size. Ideally, the elapsed time is inversely proportional to the number of processing units (CPUs, cores, GPUs).*

Definition 4.2 (Weak scalability). *Weak scalability is defined as how the solution time varies with the number of cores for a fixed problem size per core. Ideally, the elapsed time is constant for a fixed ratio between the size of the problem and the number of processing units (CPUs, cores, GPUs).*

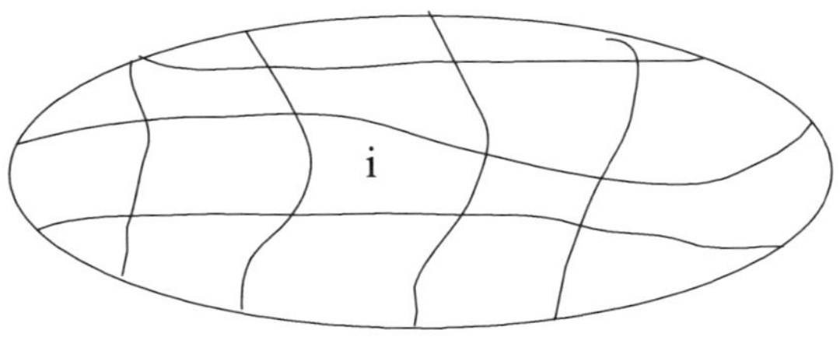

Figure 4.1. *Decomposition into many subdomains.*

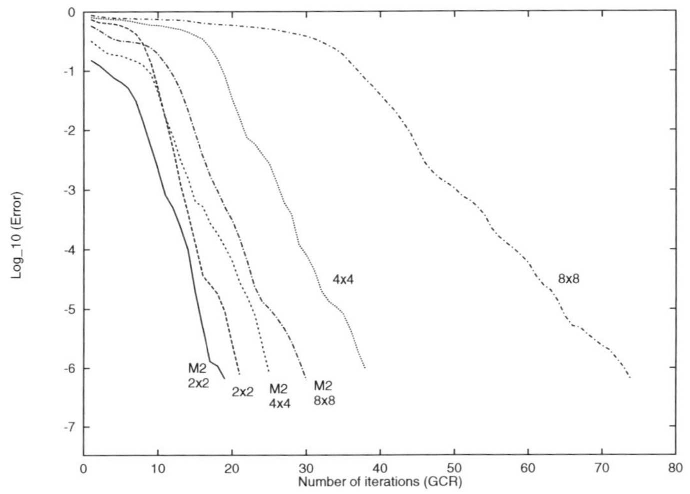

Figure 4.2. *Convergence curves with and without coarse spaces, see* [115].

A mechanism through which the *scalability* can be achieved (in this case the weak scalability) consists in the introduction of a two-level preconditioner via a coarse space correction. In two-level methods, a small problem whose size is proportional to the number of subdomains couples all subdomains at each iteration. Note that here the term *scalable* refers to the number of iterations. In real world applications, scalability is defined in terms of the total computing time. When measured by the computing time, the two-level methods are often not scalable for large-scale calculations because the coarse grid solver is not scalable. As a result, a multilevel (more than two) approach is often necessary.

To illustrate the need for a coarse space, in Figure 4.2 we consider a 2D problem decomposed into 2×2, 4×4, and 8×8 subdomains. For each domain decomposition, we have two curves: one with a one-level method which has longer and longer plateaus, and another with a coarse grid correction which is denoted by M2, for which the plateaus are much smaller. The problem of the one-level method and its cure are also well illustrated in Figure 4.3 for a domain decomposition into 64 strips, as well as in Table 4.1. The one-level method has a long plateau in the convergence, whereas with a coarse space correction convergence is quite fast. We also see that for the one-level method curves, the plateau has a size proportional to the number of subdomains in one direction. This can be understood by looking at Figure 4.4, where we plot the slow decrease of the error for a one-dimensional Poisson problem.

From the linear algebra point of view, stagnation in the convergence of one-level Schwarz methods corresponds to a few very low eigenvalues in the spectrum of the preconditioned problem. Using the preconditioners M^{-1}_{ASM} or M^{-1}_{RAS}, we can remove the

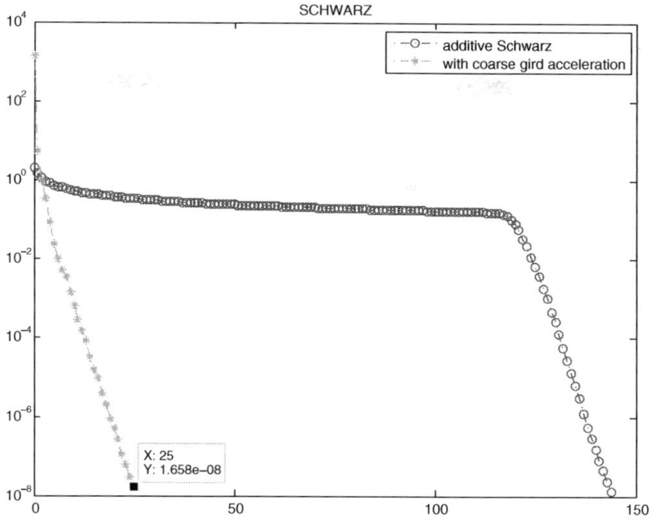

Figure 4.3. *Convergence curves with and without a coarse space correction for a decomposition into 64 strips.*

Table 4.1. *Iteration counts for a Poisson problem on a domain decomposed into strips. The number of unknowns is proportional to the number of subdomains (weak scalability).*

N subdomains	Schwarz	With coarse grid
4	18	25
8	37	22
16	54	24
32	84	25
64	144	25

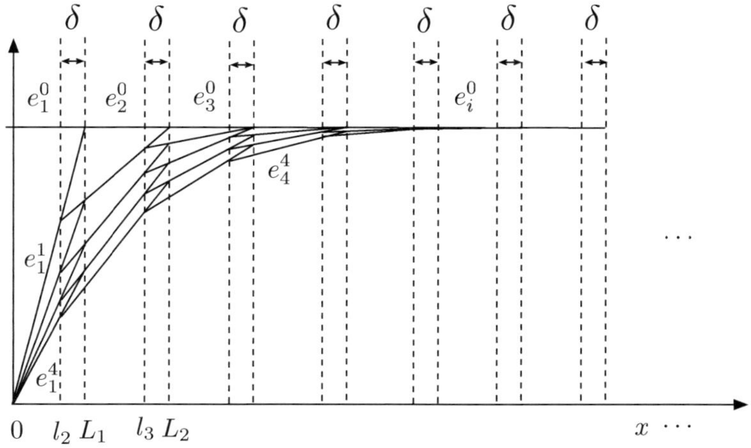

Figure 4.4. *Convergence of the error for the one-level Schwarz method in the* 1*D case.*

influence of very large eigenvalues of the coefficient matrix, which correspond to high-frequency modes. It has been proved that for an SPD (symmetric positive definite) matrix, the largest eigenvalue of the preconditioned system by M_{ASM}^{-1} is bounded by the number of colors needed for the overlapping decomposition such that different colors are used for adjacent subdomains; see [185] or [169], for instance. But the small eigenvalues still exist and hamper the convergence. These small eigenvalues correspond to low-frequency modes and represent certain global information with which we have to deal efficiently.

A classical remedy consists in the introduction of a *coarse grid* or *coarse space correction* that couples all subdomains at each iteration of the iterative method. This is closely related to deflation technique classical in linear algebra; see Tang et al.'s paper [184] and references therein. This is connected as well to augmented or recycled Krylov space methods; see, e.g., [76, 154, 168] or [24] and references therein.

Suppose we have identified the modes corresponding to the slow convergence of the iterative method used to solve the linear system

$$A\mathbf{x} = \mathbf{b}$$

with a preconditioner M, in our case a domain decomposition method. That is, we have some a priori knowledge of the small eigenvalues of the preconditioned system $M^{-1}A$. For a Poisson problem, these slow modes correspond to constant functions that are in the null space (kernel) of the Laplace operators. For a homogeneous elasticity problem, they correspond to the rigid body motions.

Let us call Z the rectangular matrix whose columns correspond to these slow modes. There are algebraic ways of incorporating this information to accelerate the method. We give here the presentation that is classical in domain decomposition; see, e.g., [137] or [185]. In the case where A is SPD, the starting point is to consider the minimization problem

$$\min_{\beta} \|A(\mathbf{y} + Z\beta) - \mathbf{b}\|_{A^{-1}}\,.$$

It corresponds to finding the best correction to an approximate solution $\mathbf{y}$ by a vector $Z\beta$ in the vector space spanned by the n_c columns of Z. This problem is equivalent to

$$\min_{\beta \in \mathbb{R}^{n_c}} 2(A\mathbf{y} - \mathbf{b}, Z\beta)_2 + (AZ\beta, Z\beta)_2$$

whose solution is

$$\beta = (Z^T A Z)^{-1} Z^T (\mathbf{b} - A\mathbf{y})\,.$$

Thus, the correction term is

$$Z\beta = Z(Z^T A Z)^{-1} Z^T \underbrace{(\mathbf{b} - A\mathbf{y})}_{\mathbf{r}}\,.$$

This kind of correction is called a Galerkin correction. Let $R_0 := Z^T$ and $\mathbf{r} = \mathbf{b} - A\mathbf{y}$ be the residual associated to the approximate solution $\mathbf{y}$. We have just seen that the best correction that belongs to the vector space spanned by the columns of Z reads as

$$Z\beta = R_0^T \beta = R_0^T (R_0 A R_0^T)^{-1} R_0 \mathbf{r}\,.$$

When using such an approach with an additive Schwarz method (ASM), as defined in (1.29), it is natural to introduce an additive correction to the ASM.

Definition 4.3 (Two-level additive Schwarz preconditioner). *The two-level additive Schwarz preconditioner is defined as*

$$\boxed{M_{ASM,2}^{-1} := R_0^T (R_0 A R_0^T)^{-1} R_0 + \sum_{i=1}^{N} R_i^T (R_i A R_i^T)^{-1} R_i,} \tag{4.1}$$

where the R_i's $(1 \le i \le N)$ are the restriction operators to the overlapping subdomains and $R_0 = Z^T$.

Remark 4.1.1. *This definition suggests the following remarks:*

- *The structure of the two level preconditioner $M_{ASM,2}^{-1}$ is the same as that of the one-level method.*
- *Compared to the one-level Schwarz method, where only local subproblems have to be solved in parallel, the two-level method adds the solving of a linear system of matrix $R_0 A R_0^T$ which is global in nature. This global problem is called the coarse problem.*
- *The coarse problem couples all subdomains at each iteration, but its matrix is a small $O(n_c \times n_c)$ square matrix and the extra cost is negligible compared to the gain provided that the number of subdomains is not too large.*

In Table 4.1 we display the iteration counts for a decomposition of the domain in an increasing number of subdomains. In Figure 4.3, we see that without a coarse grid correction, the convergence curve of the one-level Schwarz method has a very long plateau that can be bypassed by a two-level method.

4.2 ▪ Nicolaides coarse space

We give here a precise definition of the rectangular matrix Z for a Poisson problem. This construction was introduced by Nicolaides [151]. We take Z so that it has a domain decomposition structure. Z is defined by vectors which have local support in the subdomains and so that the constant function $\mathbf{1}$ belongs to the vector space spanned by the columns of Z.

Recall first that we have a discrete partition of unity (see section 1.3) in the following sense. Let $\mathcal{N}_i$ be subsets of indices and D_i be diagonal matrices, $1 \le i \le N$,

$$D_i : \mathbb{R}^{\#\mathcal{N}_i} \longmapsto \mathbb{R}^{\#\mathcal{N}_i}, \tag{4.2}$$

so that we have

$$\sum_{i=1}^{N} R_i^T D_i R_i = \mathrm{Id}.$$

With these ingredients we are ready to provide the definition of what we will hereafter call the *classical* or *Nicolaides coarse space*.

Definition 4.4 (Nicolaides coarse space). *We define Z as the matrix whose ith column is*

$$\boxed{Z_i := R_i^T D_i R_i \mathbf{1} \text{ for } 1 \le i \le N,} \tag{4.3}$$

where **1** *is the vector of dimension* $\mathcal{N}$ *full of ones. The global structure of* Z *is thus the following:*

$$Z = \begin{bmatrix} D_1R_1\mathbf{1} & 0 & \cdots & 0 \\ 0 & D_2R_2\mathbf{1} & \ddots & \\ \vdots & \ddots & \ddots & 0 \\ 0 & 0 & 0 & D_NR_N\mathbf{1} \end{bmatrix}. \tag{4.4}$$

The numerical results of Figure 4.3 and Table 4.1 have been obtained by a two-level Schwarz preconditioner (4.1) where $R_0 = Z^T$ with Z being defined by (4.4).

4.2.1 ▪ Nicolaides coarse space using FreeFem++

We provide FreeFem++ scripts that will illustrate numerically the performance of the Nicolaides coarse space. We first need to build matrix Z whose expression is given in (4.4),

```
// Build coarse space
for(int i=0;i<npart;++i)
{
  Z[i]=1.;
  // Z[i][] = Z[i][].*intern[];
  real[int] zit = Rih[i]*Z[i][];
  real[int] zitemp = Dih[i]*zit;
  Z[i][]=Rih[i]'*zitemp;
}
```

Listing 4.1. `./FreefemCommon/coarsespace.idp`

then the coarse matrix $A_0 = Z^T A Z = R_0 A R_0^T$,

```
real[int,int] Ef(npart,npart); // E = Z^T*A*Z
for(int i=0;i<npart;++i)
{
    real[int] vaux = A(Z[i][]);
    for(int j=0;j<npart;++j)
    {
    Ef(j,i) = Z[j][]'*vaux;
    }
}
matrix E;
E = Ef;
set(E,solver=UMFPACK);
```

Listing 4.2. `./FreefemCommon/coarsespace.idp`

then the coarse solver $Q = R_0^T (R_0 A R_0^T)^{-1} R_0$,

```
func real[int] Q(real[int] &l) // Q = Z*E^-1*Z^T
{
    real[int] res(l.n);
    res=0.;
    real[int] vaux(npart);
    for(int i=0;i<npart;++i)
```

```
  {
    vaux[i]=Z[i][]'*l;
  }
  real[int] zaux=E^-1*vaux; // zaux=E^-1*Z^T*l
  for(int i=0;i<npart;++i) // Z*zaux
  {
    res +=zaux[i]*Z[i][];
  }
  return res;
}
```

Listing 4.3. `./FreefemCommon/coarsespace.idp`

and finally the projector outside the coarse space $P = I - QA$ and its transpose $P^T = I - AQ$:

```
func real[int] P(real[int] &l) // P = I − A*Q
{
  real[int] res=Q(l);
  real[int] res2=A(res);
  res2 −= l;
  res2 *= −1.;
  return res2;
}
func real[int] PT(real[int] &l) // P^T = I−Q*A
{
  real[int] res=A(l);
  real[int] res2=Q(res);
  res2 −= l;
  res2 *= −1.;
  return res2;
}
```

Listing 4.4. `./FreefemCommon/coarsespace.idp`

Based on the previous quantities, we can build two versions of the additive Schwarz preconditioner: first the classical one, called `AS2` corresponding to formula (4.1), and then the alternative hybrid version $Q + P^T M_{AS,1}^{-1} P$, detailed in section 7.5.

```
func real[int] AS2(real[int] & r){
    real[int] z = Q(r);
    real[int] aux = AS(r);
    z += aux;
     return z;
}
func real[int] BNN(real[int] &u) // precond BNN
{
    real[int] aux1 = Q(u);
    real[int] aux2 = P(u);
    real[int] aux3 = AS(aux2);
    aux2 = PT(aux3);
    aux2 += aux1;
  return aux2;
}
```

Listing 4.5. `./COARSEGRID/FreefemProgram/PCG-CS.idp`

After that we can use one of these inside a preconditioned conjugate gradient. Finally, the main program script is given by `AS2-PCG.edp`, whose first part is identical to that in the case of the one-level method.

```
include "../../FreefemCommon/matvecAS.idp"
include "../../FreefemCommon/coarsespace.idp"
include "PCG-CS.idp"
Vh un = 0, sol; // initial guess un and final solution
cout << "Schwarz Dirichlet algorithm" << endl;
sol[] = myPCG(un[], tol, maxit); // PCG with initial guess un
plot(sol,cmm=" Final solution", wait=1,dim=3,fill=1,value=1);
Vh er = sol-uglob;
cout << " Final error: " << er[].linfty << endl;
```

Listing 4.6. `./COARSEGRID/FreefemProgram/AS2-PCG.edp`

An identical task can be performed for the nonsymmetric preconditioner, by simply replacing `AS` by `RAS` in the two versions of the preconditioners. After that we can use one of these inside BiCGSTAB or GMRES.

```
func real[int] RAS2(real[int] & r){
    real[int] z = Q(r);
    real[int] aux = RAS(r);
    z += aux;
     return z;
}
func real[int] BNN(real[int] &u) // precond BNN
{
    real[int] aux1 = Q(u);
    real[int] aux2 = P(u);
    real[int] aux3 = RAS(aux2);
    aux2 = PT(aux3);
    aux2 += aux1;
  return aux2;
}
```

Listing 4.7. `./COARSEGRID/FreefemProgram/BiCGstab-CS.idp`

Figure 4.5. *Decomposition into subdomains.*

We perform now the following numerical experiment: consider a uniform decomposition into $n \times n$ square subdomains, cf. Figure 4.5 where each domain has 20×20 discretization points with two layers of overlap. We see in Figure 4.6 that in the absence of a coarse space, the iteration counts of a Schwarz method used as a preconditioner in a Krylov method depends linearly on the number of subdomains in one direction. By introducing the Nicolaides coarse space we remove this dependence on the number of subdomains.

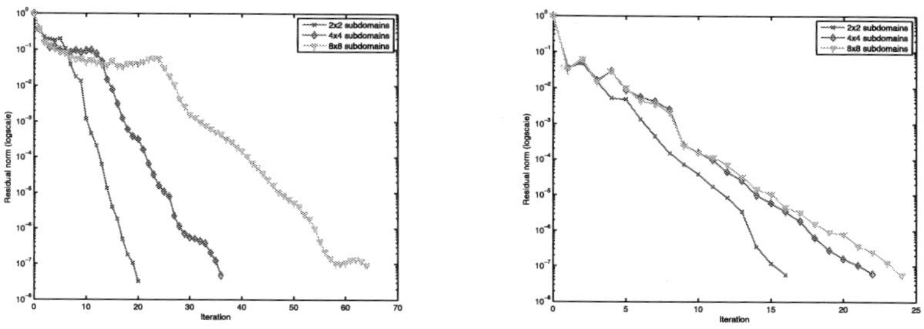

Figure 4.6. *Convergence history without (left) and with (right) a coarse space for various domain decompositions.*

Chapter 5

Theory of two-level additive Schwarz methods

In this chapter we first introduce the spectral coarse space which generalizes the idea of Nicolaides coarse space. In order to quantify the action of this new preconditioner we need to estimate the condition number of the preconditioned matrix $M_{ASM,2}^{-1}A$. This requires a theoretical framework allowing a functional analysis of the condition number of the two-level Schwarz method with this spectral coarse space.

5.1 ▪ Introduction of a spectral coarse space

In this section we introduce a more general coarse space construction that reduces to that of Nicolaides for a particular choice of a threshold parameter. This new construction answers the natural question of how to enrich the coarse space by allowing it to have a size larger than the number of subdomains. This enrichment is mainly motivated by the complexity of the problem to be solved, which is often a result of highly heterogeneous coefficients. We will try to keep it for the moment as elementary as possible without any theoretical convergence proof in this section.

Classical coarse spaces are known to give good results when the jumps in the coefficients are across subdomain interfaces (see, e.g., [64, 133, 48, 49]), or are inside the subdomains and not near their boundaries (see [160, 159]). However, when the discontinuities are along subdomain interfaces, classical methods no longer work. In [147], we proposed the construction of a coarse subspace for a scalar elliptic positive definite operator, which leads to a two-level method that is robust with respect to heterogeneous coefficients for fairly arbitrary domain decompositions, e.g., as provided by an automatic graph partitioner such as METIS or SCOTCH [119, 27]. This method was extensively studied from the numerical point of view in [148]. The main idea behind the construction of the coarse space is the computation of the low-frequency modes associated with a generalized eigenvalue problem based on the Dirichlet-to-Neumann (DtN) map on the boundary of each subdomain. We use the harmonic extensions of these low-frequency modes to the whole subdomain to build the coarse space. With this method, even for discontinuities along (rather than across) the subdomain interfaces (see Figure 5.1), the iteration counts remain fairly constant for arbitrarily large jumps in the coefficients, leading to a very efficient, automatic preconditioning method for scalar heterogeneous problems. As usual with domain decomposition methods, it is also well suited for parallel implementation, as illustrated in [117].

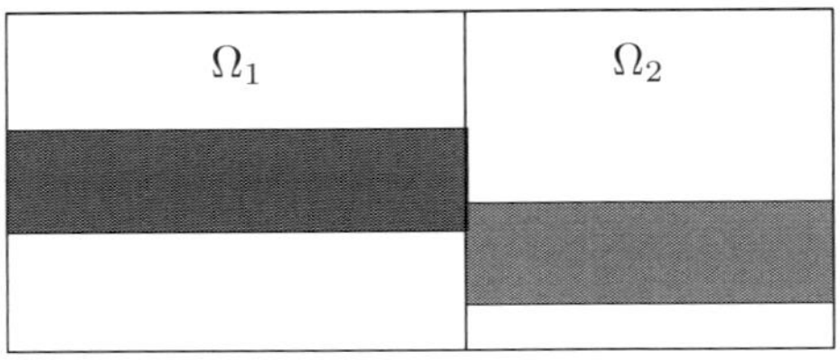

Figure 5.1. *Coefficient α varying along and across the interface.*

We now propose a construction of the coarse space that will be suitable for parallel implementation and efficient for arbitrary domain decompositions and highly heterogeneous problems such as the Darcy problem with Dirichlet boundary conditions:

$$-\mathrm{div}(\alpha \nabla u) = f \text{ in } \Omega,$$

where α can vary by several orders of magnitude, typically between 1 and 10^6.

We still choose Z such that it has the form

$$Z = \begin{bmatrix} W^1 & 0 & \cdots & 0 \\ \vdots & W^2 & \cdots & 0 \\ \vdots & \vdots & \cdots & \vdots \\ 0 & 0 & \cdots & W^N \end{bmatrix}, \tag{5.1}$$

where N is the number of overlapping subdomains. But W^i is now a rectangular matrix whose columns are based on the harmonic extensions of the eigenvectors corresponding to small eigenvalues of the DtN map in each subdomain Ω_i multiplied by the local component of the partition of unity; see Definition 5.13. Note that the sparsity of the coarse operator $E = Z^T A Z$ results from the sparsity of Z. The nonzero components of E correspond to adjacent subdomains.

More precisely, let us consider first at the continuous level the Dirichlet-to-Neumann map DtN_{Ω_i}.

Definition 5.1 (Dirichlet-to-Neumann map). *For any function defined on the interface $u_{\Gamma_i} : \Gamma_i \mapsto \mathbb{R}$, we consider the Dirichlet-to-Neumann map*

$$\mathrm{DtN}_{\Omega_i}(u_{\Gamma_i}) = \left. \frac{\partial v}{\partial \mathbf{n}_i} \right|_{\Gamma_i},$$

where $\Gamma_i := \partial\Omega_i \setminus \partial\Omega$ and v satisfies

$$\begin{cases} -\mathrm{div}(\alpha \nabla v) = 0 & \text{in } \Omega_i, \\ v = u_{\Gamma_i} & \text{on } \Gamma_i, \\ v = 0 & \text{on } \partial\Omega_i \cap \partial\Omega. \end{cases} \tag{5.2}$$

If the subdomain is not a floating one (i.e., $\partial\Omega_i \cap \partial\Omega \neq \emptyset$), we use on $\partial\Omega_i \cap \partial\Omega$ the boundary condition from the original problem $v = 0$.

To construct the coarse grid subspace, we use some low-frequency modes of the DtN operator by solving the following eigenvalue problem:

$$\mathrm{DtN}_{\Omega_i}(u) = \lambda u. \tag{5.3}$$

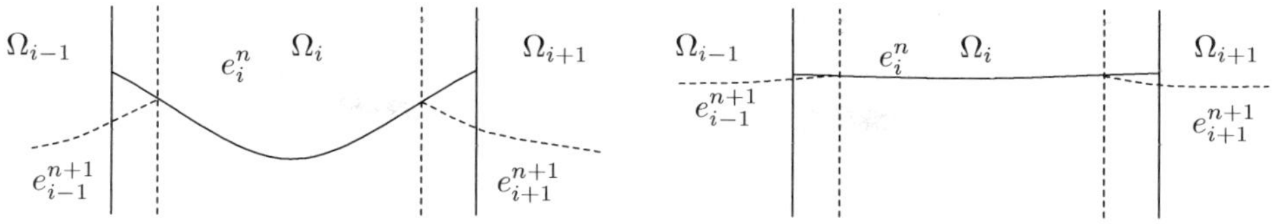

Figure 5.2. *Fast or slow convergence of the Schwarz algorithm.*

In order to obtain the variational formulation of (5.3) we multiply (5.2) by a test function w, equal to zero on $\partial\Omega_i \cap \partial\Omega$:

$$\int_{\Omega_i} \alpha \nabla v \cdot \nabla w - \int_{\Gamma_i} \alpha \frac{\partial v}{\partial \mathbf{n}_i} w = 0.$$

By using the fact that

$$\alpha \frac{\partial v}{\partial \mathbf{n}_i} = \lambda \alpha v \text{ on } \Gamma_i \text{ and } w = 0 \text{ on } \partial\Omega_i \cap \partial\Omega$$

we get the variational formulation of the eigenvalue problem of the local Dirichlet-to-Neumann map:

$$\boxed{\text{Find } (v,\lambda) \text{ such that } \int_{\Omega_i} \alpha \nabla v \cdot \nabla w = \lambda \int_{\partial\Omega_i} \alpha v\, w \quad \forall w.} \tag{5.4}$$

We now explain our choice of this coarse space based on the DtN map. We write the original Schwarz method at the continuous level, where the domain Ω is decomposed in a one-way partitioning. The error e_i^n between the current iterate at step n of the algorithm and the solution $u_{|\Omega_i}$ ($e_i^n := u_i^n - u_{|\Omega_i}$) in subdomain Ω_i at step n of the algorithm is harmonic:

$$-\text{div}(\alpha \nabla e_i^{n+1}) = 0 \text{ in } \Omega_i\,. \tag{5.5}$$

In the 1D example sketched in Figure 5.2, we see that the rate of convergence of the algorithm is related to the decay of the harmonic functions e_i^n in the vicinity of $\partial\Omega_i$ via the subdomain boundary condition. Indeed, a small value for this boundary condition leads to a smaller error in the entire subdomain thanks to the maximum principle. That is, a fast decay for this value corresponds to a large eigenvalue of the DtN map, whereas a slow decay corresponds to small eigenvalues of this map because the DtN operator is related to the normal derivative at the interface and the overlap is thin. Thus the small eigenvalues of the DtN map are responsible for the slow convergence of the algorithm and it is natural to incorporate them in the coarse grid space.

5.1.1 ▪ Spectral coarse spaces for other problems

The construction of coarse spaces is also a challenge for other classes of problems, such as compressible and incompressible flows in [3] and indefinite problems such as Helmholtz equations in [36]. In the latter, the construction of the coarse space is completely automatic and robust. The construction of the DtN coarse space inherently respects variations in the wave number, making it possible to treat heterogeneous Helmholtz problems. Moreover, it does not suffer from ill-conditioning, as does the standard approach based on plane waves (see [79, 77, 121]).

5.2 ▪ Variational setting

We consider the variational formulation of a Poisson boundary value problem with Dirichlet boundary conditions: Find $u \in H_0^1(\Omega)$ for a given polygonal (polyhedral) domain $\Omega \subset \mathbb{R}^d$ ($d=2$ or 3) and a source term $f \in L_2(\Omega)$ such that

$$\underbrace{\int_\Omega \nabla u \cdot \nabla v}_{:=a(u,v)} = \underbrace{\int_\Omega f(x)v(x)}_{:=(f,v)} \quad \forall v \in H_0^1(\Omega). \tag{5.6}$$

For any domain $D \subset \Omega$ we need the usual norm $\|\cdot\|_{L_2(D)}$ and seminorm $|\cdot|_{H^1(D)}$, as well as the L_2 inner product $(v,w)_{L_2(D)}$. To simplify notation we write

$$\|v\|_{0,D}^2 = \int_D v^2,$$

and the seminorm $|\cdot|_{a,D}$ is defined by

$$|v|_{a,D}^2 = \int_D |\nabla v|^2.$$

When $D = \Omega$ we omit the domain from the subscript and write $\|\cdot\|_0$ and $|\cdot|_a$ instead of $\|\cdot\|_{0,\Omega}$ and $|\cdot|_{a,\Omega}$, respectively. Note that the seminorm $|\cdot|_a$ becomes a norm on $H_0^1(\Omega)$.

Finally, we will also need averages and norms defined on $(d-1)$-dimensional manifolds $X \subset \mathbb{R}^d$; namely for any $v \in L_2(X)$ we define

$$\overline{v}^X := \frac{1}{|X|}\int_X u \quad \text{and} \quad \|v\|_{0,X}^2 := \int_X v^2,$$

where $|X|$ is the measure of the manifold X.

We consider a discretization of the variational problem (5.6) with $P1$ continuous, piecewise linear finite elements (FEs). To define the FE spaces and the approximate solution, we assume that we have a shape regular, simplicial triangulation $\mathcal{T}_h$ of Ω. The standard space of continuous and piecewise linear (w.r.t. $\mathcal{T}_h$) functions is then denoted by V_h and the subspace of functions from V_h that vanish on the boundary of Ω by $V_{h,0}$. Note that $V_{h,0} = V_h \cap H_0^1(\Omega)$. Denote by $\{\phi_k\}_{k\in\mathcal{N}}$ the basis of $V_{h,0}$ and by $\{x_k\}_{k\in\mathcal{N}}$ the associated nodes in Ω. Let us write the discrete FE problem that we want to solve: Find $u_h \in V_{h,0}$ such that

$$a(u_h, v_h) = (f, v_h) \quad \forall v_h \in V_{h,0}, \tag{5.7}$$

which gives the matrix form

$$A\mathbf{U} = \mathbf{F}, A_{ij} = a(\phi_j, \phi_i), F_i = (f, \phi_i) \quad \forall i, j \in \mathcal{N},$$

so that $u_h = \sum_{k\in\mathcal{N}} U_k \phi_k$, where $\mathbf{U} = (U_k)_{k\in\mathcal{N}}$.

For technical reasons, we also need the standard nodal value interpolation operator I_h from $C(\bar{\Omega})$ to $V_h(\Omega)$:

$$I_h(v) := \sum_{k\in\mathcal{N}} v(x_k)\phi_k.$$

This interpolant is known to be stable in the sense that there exists a constant $C_{I_h} > 0$ such that for all continuous, piecewise quadratic (w.r.t. $\mathcal{T}_h$) functions v, we have

$$\|I_h(v)\|_a \le C_{I_h} \|v\|_a. \tag{5.8}$$

5.3 ▪ Additive Schwarz setting

In order to automatically construct a robust two-level Schwarz preconditioner, we first partition our domain Ω into a set of nonoverlapping subdomains $\{\Omega'_j\}_{j=1}^N$ resolved by $\mathcal{T}_h$ using, for example, a graph partitioner such as METIS [119] or SCOTCH [27]. Each subdomain Ω'_j is then extended to a domain Ω_j by adding one or several layers of mesh elements in the sense of Definition 5.2, thus creating an overlapping decomposition $\{\Omega_j\}_{j=1}^N$ of Ω (see Figure 1.8).

Definition 5.2. *Given an open subset $D' \subset \Omega$ which is resolved by $\mathcal{T}_h$, the extension of D' by one layer of elements is*

$$D = \mathrm{Int}\left(\bigcup_{k \in \overline{\mathrm{dof}}(D')} \mathrm{supp}(\phi_k) \right), \quad \textit{where} \quad \overline{\mathrm{dof}}(D') := \{k \in \mathcal{N} : \mathrm{supp}(\phi_k) \cap D' \neq \emptyset\},$$

and Int(·) *denotes the interior of a domain. Extensions by more than one layer can then be defined recursively.*

The proof of the following lemma is a direct consequence of Definition 5.2.

Lemma 5.3. *For every degree of freedom k, with $k \in \mathcal{N}$, there is a subdomain Ω_j, with $1 \leq j \leq N$, such that* $\mathrm{supp}(\phi_k) \subset \overline{\Omega}_j$.

Note that in the case of $P1$ finite elements the degrees of freedom are in fact the values of the unknown function in the vertices of the mesh.

Definition 5.4 (Multiplicity of intersections). *We define*

$$\boxed{k_0 = \max_{\tau \in \mathcal{T}_h} \left(\#\{\Omega_j : 1 \leq j \leq N, \tau \subset \Omega_j\} \right)}$$

as the maximum number of subdomains to which one cell can belong.

Another ingredient of a decomposition is the following.

Definition 5.5 (Number of colors). *We define the number N_c as the minimum number of colors we can use to color a decomposition such that any two neighboring domains have different colors.*

Note that there is an obvious relation: $k_0 \leq N_c$.

Definition 5.6 (Overlapping parameter). *We define for each domain the overlapping parameter as*

$$\delta_j := \inf_{x \in \Omega_j \setminus \cup_{i \neq j} \bar{\Omega}_i} dist(x, \partial\Omega_j \setminus \partial\Omega). \tag{5.9}$$

Lemma 5.7 (Partition of unity). *With k_0 as in Definition 5.4, there exist functions $\{\chi_j\}_{j=1}^N \subset V_h(\Omega)$ such that*

- $\chi_j(x) \in [0,1] \quad \forall x \in \bar{\Omega}$,
- $\mathrm{supp}(\chi_j) \subset \bar{\Omega}_j$,
- $\sum_{j=1}^{N} \chi_j(x) = 1 \quad \forall x \in \bar{\Omega}$,
- $|\nabla \chi_j| \leq C_\chi \delta_j^{-1}$.

Proof. Let

$$d_j(x) := \begin{cases} \mathrm{dist}(x, \partial\Omega_j \setminus \partial\Omega),\, x \in \Omega_j, \\ 0, x \in \Omega \setminus \Omega_j. \end{cases}$$

Then it is enough to set

$$\chi_j(x) := I_h \left(\frac{d_j(x)}{\sum_{j=1}^{N} d_j(x)} \right).$$

The first three properties are obvious; the last one can be found in [185, p. 57]. □

Note that with the help of the partition of unity $\{\chi_j\}_{j=1}^{N}$ defined on Ω, subordinate to the overlapping decomposition $\{\Omega_j\}_{j=1}^{N}$ we can define

$$\Omega_j^{\circ} := \{x \in \Omega_j : \chi_j(x) < 1\}$$

as the boundary layer of Ω_j that is overlapped by neighboring domains.

Now, for each $j = 1, \ldots, N$, let

$$V_h(\Omega_j) := \{v|_{\Omega_j} : v \in V_h\}$$

denote the space of restrictions of functions in V_h to Ω_j. Furthermore, let

$$V_{h,0}(\Omega_j) := \{v|_{\Omega_j} : v \in V_h,\ \mathrm{supp}(v) \subset \bar{\Omega}_j\}$$

denote the space of finite element functions supported in $\bar{\Omega}_j$. By definition, the extension by zero of a function $v \in V_{h,0}(\Omega_j)$ to Ω lies again in V_h. We denote the corresponding extension operator by

$$E_j : V_{h,0}(\Omega_j) \to V_h. \tag{5.10}$$

Lemma 5.3 guarantees that $V_h = \sum_{j=1}^{N} E_j V_{h,0}(\Omega_j)$. The adjoint of E_j,

$$E_j^{\top} : V_h' \to V_{h,0}(\Omega_j)',$$

called the restriction operator, is defined by $\langle E_j^{\top} g, v\rangle = \langle g, E_j v\rangle$ (where $\langle\cdot,\cdot\rangle$ denotes the duality pairing) for $v \in V_{h,0}(\Omega_j)$, $g \in V_h'$. However, for the sake of simplicity, we will often leave out the action of E_j and view $V_{h,0}(\Omega_j)$ as a subspace of V_h.

The final ingredient is a coarse space $V_0 \subset V_h$ which will be defined later. Let $E_0 : V_0 \to V_h$ denote the natural embedding and $E_0^{\top}$ its adjoint. Then the two-level additive Schwarz preconditioner (in matrix form) reads as

$$M_{ASM,2}^{-1} = R_0^T A_0^{-1} R_0 + \sum_{j=1}^{N} R_j^T A_j^{-1} R_j,\ A_0 := R_0 A R_0^T, \text{ and } A_j := R_j A R_j^T, \tag{5.11}$$

where R_j, R_0 are the matrix representations of $E_j^{\top}$ and $E_0^{\top}$ with respect to the basis $\{\phi_k\}_{k\in\mathcal{N}}$ and the chosen basis of the coarse space V_0 which will be specified later.

As is usual for standard elliptic BVPs, A_j corresponds to the original (global) system matrix restricted to subdomain Ω_j with Dirichlet conditions on the artificial boundary $\partial\Omega_j \setminus \partial\Omega$.

On each of the local spaces $V_{h,0}(\Omega_j)$ the bilinear form $a_{\Omega_j}(\cdot,\cdot)$ is positive definite since

$$a_{\Omega_j}(v,w) = a(E_j v, E_j w) \quad \forall v,w \in V_{h,0}(\Omega_j)$$

and because $a(\cdot,\cdot)$ is coercive on V_h. For the same reason, the matrix A_j in (5.11) is invertible. An important ingredient in the definition of Schwarz algorithms are the following operators:

$$\tilde{\mathcal{P}}_j : V_h \to V_{h,0}(\Omega_j),\; a_{\Omega_j}(\tilde{\mathcal{P}}_j u, v_j) = a(u, E_j v_j) \quad \forall v_j \in V_{h,0}(\Omega_j)$$

and

$$\mathcal{P}_j = E_j \tilde{\mathcal{P}}_j : V_h \to V_h, j = 1,\ldots,N.$$

Lemma 5.8 (Properties of $\mathcal{P}_i$). *The operators $\mathcal{P}_j$ are symmetric projectors with respect to the scalar product induced by $a(\cdot,\cdot)$.*

Proof. By definition of $\tilde{\mathcal{P}}_j$, $\mathcal{P}_j$, and a_{Ω_j} we have

$$\begin{aligned} a(\mathcal{P}_j u, v) &= a(E_j \tilde{\mathcal{P}}_j u, v) = a(v, E_j \tilde{\mathcal{P}}_j u) \\ &= a_{\Omega_j}(\tilde{\mathcal{P}}_j u, \tilde{\mathcal{P}}_j v) = a(E_j \tilde{\mathcal{P}}_j u, E_j \tilde{\mathcal{P}}_j v) = a(\mathcal{P}_j u, \mathcal{P}_j v). \end{aligned} \tag{5.12}$$

Similarly we have $a(u, \mathcal{P}_j v) = a(\mathcal{P}_j v, u) = a(\mathcal{P}_j v, \mathcal{P}_j u)$, which proves that $a(\mathcal{P}_j u, v) = a(u, \mathcal{P}_j v)$, that is, $\mathcal{P}_j$ is self-adjoint w.r.t. a scalar product.

They are also projectors since by (5.12)

$$a(\mathcal{P}_j u, v) = a(\mathcal{P}_j u, \mathcal{P}_j v) = a(\mathcal{P}_j^2 u, \mathcal{P}_j v) = a(\mathcal{P}_j^2 u, v).$$

As a consequence of the coercivity of the bilinear forms a_{Ω_j} we have

$$a(\mathcal{P}_j u, u) = a(E_j \tilde{\mathcal{P}}_j u, u) = a(u, E_j \tilde{\mathcal{P}}_j u) = a_{\Omega_j}(\tilde{\mathcal{P}}_j u, \tilde{\mathcal{P}}_j u) \geq 0. \qquad \square$$

We will denote the *additive Schwarz operator* by

$$\boxed{\mathcal{P}_{ad} = \sum_{i=0}^{N} \mathcal{P}_i = \sum_{i=0}^{N} E_j \tilde{\mathcal{P}}_j.}$$

We can give an interpretation for the above from the linear algebra point of view. In this case, if we denote by $\tilde{P}_j$ and P_j the matrix counterparts of $\tilde{\mathcal{P}}_j$ and $\mathcal{P}_j$ w.r.t. the finite element basis $\{\phi_k\}_{k\in\mathcal{N}}$, we get

$$\tilde{P}_j = A_j^{-1} R_j A,\; P_j = R_j^T A_j^{-1} R_j A.$$

Taking into account (5.11), the *additive* matrix which corresponds to the "parallel" or block Jacobi version of the original Schwarz method is then given:

$$P_{ad} := M_{ASM,2}^{-1} A = \sum_{i=0}^{N} P_i. \tag{5.13}$$

A measure of the convergence is the condition number of P_{ad}, $\kappa(P_{ad})$, which can be computed using its largest and smallest eigenvalues (which are real since $\mathcal{P}_{ad}$ is self-adjoint w.r.t. to the a-inner product):

$$\lambda_{\max}(P_{ad}) = \sup_{u \in V_h} \frac{a(\mathcal{P}_{ad} u, u)}{a(u,u)}, \ \lambda_{\min}(P_{ad}) = \inf_{u \in V_h} \frac{a(\mathcal{P}_{ad} u, u)}{a(u,u)}.$$

5.4 ▪ Abstract theory for the two-level ASM

We present here the abstract framework for additive Schwarz methods (see [185, Chapter 2]). In the following we summarize the most important ingredients.

Lemma 5.9 (Upper bound of the eigenvalues of P_{ad}). *With N_c as in Definition 5.5, the largest eigenvalue of the additive Schwarz preconditioned operator $P_{ad} = M_{ASM,2}^{-1} A$ is bounded as follows:*

$$\boxed{\lambda_{\max}(P_{ad}) \le N_c + 1.}$$

Proof. We start by proving that

$$\lambda_{\max}(P_{ad}) \le N + 1.$$

By using the fact that $\mathcal{P}_i$ are a-orthogonal projectors, we get for all $u \in V_h$

$$\frac{a(\mathcal{P}_i u, u)}{\|u\|_a^2} = \frac{a(\mathcal{P}_i u, \mathcal{P}_i u)}{\|u\|_a^2} \le 1,$$

which implies

$$\lambda_{\max}(P_{ad}) = \sup_{u \in V_h} \sum_{i=0}^{N} \frac{a(\mathcal{P}_i u, u)}{\|u\|_a^2} \le \sum_{i=0}^{N} \sup_{u \in V_h} \frac{a(\mathcal{P}_i u, u)}{\|u\|_a^2} \le N + 1.$$

As for the estimate with a bound on the maximum number of colors, we proceed in the following way. By assumption, there are N_c sets of indices Θ_i, $i = 1, \ldots, N_c$, such that all subdomains Ω_j for $j \in \Theta_i$ have no intersections with one another. Then,

$$\mathcal{P}_{\Theta_i} := \sum_{j \in \Theta_i} \mathcal{P}_j, \ i = 1, \ldots, N_c$$

are again a-orthogonal projectors to which we can apply the same reasoning to reach the conclusion. Note that this result can be further improved to $\lambda_{\max}(P_{ad}) \le k_0$, where k_0 is the maximum multiplicity of the intersection between subdomains; see [185]. □

Definition 5.10 (Stable decomposition). *Given a coarse space $V_0 \subset V_h$, local subspaces $\{V_{h,0}(\Omega_j)\}_{1 \le j \le N}$, and a constant C_0, a C_0-stable decomposition of $v \in V_h$ is a family of* *functions* $\{v_j\}_{0\le j\le N}$ *that satisfies*

$$v=\sum_{j=0}^{N} E_j v_j, \qquad \text{with} \quad v_0\in V_0, v_j\in V_{h,0}(\Omega_j) \quad \text{for } j\ge 1, \tag{5.14}$$

and

$$\sum_{j=0}^{N}\left\|v_j\right\|_{a,\Omega_j}^2 \le C_0^2\|v\|_a^2 . \tag{5.15}$$

Theorem 5.11. *If every* $v\in V_h$ *admits a* C_0*-stable decomposition (with uniform* C_0*), then the smallest eigenvalue of* $M_{ASM,2}^{-1}A$ *satisfies*

$$\boxed{\lambda_{\min}(M_{ASM,2}^{-1}A)\ \ge\ C_0^{-2}.} \tag{5.16}$$

Proof. We use the projector properties of $\mathcal{P}_j$ and the definitions of $\mathcal{P}_j$ and $\tilde{\mathcal{P}}_j$ to get

$$\begin{aligned} a(\mathcal{P}_{ad}u,u) &= \sum_{j=0}^{N} a(\mathcal{P}_j u,u)=\sum_{j=0}^{N} a(\mathcal{P}_j u,\mathcal{P}_j u)\\ &=\sum_{j=0}^{N} a(E_j\tilde{\mathcal{P}}_j u,E_j\tilde{\mathcal{P}}_j u)=\sum_{j=0}^{N} a_{\Omega_j}(\tilde{\mathcal{P}}_j u,\tilde{\mathcal{P}}_j u)\\ &=\sum_{j=0}^{N}\left\|\tilde{\mathcal{P}}_j u\right\|_{a,\Omega_j}^2 . \end{aligned} \tag{5.17}$$

On the other hand, by using the same properties, the decomposition of $u=\sum_{j=0}^{N}E_j u_j$, and a repeated application of the Cauchy–Schwarz inequality we get

$$\begin{aligned} \|u\|_a^2 = a(u,u) &= \sum_{j=0}^{N} a(u,E_j u_j)=\sum_{j=0}^{N} a_{\Omega_j}(\tilde{\mathcal{P}}_j u,u_j)\\ \le \sum_{j=0}^{N}\left\|\tilde{\mathcal{P}}_j u\right\|_{a,\Omega_j}\left\|u_j\right\|_{a,\Omega_j} &\le \left(\sum_{j=0}^{N}\left\|\tilde{\mathcal{P}}_j u\right\|_{a,\Omega_j}^2\right)^{1/2}\left(\sum_{j=0}^{N}\left\|u_j\right\|_{a,\Omega_j}^2\right)^{1/2}. \end{aligned} \tag{5.18}$$

From (5.18) we get first that

$$\|u\|_a^4 \le \left(\sum_{j=0}^{N}\left\|\tilde{\mathcal{P}}_j u\right\|_{a,\Omega_j}^2\right)\left(\sum_{j=0}^{N}\left\|u_j\right\|_{a,\Omega_j}^2\right),$$

which can be combined with (5.17) to conclude that

$$\frac{a(\mathcal{P}_{ad}u,u)}{\|u\|_a^2}=\frac{\sum_{j=0}^{N}\left\|\tilde{\mathcal{P}}_j u\right\|_{a,\Omega_j}^2}{\|u\|_a^2}\ge\frac{\|u\|_a^2}{\sum_{j=0}^{N}\left\|u_j\right\|_{a,\Omega_j}^2}.$$

We can clearly see that if the decomposition $u=\sum_{j=0}^{N}E_j u_j$ is C_0^2-stable, that is, if (5.15) is verified, then the smallest eigenvalue of P_{ad} satisfies

$$\lambda_{\min}(P_{ad})=\inf_{u\in V_h}\frac{a(\mathcal{P}_{ad}u,u)}{\|u\|_a^2}\ge C_0^{-2}. \qquad \square$$

From Lemma 5.9 and (5.16) we get the final condition number estimate.

Corollary 5.12. *The condition number of the Schwarz preconditioner* (5.11) *can be bounded by*

$$\kappa(M_{ASM,2}^{-1}A) \leq C_0^2(N_c+1).$$

5.5 ▪ Definition and properties of Nicolaides and spectral coarse spaces

In the following, we will construct a C_0-stable decomposition in a specific framework, but prior to that we will provide, in an abstract setting, a sufficient and simplified condition of stability.

5.5.1 ▪ Nicolaides coarse space

The Nicolaides coarse space is made of local components associated to piecewise constant functions per subdomain. These functions are not in $V_{h,0}(\Omega_j)$ since they do not vanish on the boundary. To fix this problem, we have to distinguish between the so-called *floating* subdomains Ω_j, which do not touch the global boundary where a Dirichlet boundary condition is imposed ($\partial\Omega \cap \partial\Omega_j = \emptyset$), and subdomains that have part of their boundary on $\partial\Omega$. For floating subdomains, we will use the partition of unity, while the other subdomains will not contribute to the coarse space. More precisely, we define

$$\Pi_j^{Nico} u := \begin{cases} \left(\frac{1}{|\Omega_j|}\int_{\Omega_j} u\right) \mathbf{1}_{\Omega_j} = \bar{u}^{\Omega_j}\mathbf{1}_{\Omega_j} & \text{if } \Omega_j \text{ is floating,} \\ 0 & \text{otherwise.} \end{cases}$$

Functions $\mathbf{1}_{\Omega_j}$ are not in $V_{h,0}(\Omega_j)$, so we will use the partition of unity to build the global coarse space. Recall that I_h denotes the standard nodal value interpolation operator from $C(\bar{\Omega})$ to $V_h(\Omega)$ and define

$$\Phi_j^H := I_h(\chi_j \mathbf{1}_{\Omega_j}).$$

The coarse space is now defined as

$$\boxed{V_0 := \operatorname{span}\{\Phi_j^H : \Omega_j \text{ is a floating subdomain}\}.}$$

The use of the standard nodal value interpolation operator I_h is made necessary by the fact that $\chi_j \mathbf{1}_{\Omega_j}$ is not a $P1$ finite element function. By construction, each of the functions $\Phi_j \in V_{h,0}$, so that as required $V_0 \subset V_{h,0}$. The dimension of V_0 is the number of floating subdomains.

5.5.2 ▪ Spectral coarse space

As explained in section 5.1, the coarse space is made of local components associated to generalized eigenvalue problems in the subdomains, glued together with the partition of unity functions, except that in this section we have a constant coefficient operator (see [62] for the theory with heterogeneous coefficients).

Consider the variational generalized eigenvalue problem:

$$\text{Find } (v, \lambda) \in V_h(\Omega_j) \times \mathbb{R} \text{ such that}$$
$$\int_{\Omega_j} \nabla v \cdot \nabla w = \lambda \int_{\partial\Omega_j} v\, w \quad \forall w \in V_h(\Omega_j). \tag{5.19}$$

Let $n_{\partial\Omega_j}$ be the number of degrees of freedom on $\partial\Omega_j$. Since the right-hand side of the generalized eigenvalue problem is an integral over the boundary of the subdomain Ω_j, (5.19) has at most $n_{\partial\Omega_j}$ eigenpairs. Let $(v_\ell^{(j)}, \lambda_\ell^{(j)})_{1\le\ell\le n_{\partial\Omega_j}}$ denote these eigenpairs.

For some m_j ($1 \le m_j \le n_{\partial\Omega_j}$) the local coarse space is now defined as the span of the first m_j finite element functions $v_\ell^{(j)} \in V_h(\Omega_j)$, $1 \le \ell \le m_j$. For any $u \in V_h(\Omega_j)$, we can define the projection on $\text{span}\{v_\ell^{(j)}\}_{\ell=1}^{m_j}$ by

$$\Pi_j^{spec} u := \sum_{\ell=1}^{m_j} a_j\left(v_\ell^{(j)}, u\right) v_\ell^{(j)}, \tag{5.20}$$

where a_j is the local bilinear form associated to the Laplace operator on the local domain Ω_j,

$$a_j(v, w) := \int_{\Omega_j} \nabla v \cdot \nabla w \quad \forall v, w \in V_h(\Omega_j),$$

where the eigenvectors are normalized: $a_j(v_\ell^{(j)}, v_\ell^{(j)}) = 1$. Functions $v_\ell^{(j)}$ are not in $V_{h,0}(\Omega_j)$ so we will use the partition of unity functions χ_j to build the global coarse space.

Definition 5.13 (Spectral coarse space). *The spectral coarse space is defined as*

$$\boxed{V_0 := \text{span}\{\Phi_{j,\ell}^H : 1 \le j \le N \text{ and } 1 \le \ell \le m_j\}, \quad \text{where} \quad \Phi_{j,\ell}^H := I_h(\chi_j v_\ell^{(j)}).}$$

Remark 5.5.1. *The structure of this coarse space imposes the following remarks:*

- *As in the case of the Nicolaides coarse space, the use of the standard nodal value interpolation operator I_h is justified by the fact that $\chi_j v_\ell^{(j)}$ is not a $P1$ finite element function.*
- *The dimension of V_0 is $\sum_{j=1}^N m_j$.*
- *By construction, each of the functions $\Phi_{j,\ell} \in V_{h,0}$, so that as required $V_0 \subset V_{h,0}$.*
- *When $m_j = 1$ and the subdomain j does not touch the boundary of Ω, the lowest eigenvalue of the DtN map is zero and the corresponding eigenvector is a constant vector. Thus the Nicolaides and spectral coarse spaces are then identical.*

In the following we will illustrate some very natural properties of the projection operators defined before.

Lemma 5.14 (Stability of the local projections). *There exist constants C_P and C_{tr} independent of δ_j and of the size of the subdomains such that for any $u \in V_h(\Omega_j)$ and for*

$\Pi_j = \Pi_j^{Nico}$ *or* $\Pi_j = \Pi_j^{spec}$

$$|\Pi_j u|_{a,\Omega_j} \le |u|_{a,\Omega_j}, \tag{5.21}$$

$$\|u - \Pi_j u\|_{0,\Omega_j^\circ} \le 2\sqrt{c_j}\,\delta_j\,|u|_{a,\Omega_j}, \tag{5.22}$$

where

$$c_j := \begin{cases} C_P^2 + \left(\delta_j \lambda^{(j)}_{m_j+1}\right)^{-1} & \text{for } \Pi_j = \Pi_j^{spec}, \\ C_P^2 + C_{tr}^{-1}\left(\frac{H_j}{\delta_j}\right) & \text{for } \Pi_j = \Pi_j^{Nico}, \end{cases} \tag{5.23}$$

where H_j denotes the diameter of the subdomain Ω_j.

Proof. The results in the first part of the proof are true for both cases; that is, $\Pi_j = \Pi_j^{Nico}$ or $\Pi_j = \Pi_j^{spec}$. Stability estimate (5.21) follows immediately from the fact that Π_j^{spec} is an $a^{(j)}$-orthogonal projection and from the consideration that $|\Pi_j^{Nico} u|_{a,\Omega_j} = 0$ since $\Pi_j^{Nico} u$ is a constant function.

To prove (5.22) let us first apply Lemma 5.19, i.e.,

$$\|u - \Pi_j u\|^2_{0,\Omega_j^\circ} \le 2C_P^2\delta_j^2|u - \Pi_j u|^2_{a,\Omega_j^\circ} + 4\,\delta_j\|u - \Pi_j u\|^2_{0,\partial\Omega_j}. \tag{5.24}$$

It follows from the triangle inequality and (5.21) that

$$|u - \Pi_j u|^2_{a,\Omega_j^\circ} \le |u - \Pi_j u|^2_{a,\Omega_j} \le 2\,|u|^2_{a,\Omega_j}, \tag{5.25}$$

and so it only remains to bound $\|u - \Pi_j u\|^2_{0,\partial\Omega_j}$.

At this point we have to distinguish between the Nicolaides and spectral coarse spaces.

So if $\Pi_j = \Pi_j^{Nico}$ and Ω_j is a floating subdomain, under some regularity assumptions of the subdomains, we have the following inequality (see [185] for this type of result):

$$\|u - \Pi_j u\|^2_{0,\partial\Omega_j} \le C_{tr} H_j |u|^2_{a,\Omega_j}.$$

When Ω_j is not floating, $\Pi_j u = 0$ and the above inequality becomes a classical trace inequality.

In the case of the spectral coarse space, that is, $\Pi_j = \Pi_j^{spec}$, let us extend the set $\{v_\ell^{(j)}\}_{\ell=1}^{n_{\partial\Omega_j}}$ to an a_j-orthonormal basis of $V_h(\Omega_j)$ by adding $n_j = \dim V_{h,0}(\Omega_j)$ suitable functions $v^{(j)}_{n_{\partial\Omega_j}+\ell}$, $\ell = 1,\dots,n_j$, and write

$$u = \sum_{\ell=1}^{n_j + n_{\partial\Omega_j}} a_j\left(v_\ell^{(j)}, u\right) v_\ell^{(j)}. \tag{5.26}$$

The restriction of the functions $\{v_\ell^{(j)}\}_{\ell=1}^{n_{\partial\Omega_j}}$ to the boundary $\partial\Omega_j$ forms a complete basis of $V_h(\partial\Omega_j)$. This implies that $v^{(j)}_{n_{\partial\Omega_j}+\ell} \equiv 0$ on $\partial\Omega_j$ for all $\ell = 1,\dots,n_j$. Moreover, it follows from the definition of the eigenproblem that the functions $\{v_\ell^{(j)}\}_{\ell=1}^{n_{\partial\Omega_j}}$ are orthogonal also in the $(\cdot,\cdot)_{0,\partial\Omega_j}$ inner product. Therefore

$$\|u-\Pi_j u\|^2_{0,\partial\Omega_j} = \sum_{\ell=m_j+1}^{n_{\partial\Omega_j}} a\left(v_\ell^{(j)},u\right)^2 \|v_\ell^{(j)}\|^2_{0,\partial\Omega_j} \tag{5.27}$$

$$= \sum_{\ell=m_j+1}^{n_{\partial\Omega_j}} \frac{1}{\lambda_\ell^{(j)}} a\left(v_\ell^{(j)},u\right)^2 \tag{5.28}$$

$$\le \frac{1}{\lambda_{m_j+1}^{(j)}} \sum_{\ell=1}^{n_{\partial\Omega_j}} a\left(v_\ell^{(j)},u\right)^2 = \frac{1}{\lambda_{m_j+1}^{(j)}} |u|^2_{a,\Omega_j}$$

and the result follows from (5.24), (5.25), and (5.27). Note that this estimate does not require any regularity assumption on the subdomain. □

5.6 ▪ Convergence theory for the ASM with Nicolaides and spectral coarse spaces

We are now ready to state the existence of a stable splitting for the previously built coarse spaces.

Lemma 5.15 (Existence of a stable splitting). *For any $u \in V_{h,0}$, let $\{u_j\}_{0\le j\le N}$ be defined by*

$$\boxed{u_0 := I_h\left(\sum_{j=1}^{J} \chi_j \Pi_j u|_{\Omega_j}\right) \in V_0 \text{ and } u_j := I_h(\chi_j(u-\Pi_j u)),} \tag{5.29}$$

where Π_j is Π_j^{Nico} or Π_j^{spec}. Then $\{u_j\}_{0\le j\le N}$ form a C_0^2-stable decomposition of u in the sense of Definition 5.10 with

$$C_0^2 := \left(8 + 8\, C_\chi^2 \max_{j=1}^{N} c_j\right) k_0\, C_{I_h}\, (k_0+1) + 1,$$

where c_j depends on the choice of the coarse space and is given by formula (5.23).

Proof. Note that the proof is the same for both cases, that is, $\Pi_j = \Pi_j^{Nico}$ and $\Pi_j = \Pi_j^{spec}$. In the following, to ease the presentation when there is no confusion and it is clear from the context, we will simply denote the restriction $u|_{\Omega_j}$ of u onto Ω_j by u and write, e.g., $\Pi_j u$ instead of $\Pi_j u|_{\Omega_j}$. Conversely, where there is no confusion we will simply write u_j instead of its extension to Ω, $E_j u_j$.

Let us first show (5.14), that is, $u = \sum_{j=0}^{N} u_j$:

$$\sum_{j=1}^{N} u_j = \sum_{j=1}^{N} I_h(\chi_j(u-\Pi_j u)) = I_h\left(\sum_{j=1}^{N}\chi_j u\right) - \sum_{j=1}^{J} I_h\left(\chi_j \Pi_j u\right) = u - u_0 .$$

We turn now to the second part, that is, to the proof of (5.15). From a simple application of the triangle inequality it follows that

$$\sum_{j=0}^{N} \|u_j\|_a^2 \le \|u-u_0\|_a^2 + \|u\|_a^2 + \sum_{j=1}^{N} \|u_j\|_a^2. \tag{5.30}$$

Since the interpolant I_h is stable with respect to the a-norm (see (5.8)) and since each cell is overlapped by at most k_0 domains, we have

$$\begin{aligned} \|u-u_0\|_a^2 &= \left\| I_h\left(\sum_{j=1}^{J} \chi_j(u-\Pi_j u)\right)\right\|_a^2 \le C_{I_h} \left\| \sum_{j=1}^{N} \chi_j(u-\Pi_j u)\right\|_a^2 \\ &\le C_{I_h} k_0 \sum_{j=1}^{N} \|\chi_j(u-\Pi_j u)\|_a^2 . \end{aligned}$$

Substituting this into (5.30) and using the definition of u_j as well as the a-stability of the interpolant I_h we get

$$\sum_{j=0}^{N} \|u_j\|_a^2 \le C_{I_h}(k_0+1)\sum_{j=1}^{N} \|\chi_j(u-\Pi_j u)\|_a^2 + \|u\|_a^2. \tag{5.31}$$

Note that $\mathrm{supp}\{\chi_j\} = \overline{\Omega}_j$ and $\mathrm{supp}\{\nabla\chi_j\} = \overline{\Omega}_j^{\circ}$. Thus, using the triangle inequality and the product rule,

$$\begin{aligned} \sum_{j=1}^{N} \|\chi_j(u-\Pi_j u)\|_a^2 &\le 2\sum_{j=1}^{N} \|\chi_j\|_\infty^2 |u-\Pi_j u|_{a,\Omega_j}^2 + \|\nabla\chi_j\|_\infty^2 \|u-\Pi_j u\|_{0,\Omega_j^\circ}^2 \\ &\le \sum_{j=1}^{N} 4|u|_{a,\Omega_j}^2 + 4|\Pi_j u|_{a,\Omega_j}^2 + 2C_\chi^2 \delta_j^{-2} \|u-\Pi_j u\|_{0,\Omega_j^\circ}^2 . \end{aligned} \tag{5.32}$$

Note that in the last part we used the last property of the partition of unity stated in Lemma 5.7.

We have

$$\begin{aligned} \sum_{j=1}^{N} \|\chi_j(u-\Pi_j u)\|_a^2 &\le \sum_{j=1}^{N} 4|u|_{a,\Omega_j}^2 + 4|\Pi_j u|_{a,\Omega_j}^2 + 2C_\chi^2 \delta_j^{-2} \|u-\Pi_j u\|_{0,\Omega_j^\circ}^2 \\ &\le \sum_{j=1}^{N} |u|_{a,\Omega_j}^2 (8+8\,C_\chi^2\, c_j) \\ &\le \left(8+8\,C_\chi^2 \max_{j=1}^{N} c_j\right) \sum_{j=1}^{N} |u|_{a,\Omega_j}^2 \\ &\le \left(8+8\,C_\chi^2 \max_{j=1}^{N} c_j\right) k_0\, |u|_a^2. \end{aligned} \tag{5.33}$$

The last inequality comes from the assumption that each point $x \in \Omega$ is contained in at most k_0 subdomains.

From (5.31) and (5.33) we conclude that

$$\sum_{j=0}^{N} \|u_j\|_a^2 \le \left[\left(8+8\,C_\chi^2 \max_{j=1}^{N} c_j\right) k_0\, C_{I_h}\,(k_0+1)+1\right] \|u\|_a^2,$$

which ends the proof and yields the following formula for the constant C_0:

$$C_0^2 := \left(8 + 8\, C_\chi^2 \max_{j=1}^{N} c_j\right) k_0\, C_{I_h}\,(k_0+1)+1. \qquad \square \tag{5.34}$$

From the abstract Schwarz theory, we have the condition number estimates in the two cases (see Corollary 5.12).

Theorem 5.16 (Convergence of the ASM with the Nicolaides coarse space). *The condition number of the two-level Schwarz algorithm with the Nicolaides coarse space can be bounded by*

$$\boxed{\begin{aligned}&\kappa(M^{-1}_{ASM,2}A)\\ &\le \left[\left(8+8\,C_\chi^2 \max_{j=1}^{N}\left(C_P^2 + C_{tr}^{-1}\left(\frac{H_j}{\delta_j}\right)\right)\right)k_0 C_{I_h}(k_0+1)+1\right](N_c+1).\end{aligned}}$$

Theorem 5.17 (Convergence of the ASM with a spectral coarse space). *The condition number of the two-level Schwarz algorithm with a coarse space based on local DtN maps can be bounded by*

$$\boxed{\begin{aligned}&\kappa(M^{-1}_{ASM,2}A)\\ &\le \left[\left(8+8\,C_\chi^2 \max_{j=1}^{N}\left(C_P^2 + \left(\frac{1}{\delta_j \lambda_{m_j+1}}\right)\right)\right)k_0 C_{I_h}(k_0+1)+1\right](N_c+1).\end{aligned}}$$

Remark 5.6.1. *Note that, by choosing the number m_j of modes per subdomain in the coarse space such that $\lambda^{(j)}_{m_j+1} \ge H_j^{-1}$, the preconditioned problem using the coarse space verifies*

$$\boxed{\begin{aligned}&\kappa(M^{-1}_{ASM,2}A)\\ &\le \left[\left(8+8\,C_\chi^2 \max_{j=1}^{N}\left(C_P^2 + \left(\frac{H_j}{\delta_j}\right)\right)\right)k_0 C_{I_h}(k_0+1)+1\right](N_c+1).\end{aligned}}$$

Hence, we have the same type of estimate as for the Nicolaides coarse space. An interesting observation is that the bound depends only in an additive way on the constant C_P and on the ratio of the subdomain diameter to overlap.

5.7 ▪ Functional analysis results

We prove Lemma 5.19 that was used to establish estimate (5.22). We start by stating an assumption on the overlapping subregions.

Assumption 5.7.1. We assume that each Ω_j°, $j=1,\dots,N$, can be subdivided into regions D_{jk}, $k=1,\dots,K_j$ (see Figure 5.3), such that $\mathrm{diam}(D_{jk}) \approx \delta_j$ and $|D_{jk}| \approx \delta_j^d$, and such

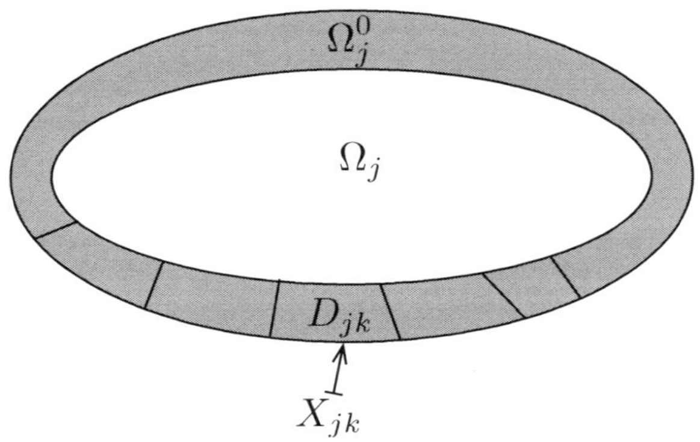

Figure 5.3. *Subdivision of the overlapping region.*

that, for each k, the $(d-1)$-dimensional manifold $X_{jk} := \partial\Omega_j \cap \overline{D}_{jk}$ has the following properties:

(i) $|X_{jk}| \approx \delta_j^{d-1}$;

(ii) $\frac{|D_{jk}|}{|X_{jk}|} \leq 2\delta_j$.

We assume as well that the triangulation $\mathcal{T}_h$ resolves each of the regions D_{jk} and each of the manifolds X_{jk}.

In [161], the following uniform Poincaré inequality is proved.

Lemma 5.18. *There exists a uniform constant $C_P > 0$ such that the following Poincaré-type inequality holds for all $j = 1, \ldots, N$ and $k = 1, \ldots, K_j$:*

$$\|v - \overline{v}^{X_{jk}}\|_{0,D_{jk}} \leq C_P\, \delta_j\, |v|_{a,D_{jk}} \qquad \forall\, v \in V_h(D_{jk}).$$

The constant C_P is independent of h and δ_j.

We then prove the following result.

Lemma 5.19. *We have*

$$\|u\|_{0,\Omega_j^\circ}^2 \leq 2C_P^2\delta_j^2|u|_{a,\Omega_j^\circ}^2 + 4\delta_j\|u\|_{0,\partial\Omega_j}^2 \qquad \forall\, u \in V_h(\Omega_j^\circ).$$

Proof. It follows from Lemma 5.18, as well as from the triangle and Cauchy–Schwarz inequalities, that

$$\begin{aligned}
\tfrac{1}{2}\|u\|_{0,D_{jk}}^2 &\leq \|u - \overline{u}^{X_{jk}}\|_{0,D_{jk}}^2 + \|\overline{u}^{X_{jk}}\|_{0,D_{jk}}^2 \\
&\leq C_P^2\delta_j^2|u|_{a,D_{jk}}^2 + \frac{|D_{jk}|}{|X_{jk}|^2}\left(\int_{X_{jk}} u\right)^2 \\
&\leq C_P^2\delta_j^2|u|_{a,D_{jk}}^2 + \frac{|D_{jk}|}{|X_{jk}|}\int_{X_{jk}} u^2 \\
&\leq C_P^2\delta_j^2|u|_{a,D_{jk}}^2 + 2\delta_j\,\|u\|_{0,X_{jk}}^2 .
\end{aligned}$$

The final result follows by summing over $k = 1, \ldots, K_j$. $\square$

5.8 ▪ Theory of spectral coarse spaces for scalar heterogeneous problems

The analysis of the spectral coarse space introduced in [147] for the highly heterogeneous Darcy equation

$$-\text{div}(\alpha \nabla u) = f \text{ in } \Omega,$$

where α can vary by several orders of magnitude, typically between 1 and 10^6, was analyzed in [62]. A similar but slightly different approach is developed in [84]. These results generalize the classical estimates of overlapping Schwarz methods to the case where the coarse space is richer than just the kernel of the local operators (which is the set of constant functions) [151], or other classical coarse spaces [185]. It is particularly well suited to the small overlap case. In the case of generous overlap, our analysis relies on an assumption on the coefficient distribution that may not be satisfied for small-scale coefficient variation. This assumption does not seem to be necessary in practice (see also [148] for more extensive numerical results).

The idea of a coarse space based on spectral information that allows one to achieve any a priori chosen target convergence rate was developed and implemented in the spectral AMGe method in [25]. More recently, in the framework of two-level overlapping Schwarz, [84, 83, 147, 148, 70] also build coarse spaces for problems with highly heterogeneous coefficients by solving local eigenproblems. However, compared to earlier works in the AMG context, the recent papers all focus on generalized eigenvalue problems. We can distinguish three sets of methods that all differ by the choice of the bilinear form on the right-hand side of the generalized eigenproblem. In [84, 83], the right-hand side is the local mass matrix, or a "homogenized" version obtained by using a multiscale partition of unity. In [147, 148] the right-hand side corresponds to an L_2-product on the subdomain boundary, so that the problem can be reduced to a generalized eigenproblem for the DtN operator on the subdomain boundary. The latest set of papers, including [70], uses yet another type of bilinear form on the right-hand side, inspired by theoretical considerations. In [173], the construction of energy-minimizing coarse spaces that obey certain functional constraints is used to build robust coarse spaces for elliptic problems with large variations in the coefficients.

All these approaches have their advantages and disadvantages, which depend on many factors, in particular the type of coefficient variation and the size of the overlap. When there are very small scale coefficient variations, many of the above approaches lead to rather large (and therefore costly) coarse spaces, and it is still an open theoretical question how large the coarse space will have to become in each case to achieve robustness for an arbitrary coefficient variation.

Chapter 6

Neumann–Neumann and FETI algorithms

The last decade has shown that Neumann–Neumann [14, 132, 125] and FETI [81] algorithms, as well as their variants such as BDDC [47] and FETI-DP [78] algorithms, are very efficient domain decomposition methods. Most of the early theoretical and numerical work has been carried out for symmetric positive definite second-order problems; see, for example, [39, 134, 81, 135]. Then, the method was extended to other problems, such as the advection-diffusion equations [95, 1], plate and shell problems [182], and the Stokes equations [155, 183].

These methods require pseudoinverses for local Neumann solves which can be ill-posed either in the formulation of the domain decomposition problem or in the domain decomposition preconditioner. FETI methods are based on the introduction of Lagrange multipliers on the interfaces to ensure a weak continuity of the solution. We present here an alternative formulation of the Neumann–Neumann algorithm with the purpose of simplifying the numerical implementation. This is made at the expense of minor modifications to the original algorithm while keeping its main features. We start with a historical and heuristic presentation of these methods.

6.1 ▪ Direct and hybrid substructured solvers

Neumann–Neumann and FETI methods originated from parallel direct solvers. For a general matrix, direct methods consist in performing its Gauss elimination into a lower triangular matrix and an upper triangular one, a.k.a. LU decomposition. For a symmetric positive definite matrix, a Cholesky decomposition into a lower triangular matrix times its transpose is cheaper (a.k.a. LL^T decomposition). In order to benefit from the sparsity of the matrix arising from a finite element discretization of a partial differential equation, a variant of Gaussian elimination, the frontal method, that automatically avoids a large number of operations involving zero terms was developed. A frontal solver builds an LU or Cholesky decomposition of a sparse matrix given as the assembly of element matrices by eliminating equations only on a subset of elements at a time. This subset is called *the front,* and it is essentially the transition region between the part of the system already finished and the part not touched yet. These methods are basically sequential since the unknowns are processed one after another or one front after another. In order to benefit from multicore processors, a multifrontal solver (see Duff and Reid [69]) is an improvement on the frontal solver that uses several independent fronts at the same time. The fronts can be worked on by different processors, which enables parallel computing.

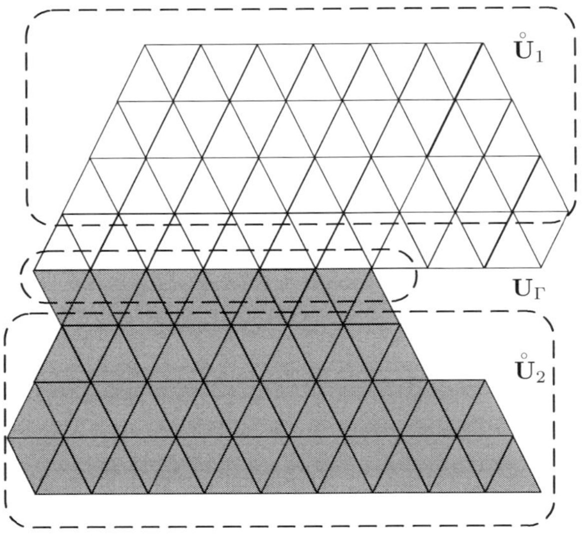

Figure 6.1. *Degrees of freedom partition for a multifrontal method.*

In order to simplify the presentation of a multifrontal solver, we consider the two-subdomain case for a matrix arising from a finite element discretization. As we have repeatedly shown in the previous chapters, a variational problem discretized by a finite element method yields a linear system of the form $A\mathbf{U} = \mathbf{F}$, where $\mathbf{F}$ is a given right-hand side and $\mathbf{U}$ is the set of unknowns.

The set of degrees of freedom $\mathcal{N}$ is partitioned into interior degrees of freedom (dofs) $\mathcal{N}_1$ and $\mathcal{N}_2$ and interface dofs $\mathcal{N}_\Gamma$ (separator); see section 6.4 for a complete definition of this intuitive notion and Figure 6.1 as an illustration of the two-subdomain case.

The vector of unknowns $\mathbf{U}$ (resp., $\mathbf{F}$) is accordingly partitioned into the interior unknowns $\overset{\circ}{\mathbf{U}}_1$ and $\overset{\circ}{\mathbf{U}}_2$ (resp., $\overset{\circ}{\mathbf{F}}_1$, $\overset{\circ}{\mathbf{F}}_2$), and into the interface unknowns $\mathbf{U}_\Gamma$ (resp., $\mathbf{F}_\Gamma$). Numbering the interface equations last leads to a block decomposition of the linear system which has the shape of an arrow (pointing down to the right):

$$\begin{pmatrix} A_{11} & 0 & A_{1\Gamma} \\ 0 & A_{22} & A_{2\Gamma} \\ A_{\Gamma 1} & A_{\Gamma 2} & A_{\Gamma\Gamma} \end{pmatrix} \begin{pmatrix} \overset{\circ}{\mathbf{U}}_1 \\ \overset{\circ}{\mathbf{U}}_2 \\ \mathbf{U}_\Gamma \end{pmatrix} = \begin{pmatrix} \overset{\circ}{\mathbf{F}}_1 \\ \overset{\circ}{\mathbf{F}}_2 \\ \mathbf{F}_\Gamma \end{pmatrix}. \tag{6.1}$$

A simple computation shows that we have a block factorization of matrix A:

$$A = \begin{pmatrix} I & & \\ 0 & I & \\ A_{\Gamma 1}A_{11}^{-1} & A_{\Gamma 2}A_{22}^{-1} & I \end{pmatrix} \begin{pmatrix} A_{11} & & \\ & A_{22} & \\ & & \mathbb{S} \end{pmatrix} \begin{pmatrix} I & 0 & A_{11}^{-1}A_{1\Gamma} \\ & I & A_{22}^{-1}A_{2\Gamma} \\ & & I \end{pmatrix}. \tag{6.2}$$

Here, the matrix

$$\boxed{\mathbb{S} := A_{\Gamma\Gamma} - A_{\Gamma 1}A_{11}^{-1}A_{1\Gamma} - A_{\Gamma 2}A_{22}^{-1}A_{2\Gamma}} \tag{6.3}$$

is a Schur complement matrix and it is dense. It corresponds to an elimination of the interior unknowns $\overset{\circ}{\mathbf{U}}_i$, $i = 1, 2$.

From (6.2) we can see that the inverse of A can be easily computed from its factorization

$$A^{-1} = \begin{pmatrix} I & 0 & -A_{11}^{-1}A_{1\Gamma} \\ & I & -A_{22}^{-1}A_{2\Gamma} \\ & & I \end{pmatrix} \begin{pmatrix} A_{11}^{-1} & & \\ & A_{22}^{-1} & \\ & & \mathbb{S}^{-1} \end{pmatrix} \begin{pmatrix} I & & \\ 0 & I & \\ -A_{\Gamma 1}A_{11}^{-1} & -A_{\Gamma 2}A_{22}^{-1} & I \end{pmatrix}. \tag{6.4}$$

Note that in practice, the three matrices A_{11}^{-1}, A_{22}^{-1}, and $\mathbb{S}^{-1}$ are never evaluated explicitly. From the above formula, it appears that applying A^{-1} to the right-hand side $\mathbf{F}$ is equivalent to solving linear systems for the above three matrices, which can be done by factorizing them. The parallelism comes from the fact that matrices A_{11} and A_{22} can be factorized concurrently. We say that we advance two fronts in parallel. Once this is done, matrix $\mathbb{S}$ can be computed and then factorized.

By considering more than two fronts and recursive variants, it is possible to have numerical efficiency for one or possibly two dozen cores and problems of one or two million unknowns in two dimensions and hundreds of thousands of degrees of freedom in three dimensions. Although these figures improve over time, it is generally accepted that *purely direct solvers cannot scale well on large parallel platforms*. The bottleneck comes from the fact that $\mathbb{S}$ is a full matrix which is costly to compute and factorize. This is unfortunate since direct solvers have the advantage over iterative solvers of being very predictable, reliable, and easy to integrate via third-party libraries. There is thus a great amount of effort involved in developing hybrid direct/iterative solvers that try to take advantage of the two worlds.

A first approach consists in avoiding the factorization of the Schur complement matrix $\mathbb{S}$ by replacing every application of its inverse to some vector $\mathbf{b}_\Gamma \in \mathbb{R}^{\#\mathcal{N}_\Gamma}$ by an iterative solve of a Schur complement equation

$$\boxed{\mathbf{x}_\Gamma = \mathbb{S}^{-1}\mathbf{b}_\Gamma \Leftrightarrow (A_{\Gamma\Gamma} - A_{\Gamma 1}A_{11}^{-1}A_{1\Gamma} - A_{\Gamma 2}A_{22}^{-1}A_{2\Gamma})\mathbf{x}_\Gamma = \mathbf{b}_\Gamma} \tag{6.5}$$

with a conjugate gradient (CG) algorithm (see Chapter 3 on Krylov methods). An extra advantage of using Krylov methods is that they only require matrix-vector products $\mathbb{S}$ times some vector $\mathbf{r}_\Gamma$ which can be performed mostly in parallel without computing the entries of the Schur complement matrix. Both the computation and the factorization $\mathbb{S}$ are avoided. This way, the method combines direct solvers in the subdomain with an iterative solver for the interface between subdomains. This is an example of a hybrid direct/iterative solver.

We are now left with the efficiency of the CG method, which means there is a need for a preconditioner. Finding a good parallel preconditioner for problem (6.5) is not easy at the algebraic level. In the next section we will consider substructuring at the continuous level, which will enable us to propose the Neumann–Neumann/FETI preconditioners at the continuous level in section 6.2.2 and then define its algebraic counterpart in section 6.3.

6.2 ▪ Two-subdomain case at the continuous level

As an example, we consider a Poisson problem defined on a domain Ω:

$$\text{Find } u : \Omega \mapsto \mathbb{R} \text{ such that} \quad \begin{cases} -\Delta u = f & \text{in } \Omega, \\ u = 0 & \text{on } \partial\Omega. \end{cases} \tag{6.6}$$

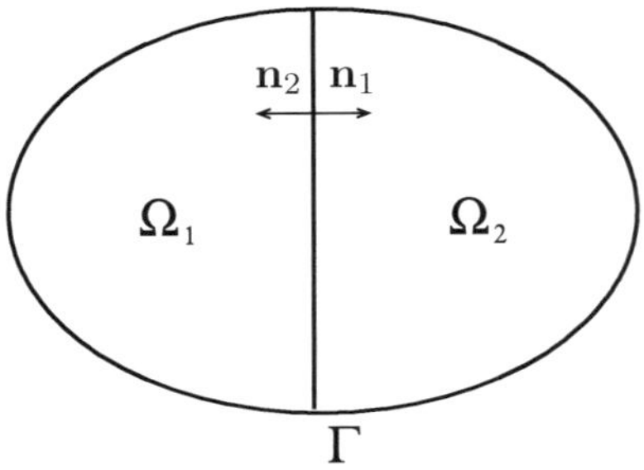

Figure 6.2. *Nonoverlapping decomposition into two subdomains.*

In order to simplify the presentation, we forget about the boundary condition on $\partial\Omega$. Suppose that the domain Ω is decomposed into two nonoverlapping subdomains Ω_1 and Ω_2. Let the interface between the subdomains be $\Gamma = \bar{\Omega}_1 \cap \bar{\Omega}_2$, and let $(\mathbf{n}_i)_{i=1,2}$ be the outward normals to the artificial boundary Γ corresponding to $\Omega_{1,2}$ as depicted in Figure 6.2.

It is known that u_1 (resp., u_2) is the restriction of the solution to (6.6) to subdomain Ω_1 (resp., Ω_2) if and only if

$$\begin{cases} -\Delta u_i = f & \text{in } \Omega_i,\ i = 1,2, \\ u_1 = u_2 & \text{on } \Gamma, \\ \dfrac{\partial u_1}{\partial \mathbf{n}_1} + \dfrac{\partial u_2}{\partial \mathbf{n}_2} = 0 & \text{on } \Gamma. \end{cases} \tag{6.7}$$

In other words, the local functions must satisfy the Poisson equation in the subdomains and match at the interface. Since the Poisson equation is a scalar second-order partial differential equation, we have two continuity equations on the interface: one for the unknown and another for the normal derivatives. Recall that for overlapping subdomains, a matching condition of the unknown on the interfaces is sufficient; see Chapter 1 on Schwarz methods.

6.2.1 ▪ Iterative Neumann–Neumann and FETI algorithms

In this section we will present iterative methods for solving (6.7). They are naturally derived from (6.7) by relaxing and then correcting the continuity relations on the interface. When the flux continuity is relaxed, we obtain what is called a Neumann–Neumann method, and when the continuity on the unknown is relaxed, we obtain an iterative version of the FETI algorithm. Both Neumann–Neumann and FETI methods will be investigated at the algebraic level in section 6.3.

Definition 6.1 (Iterative Neumann–Neumann algorithm). *Let u_i^n denote an approximation to the solution u in a subdomain Ω_i, $i = 1,2$, at the nth iteration. Starting from an initial guess u_i^0, the Neumann–Neumann iteration computes the approximation (u_1^{n+1}, u_2^{n+1}) from (u_1^n, u_2^n) by solving a local Dirichlet problem with continuous data at the interface,*

$$\begin{cases} -\Delta\left(u_i^{n+\frac{1}{2}}\right) = f & \text{in } \Omega_i, \\ u_i^{n+\frac{1}{2}} = \dfrac{1}{2}(u_j^n + u_i^n) & \text{on } \Gamma, \end{cases} \tag{6.8}$$

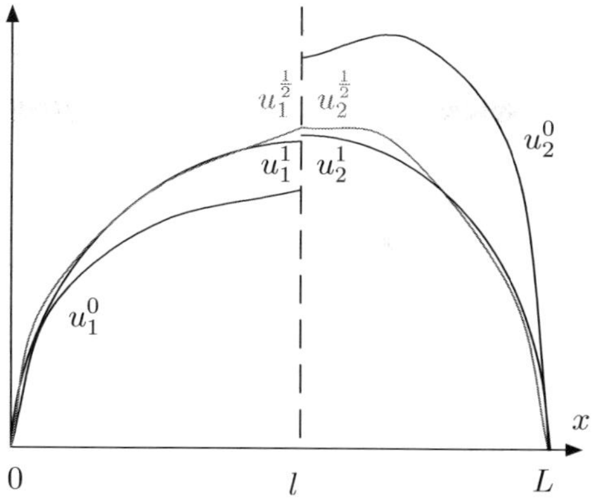

Figure 6.3. *First iteration of a Neumann–Neumann algorithm.*

followed by a correction for the jump of the fluxes on the interface

$$\begin{cases} -\Delta(e_i^{n+1}) = 0 & \text{in } \Omega_i, \\ \dfrac{\partial e_i^{n+1}}{\partial \mathbf{n}_i} = -\dfrac{1}{2}\left(\dfrac{\partial u_1^{n+\frac{1}{2}}}{\partial \mathbf{n}_1} + \dfrac{\partial u_2^{n+\frac{1}{2}}}{\partial \mathbf{n}_2}\right) & \text{on } \Gamma. \end{cases} \tag{6.9}$$

The next iteration is then given by, see Figure 6.3

$$u_i^{n+1} = u_i^{n+\frac{1}{2}} + e_i^{n+1},\ i = 1,2. \tag{6.10}$$

The rationale for the first step (6.8) is to satisfy the Poisson equation in the subdomains while ensuring the continuity of the solution on the interface Γ. After this step, the solution is continuous on the interfaces, but the fluxes may not match. The jump on the fluxes is corrected after (6.10). On Γ we have

$$\begin{aligned} &\frac{\partial u_1^{n+1}}{\partial \mathbf{n}_1} + \frac{\partial u_2^{n+1}}{\partial \mathbf{n}_2} \\ &= \frac{\partial}{\partial \mathbf{n}_1}(u_1^{n+\frac{1}{2}} + e_1^{n+1}) + \frac{\partial}{\partial \mathbf{n}_2}(u_2^{n+\frac{1}{2}} + e_2^{n+1}) \\ &= \frac{\partial u_i^{n+\frac{1}{2}}}{\partial \mathbf{n}_i} - \frac{1}{2}\left(\frac{\partial u_1^{n+\frac{1}{2}}}{\partial \mathbf{n}_1} + \frac{\partial u_2^{n+\frac{1}{2}}}{\partial \mathbf{n}_2}\right) + \frac{\partial u_2^{n+\frac{1}{2}}}{\partial \mathbf{n}_2} - \frac{1}{2}\left(\frac{\partial u_1^{n+\frac{1}{2}}}{\partial \mathbf{n}_1} + \frac{\partial u_2^{n+\frac{1}{2}}}{\partial \mathbf{n}_2}\right) = 0. \end{aligned}$$

Then, functions u_i^{n+1}, $i = 1,2$, may not match on the interface. In order to correct this misfit, correction (6.8) is applied again and so on.

Definition 6.2 (Iterative FETI algorithm). *Let u_i^n denote an approximation to the solution u in a subdomain Ω_i, $i = 1,2$, at the nth iteration. Starting from the initial guess u_i^0, the FETI iteration computes the approximation (u_1^{n+1}, u_2^{n+1}) from (u_1^n, u_2^n) by first correcting the jump of the normal derivatives on the interface,*

$$\begin{cases} -\Delta\left(u_i^{n+\frac{1}{2}}\right) = f & \text{in } \Omega_i, \\ \dfrac{\partial u_i^{n+\frac{1}{2}}}{\partial \mathbf{n}_i} = \dfrac{\partial u_i^n}{\partial \mathbf{n}_i} - \dfrac{1}{2}\left(\dfrac{\partial u_1^n}{\partial \mathbf{n}_1} + \dfrac{\partial u_2^n}{\partial \mathbf{n}_2}\right) & \text{on } \Gamma, \end{cases} \tag{6.11}$$

and then the discontinuity of the solution:

$$\begin{cases} -\Delta(e_i^{n+1}) = 0 & \text{in } \Omega_i, \\ e_i^{n+1} = \frac{1}{2}(u_j^{n+\frac{1}{2}} - u_i^{n+\frac{1}{2}}) & \text{on } \Gamma \text{ for } j \neq i\,. \end{cases} \tag{6.12}$$

The next iteration is then given by

$$u_i^{n+1} = u_i^{n+\frac{1}{2}} + e_i^{n+1}, i = 1,2. \tag{6.13}$$

6.2.2 ▪ Substructured reformulations

It consists in reformulating the algorithms from Definitions 6.1 and 6.2 in terms of the interface unknowns on Γ and the operators acting on these interface unknowns.

Definition 6.3 (Interface operators for the Neumann–Neumann algorithm). *Let the operators $\mathcal{S}_1, \mathcal{S}_2$ be defined as*

$$\mathcal{S}_i(u_\Gamma, f) := \frac{\partial u_i}{\partial \mathbf{n}_i}, \; i = 1,2, \tag{6.14}$$

where the u_i solve the local problems

$$\begin{cases} -\Delta u_i = f & \text{in } \Omega_i, \\ u_i = u_\Gamma & \text{on } \Gamma. \end{cases} \tag{6.15}$$

Let also $\mathcal{S}$ be defined as the operator

$$\boxed{\mathcal{S}(u_\Gamma, f) := \mathcal{S}_1(u_\Gamma, f) + \mathcal{S}_2(u_\Gamma, f) = \frac{\partial u_1}{\partial \mathbf{n}_1} + \frac{\partial u_2}{\partial \mathbf{n}_2}} \tag{6.16}$$

which quantifies the jump of the normal derivative at the interface. Define also operator $\mathcal{T}$ by

$$\boxed{\mathcal{T}(g_\Gamma) := \frac{1}{2}\left(\mathcal{S}_1^{-1}(g_\Gamma, 0) + \mathcal{S}_2^{-1}(g_\Gamma, 0)\right)\frac{1}{2} = \frac{1}{4}(e_{1|\Gamma} + e_{2|\Gamma}),} \tag{6.17}$$

where the $e_{i|\Gamma} := \mathcal{S}_i^{-1}(g_\Gamma, 0)$ are obtained by solving the Neumann local problems

$$\begin{cases} -\Delta e_i = 0 & \text{in } \Omega_i, \\ \dfrac{\partial e_i}{\partial \mathbf{n}_i} = g_\Gamma & \text{on } \Gamma\,. \end{cases} \tag{6.18}$$

With this notation, steps (6.9)–(6.10) can be rewritten in terms of the interface unknowns.

Lemma 6.4 (Neumann–Neumann interface iterative version). *The substructured counterpart of the Neumann–Neumann algorithm from Definition 6.1 will compute u_Γ^{n+1} from u_Γ^n by the following iteration:*

$$\boxed{u_\Gamma^{n+1} = u_\Gamma^n - (\mathcal{T} \circ \mathcal{S})(u_\Gamma^n, f).} \tag{6.19}$$

This defines what we call the iterative interface version of the Neumann–Neumann algorithm.

In order to accelerate this fixed-point method by a Krylov method, we identify this equation with formula (3.3) of Chapter 3 on Krylov methods. It is then more efficient to use a Krylov method to solve the substructured problem.

Definition 6.5 (Substructuring formulation for the Neumann–Neumann method). *The substructured formulation of the Neumann–Neumann iteration consists in solving, typically by a PCG algorithm, the interface problem*

$$\boxed{\begin{array}{l} \text{Find } u_\Gamma : \Gamma \to \mathbb{R} \text{ such that} \\ \mathcal{S}(.,0)(u_\Gamma) = -\mathcal{S}(0,f) \end{array}} \tag{6.20}$$

preconditioned by the linear operator $\mathcal{T}$.

It can be checked that both operators $\mathcal{S}(.,0)$ and $\mathcal{T}$ are symmetric positive definite and thus the preconditioned conjugate gradient algorithm (PCG) can be used. Moreover, the action of both operators $\mathcal{S}(.,0)$ and $\mathcal{T}$ can be computed mostly in parallel. This very popular preconditioner for the operator $\mathcal{S}(0,.)$ was proposed in [14].

Remark 6.2.1. *There are at least three points of view that indicate the relevance of this simply derived preconditioner:*

- **Symmetry considerations.** *In the special case where* Γ *is a symmetry axis for domain* Ω *(see Figure 6.2), we have* $\mathcal{S}_1^{-1}(.,0) = \mathcal{S}_2^{-1}(.,0)$*, and thus* $\mathcal{T}$ *is an exact preconditioner for* $\mathcal{S}(.,0)$ *since it is easy to check that* $(\mathcal{T} \circ \mathcal{S})(.,0) = I_d$.
- **Mechanical interpretation.** *The unknown* u *may be the temperature in domain* Ω *that obeys a Poisson equation with a source term* f*. Operator* $\mathcal{S}(.,0)$ *acts on temperature on the interface and returns a heat flux across the interface. A good preconditioner should act on a heat flux and return a temperature. If the Poisson equation is replaced by the system of elasticity, temperature is replaced by displacement and the normal flux by normal stress to the interface (or a force).*
- **Functional analysis considerations.** *A classical setting for the interface operator* $\mathcal{S}(.,0)$ *is to define it as a continuous linear operator from the Sobolev space* $H_{00}^{1/2}(\Gamma)$ *with values in* $H^{-1/2}(\Gamma)$*. Operator* $\mathcal{T}$ *is a continuous linear operator from* $H^{-1/2}(\Gamma)$ *with values in* $H_{00}^{1/2}(\Gamma)$*. Thus, the preconditioned operator* $\mathcal{T} \circ \mathcal{S}(.,0)$ *is continuous from* $H_{00}^{1/2}(\Gamma)$ *into itself. Any meaningful discretization of this product of operators is then expected to lead to a condition number independent of the mesh size.*

It is possible to consider the Neumann–Neumann preconditioner the other way round. The unknown is the flux at the interface between the subdomains. In this case we will obtain the FETI iterative algorithm. Note that due to the opposite signs of the outward normal $\mathbf{n}_i$ to subdomain Ω_i, the definition of the normal flux is arbitrary up to a flip of the sign. In this case we will first have to define interface operators acting on fluxes (or dual variables) which will be preconditioned by operators acting on "displacements" (or the primal variables).

Definition 6.6 (Interface operator for the FETI algorithm). *Let* λ *be a function that lives on the interface* Γ *and let* $\mathcal{T}_{feti}$ *be a function of* f *and* λ *defined by*

$$\boxed{\mathcal{T}_{feti}(\lambda, f) := v_2 - v_1\,,} \tag{6.21}$$

where the v_i are solutions of the Neumann BVPs ($i = 1,2$):

$$\begin{cases} -\Delta v_i = f & \text{in } \Omega_i, \\ \dfrac{\partial v_i}{\partial \mathbf{n}_i} = (-1)^i \lambda & \text{on } \Gamma. \end{cases} \tag{6.22}$$

This function quantifies the jump of the solutions across the interface.

Note that we have $\mathcal{T}_{feti}(\lambda, 0) = 4\,\mathcal{T}(\lambda)$ (defined by (6.17)), which means the operator

$$\boxed{\mathcal{S}_{feti} = \frac{1}{2}\mathcal{S}(.,0)\frac{1}{2}} \tag{6.23}$$

is a good preconditioner for $\mathcal{T}_{feti}$. By using a similar argument as in the case of the Neumann–Neumann algorithm, we can state the following result.

Lemma 6.7 (FETI interface iterative version). *The substructured counterpart of the FETI algorithm from Definition 6.2 will compute λ_Γ^{n+1} from λ_Γ^n by the following iteration:*

$$\boxed{\lambda_\Gamma^{n+1} = \lambda_\Gamma^n - \left(\mathcal{S}_{feti} \circ \mathcal{T}_{feti}\right)(\lambda_\Gamma^n, f).} \tag{6.24}$$

This defines what we call the iterative interface version of the FETI algorithm.

In order to accelerate this fixed-point method by a Krylov method, we identify this equation with formula (3.3) of Chapter 3. It is then more efficient to use a Krylov method to solve the substructured problem.

Definition 6.8 (Substructuring formulation for the FETI method). *The substructured formulation of the FETI iteration consists in solving the interface problem*

$$\boxed{\begin{array}{c} \text{Find } \lambda : \Gamma \mapsto \mathbb{R} \text{ such that} \\ \mathcal{T}_{feti}(\lambda, 0) = -\mathcal{T}_{feti}(0, f) \end{array}} \tag{6.25}$$

preconditioned by the operator $\mathcal{S}_{feti}$ which involves the solving in parallel of a Dirichlet boundary value problem in each subdomain.

Problem (6.25) is solved by a PCG algorithm with (6.23) as a preconditioner. This approach has been developed in [80, 37, 123].

6.2.3 ▪ FETI as an optimization problem

As explained in the following, (6.25) can be interpreted as the dual problem of a minimization problem under equality constraints. First note that the original boundary value problem (6.6) is equivalent to the minimization problem

$$\boxed{\min_{u \in H^1(\Omega)} J_\Omega(u),\; J_\Omega(u) := \frac{1}{2}\int_\Omega |\nabla u|^2 - \int_\Omega f\, u,} \tag{6.26}$$

where $H^1(\Omega)$ is the Sobolev space of functions that are square integrable and their derivatives are square integrable as well. In order to introduce domain decomposition methods, we make use of a functional analysis result that proves that space $H^1(\Omega)$ is isomorphic to the "domain decomposed space" (local H^1 functions with Dirichlet traces continuous at the interface):

$$\mathcal{H}^1(\Omega_1,\Omega_2) := \{(u_1,u_2)\in H^1(\Omega_1)\times H^1(\Omega_2)/\ u_1 = u_2 \text{ on } \Gamma\};$$

see, e.g., [15] or [31]. This allows the "splitting" of the functional of optimization problem (6.26) into local contributions constrained by the continuity condition at the interface. We thus have the following.

Lemma 6.9 (Dual optimization problem). *Minimization problem* (6.26) *is equivalent to*

$$\begin{aligned}
&\min_{(u_1,u_2)\in\mathcal{H}^1(\Omega_1,\Omega_2)} J(u_1,u_2) \\
&:= \min_{(u_1,u_2)\in\mathcal{H}^1(\Omega_1,\Omega_2)} J_{\Omega_1}(u_1)+J_{\Omega}(u_2) \\
&= \min_{\substack{u_1\in H^1(\Omega_1)\\ u_2\in H^1(\Omega_2)\\ u_1=u_2, on\ \Gamma}} \frac{1}{2}\int_{\Omega_1}|\nabla u_1|^2+\frac{1}{2}\int_{\Omega_2}|\nabla u_2|^2-\int_{\Omega_1} f\,u_1-\int_{\Omega_2} f\,u_2.
\end{aligned} \tag{6.27}$$

Note that a Lagrangian function of this constrained optimization problem can be written as

$$\mathcal{L}(u_1,u_2,\lambda) := J_{\Omega_1}(u_1)+J_{\Omega}(u_2)+\int_{\Gamma}\lambda(u_1-u_2).$$

By computing the differential of $\mathcal{L}$ with respect to the u_i's we get that the critical points satisfy weakly the local boundary value problem on the one hand,

$$\begin{cases} -\Delta u_i = f & \text{in } \Omega_1 \\ \dfrac{\partial u_i}{\partial \mathbf{n}_i} = (-1)^i\lambda & \text{on } \Gamma, \end{cases} \tag{6.28}$$

and the continuity condition

$$u_1 = u_2 \tag{6.29}$$

on the other hand. We see that (6.28) and (6.29) can be rewritten equivalently by using the relations (6.21) and (6.22) from Definition 6.6,

$$\mathcal{T}_{feti}(\lambda,f)=0,$$

which is equivalent to the substructured formulation (6.25).

6.3 ▪ Two-subdomain case at the algebraic level

The goal of this section is to write the algebraic counterpart of the algorithms defined in the previous section at the continuous level. This way, we will be able to propose a good preconditioner for the algebraic Schur complement equation (6.5). This will be done in a finite element framework.

As we have seen in the beginning of this chapter, a variational problem discretized by the finite element method yields a linear system of the form $A\mathbf{U} = \mathbf{F}$, where $\mathbf{F}$ is a given right-hand side and $\mathbf{U}$ is the set of unknowns.

The set of degrees of freedom $\mathcal{N}$ is decomposed into interior dofs $\mathcal{N}_1$ and $\mathcal{N}_2$ and interface dofs $\mathcal{N}_\Gamma$; see section 6.4 for a complete definition of this intuitive notion and Figure 6.1 as an illustration of the two-subdomain case.

The vector of unknowns $\mathbf{U}$ (resp., $\mathbf{F}$) is decomposed into interior unknowns $\mathring{\mathbf{U}}_1$ and $\mathring{\mathbf{U}}_2$ (resp., $\mathring{\mathbf{F}}_1$, $\mathring{\mathbf{F}}_2$), and into interface unknowns $\mathbf{U}_\Gamma$ (resp., $\mathbf{F}_\Gamma$). By numbering the interface equations last, we obtain a block decomposition of the linear system which has the shape of an arrow (pointing down to the right):

$$\begin{pmatrix} A_{11} & 0 & A_{1\Gamma} \\ 0 & A_{22} & A_{2\Gamma} \\ A_{\Gamma 1} & A_{\Gamma 2} & A_{\Gamma\Gamma} \end{pmatrix} \begin{pmatrix} \mathring{\mathbf{U}}_1 \\ \mathring{\mathbf{U}}_2 \\ \mathbf{U}_\Gamma \end{pmatrix} = \begin{pmatrix} \mathring{\mathbf{F}}_1 \\ \mathring{\mathbf{F}}_2 \\ \mathbf{F}_\Gamma \end{pmatrix}. \tag{6.30}$$

The algebraic counterpart of the continuous substructured problem (6.20) amounts simply to an elimination of the interior unknowns $\mathring{\mathbf{U}}_i$, $i = 1, 2$, in (6.30) which leads to a Schur complement equation:

$$\boxed{\begin{aligned} S(\mathbf{U}_\Gamma, \mathbf{F}) := & \left(A_{\Gamma\Gamma} - A_{\Gamma 1} A_{11}^{-1} A_{1\Gamma} - A_{\Gamma 2} A_{22}^{-1} A_{2\Gamma}\right)(\mathbf{U}_\Gamma) \\ & - (\mathbf{F}_\Gamma - A_{\Gamma 1} A_{11}^{-1} \mathring{\mathbf{F}}_1 - A_{\Gamma 2} A_{22}^{-1} \mathring{\mathbf{F}}_2) = 0. \end{aligned}} \tag{6.31}$$

The natural algebraic form of operator $\mathcal{S}$ (6.16) is thus defined by S in the above formula. In order to find the algebraic counterparts of the continuous preconditioner $\mathcal{T}$ (6.17) and thus of the continuous operators $\mathcal{S}_i, i = 1, 2$, in (6.14), we need somehow to split the operator into two contributions coming from the subdomains.

For this purpose, we first of all note that there is a natural decomposition of $A_{\Gamma\Gamma} = (a_{\Gamma\Gamma}^{kl})_{k,l \in \mathcal{N}_\Gamma}$ into its contribution coming from each subdomain $A_{\Gamma\Gamma} = A_{\Gamma\Gamma}^{(1)} + A_{\Gamma\Gamma}^{(2)}$. More precisely, let k, l be the indices of two degrees of freedom on the interface associated with the two basis functions ϕ^k and ϕ^l. The corresponding entry $a_{\Gamma\Gamma}^{kl}$ can be decomposed into a sum $a_{\Gamma\Gamma}^{kl} = a_{\Gamma\Gamma}^{(1),kl} + a_{\Gamma\Gamma}^{(2),kl}$, where

$$a_{\Gamma\Gamma}^{(i),kl} = \int_{\Omega_i} \nabla\phi^k \nabla\phi^l, \quad i = 1, 2.$$

In the same spirit, we define for $i = 1, 2$ and $k \in \mathcal{N}_\Gamma$

$$f_\Gamma^{(i),k} := \int_{\Omega_i} f\, \phi^k$$

and $\mathbf{F}_\Gamma^{(i)} := (f_\Gamma^{(i),k})_{k \in \mathcal{N}_\Gamma}$. In order to make things simple, we have only considered the Poisson problem. Other partial differential equations, such as the elasticity equations, can be handled in the same manner.

Finally, we have a decomposition of matrix $A_{\Gamma\Gamma}$ into

$$A_{\Gamma\Gamma} = A_{\Gamma\Gamma}^{(1)} + A_{\Gamma\Gamma}^{(2)} \tag{6.32}$$

and of $\mathbf{F}_\Gamma$ into

$$\mathbf{F}_\Gamma = \mathbf{F}_\Gamma^{(1)} + \mathbf{F}_\Gamma^{(2)}.$$

We can infer that for each subdomain $i = 1,2$, the local operators

$$\boxed{\mathrm{S}_i(\mathbf{U}_\Gamma,\mathbf{F}) := \left(A^{(i)}_{\Gamma\Gamma} - A_{\Gamma i}A^{-1}_{ii}A_{i\Gamma}\right)(\mathbf{U}_\Gamma) - \left(\mathbf{F}^{(i)}_\Gamma - A_{\Gamma i}A^{-1}_{ii}\,\mathring{\mathbf{F}}_i\right)}$$

are discretizations of the continuous operators $\mathscr{S}_i$, $i = 1,2$, and we let $\mathbb{S}_i := \mathrm{S}_i(.,0)$ be local Schur complements. Note that from (6.32) we have as in the continuous case that

$$\mathrm{S}(\mathbf{U}_\Gamma,\mathbf{F}) = \mathrm{S}_1(\mathbf{U}_\Gamma,\mathbf{F}) + \mathrm{S}_2(\mathbf{U}_\Gamma,\mathbf{F}) \text{ and } \mathbb{S} = \mathbb{S}_1 + \mathbb{S}_2.$$

Coming back to the substructuring, we obtain now a very natural definition.

Definition 6.10 (Substructuring formulation for the Neumann–Neumann method). *Since $\mathbb{S} = S(\cdot,0)$, the substructured formulation reads as*

$$\boxed{\begin{array}{c}\textit{Find } \mathbf{U}_\Gamma \in \mathbb{R}^{\#\mathcal{N}_\Gamma}, \textit{ such that}\\ \mathbb{S}(\mathbf{U}_\Gamma) = -\mathrm{S}(0,\mathbf{F}).\end{array}} \tag{6.33}$$

The preconditioner $\mathbb{T}$ analogous to (6.17) is the weighted sum

$$\mathbb{T} := \frac{1}{2}(\mathrm{S}_1(.,0)^{-1} + \mathrm{S}_2(.,0)^{-1})\frac{1}{2}. \tag{6.34}$$

The action of $\mathrm{S}_i(.,0)^{-1}$ on some interface vector g_Γ, $v_{i,\Gamma} = \mathrm{S}_i(.,0)^{-1}g_\Gamma$ can be computed by solving the discretized Neumann problem related to (6.18):

$$\begin{pmatrix} A_{ii} & A_{i\Gamma} \\ A_{\Gamma i} & A^{(i)}_{\Gamma\Gamma} \end{pmatrix}\begin{pmatrix} \mathring{v}_i \\ v_{i,\Gamma} \end{pmatrix} = \begin{pmatrix} 0 \\ g_\Gamma \end{pmatrix}. \tag{6.35}$$

To sum up, the Neumann–Neumann method consists in solving for $\mathbf{U}_\Gamma$ the Schur complement equation (6.31) or (6.33) by the PCG method using operator $\mathbb{T}$ (6.34) as a preconditioner. At convergence, the algorithm will provide the interface values. Interior values in each subdomain can be computed by solving local Dirichlet problems:

$$A_{ii}\,\mathring{\mathbf{U}}_i = \mathring{F}_i - A_{i\Gamma}\mathbf{U}_\Gamma. \tag{6.36}$$

Parallelism is natural since all these steps can mostly be computed concurrently on two processors. Matrix-vector products with operators $\mathbb{S}$ and $\mathbb{T}$ are both defined as sums of local computations and are therefore essentially parallel with a very high ratio of local computations over data transfer between processors. As for the step defined by (6.36), it is purely parallel.

As in the continuous case, it is possible to invert the role of the Neumann and Dirichlet problems and thus obtain the FETI algorithm.

Definition 6.11 (FETI interface operator). *Let the operator*

$$\mathbb{T}_{feti} : R^{\#\mathcal{N}_\Gamma} \times R^{\#\mathcal{N}} \longrightarrow R^{\#\mathcal{N}_\Gamma}$$

be defined for some $\lambda \in R^{\#\mathcal{N}_\Gamma}$ and $\mathbf{F} \in R^{\#\mathcal{N}}$ as

$$\mathbb{T}_{feti}(\lambda,\mathbf{F}) := \mathbf{V}_{2,\Gamma} - \mathbf{V}_{1,\Gamma},$$

where $(\mathring{\mathbf{V}}_i, \mathbf{V}_{i,\Gamma})$ *are the solutions of*

$$\begin{pmatrix} A_{ii} & A_{i\Gamma} \\ A_{\Gamma i} & A_{\Gamma\Gamma}^{(i)} \end{pmatrix} \begin{pmatrix} \mathring{\mathbf{V}}_i \\ \mathbf{V}_{i,\Gamma} \end{pmatrix} = \begin{pmatrix} \mathbf{F}_i \\ \mathbf{F}_\Gamma^{(i)} + (-1)^i \lambda \end{pmatrix}. \tag{6.37}$$

It can be checked that $\mathbb{T}_{feti}(\lambda, 0) = 4\,\mathbb{T}$. We can infer from this that since $\mathbb{T}$ is a good preconditioner for $\mathbb{S}$, $\frac{1}{4}\mathbb{S}$ is a good preconditioner for $\mathbb{T}_{feti}$. This leads to the following definition of the discrete FETI substructured formulation.

Definition 6.12 (Substructuring formulation for the FETI method). *The substructured formulation of the FETI method reads as*

$$\boxed{\begin{array}{c} \textit{Find } \lambda \in R^{\#\mathcal{N}_\Gamma} \textit{ such that} \\ \mathbb{T}_{feti}(\lambda, 0) = -\mathbb{T}_{feti}(0, \mathbf{F}). \end{array}}$$

This problem is solved by the PCG algorithm, using the operator

$$\frac{1}{2}\mathbb{S}\frac{1}{2}$$

as a preconditioner.

This approach has been developed in [80, 37, 123].

6.3.1 ▪ Link with approximate factorization

Following [175], we will establish a connection between the Neumann–Neumann algorithm and a block approximate factorization of the original system (6.30). Recall first that an exact block factorization of matrix A is given by (6.2) so that its inverse is given by the formula in (6.4):

$$A^{-1} = \begin{pmatrix} I & 0 & -A_{11}^{-1}A_{1\Gamma} \\ & I & -A_{22}^{-1}A_{2\Gamma} \\ & & I \end{pmatrix} \begin{pmatrix} A_{11}^{-1} & & \\ & A_{22}^{-1} & \\ & & \mathbb{S}^{-1} \end{pmatrix} \begin{pmatrix} I & & \\ 0 & I & \\ -A_{\Gamma 1}A_{11}^{-1} & -A_{\Gamma 2}A_{22}^{-1} & I \end{pmatrix}.$$

This suggests the approximation $M^{-1} \simeq A^{-1}$ of the inverse, which naturally provides a preconditioner

$$M^{-1} := \begin{pmatrix} I & 0 & -A_{11}^{-1}A_{1\Gamma} \\ & I & -A_{22}^{-1}A_{2\Gamma} \\ & & I \end{pmatrix} \begin{pmatrix} A_{11}^{-1} & & \\ & A_{22}^{-1} & \\ & & \mathbb{T} \end{pmatrix} \begin{pmatrix} I & & \\ 0 & I & \\ -A_{\Gamma 1}A_{11}^{-1} & -A_{\Gamma 2}A_{22}^{-1} & I \end{pmatrix}, \tag{6.38}$$

where $\mathbb{T}$ is given by (6.34). A direct computation shows that the preconditioned system $M^{-1}A$ has the following form:

$$M^{-1}A = \begin{pmatrix} I & 0 & 0 \\ 0 & I & 0 \\ 0 & 0 & \mathbb{T}\mathbb{S} \end{pmatrix}.$$

Due to the properties of matrices $\mathbb{S}$ and $\mathbb{T}$ we can state some remarkable features.

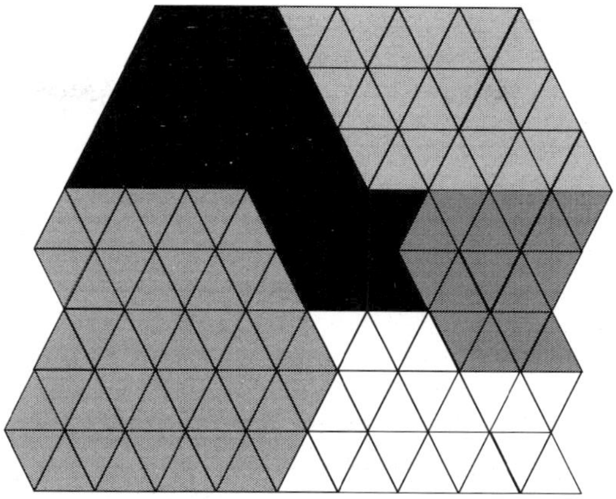

Figure 6.4. *Nonoverlapping geometric partition.*

Remark 6.3.1. *This preconditioner has a few properties:*

- *In the case of exact solves in the subdomains, that is, for the computation of the exact factorization of* A_{ii}*, the application of this preconditioner is equivalent to that of the Schur complement approach, except that it is performed at the global level.*

- *Inexact solvers can be used in the subdomains since the application of* M^{-1} *requires the repeated application of* A_{ii}^{-1}*.*

- *The preconditioner is symmetric and this symmetry can be preserved in the case of inexact solves, provided that the same inexact solves are used in the first and the last terms of* (6.38).

- *The bad news is that even if we have the spectral equivalence of the preconditioner with the diagonal blocks* A_{ii}^{-1}, M^{-1} *is not necessarily spectrally equivalent to matrix* A*, whereas with overlapping Schwarz methods, when the local subdomain solvers are replaced by a spectrally equivalent solver, the convergence behavior of the ASM is asymptotically equivalent to that of the ASM with exact solvers; see* [175].

6.4 ▪ Many-subdomain case

In a finite element setting, the computational domain is the union of elements of the finite element mesh $\mathcal{T}_h$. A geometric partition of the computational domain is natural. Here again, graph partitioning can be used by first modeling the finite element mesh by a graph and then partitioning it into N parts; see Figure 6.4.

As in section 1.3.2, we define a partition of unity for the nonoverlapping decomposition of the domain Ω into subdomains Ω_i, $i = 1,\ldots,N$. We recall that it corresponds to an overlapping decomposition of the set of degrees of freedom $\mathcal{N}$ since some basis functions have a support that intersects two or more subdomains. Let $\{\phi_k\}_{k\in\mathcal{N}}$ be a basis of the finite element space. For $1 \le i \le N$, we define

$$\boxed{\mathcal{N}_i := \{k \in \mathcal{N} : \text{supp}(\phi_k) \cap \Omega_i \neq \emptyset\}}$$

as the set of degrees of freedom whose associated basis function intersects subdomain Ω_i. For each degree of freedom $k \in \mathcal{N}$, let

$$\boxed{\mu_k := \#\{j : 1 \le j \le N \text{ and } k \in \mathcal{N}_j\}}$$

be the number of subdomains that "interact" with function ϕ_k. For example, for a $P1$ finite element method and an interface point shared by two subdomains, we have $\mu_k = 2$. If the point is a cross point shared by four subdomains, we have $\mu_k = 4$.

The global set of dofs can be decomposed into unknowns interior to subdomains Ω_i, $1 \le i \le N$,

$$\boxed{\mathring{\mathcal{N}}_i := \{k \in \mathcal{N}_i / \ \mu_k = 1\}}$$

and the set of interface unknowns, also called the *skeleton*, can be defined as

$$\boxed{\mathcal{N}_\Gamma := \{k \in \mathcal{N} / \ \mu_k > 1\}.}$$

Let $\mathring{R}_i$ and R_Γ be the restriction Boolean matrices from the global set $\mathcal{N}$ to subsets $\mathring{\mathcal{N}}_i$ and $\mathcal{N}_\Gamma$. For $\mathbf{U} \in \mathbb{R}^{\#\mathcal{N}}$, let $\mathring{\mathbf{U}}_i := \mathring{R}_i \ \mathbf{U} = (U_k)_{k \in \mathring{\mathcal{N}}_i}$ be the set of interior degrees of freedom of subdomain Ω_i and let $U_\Gamma := R_\Gamma \mathbf{U} = (U_k)_{k \in \mathcal{N}_\Gamma}$ denote the set of interface degrees of freedom. Note that the sets $(\mathring{\mathcal{N}}_i)_{1 \le i \le N}$ and $\mathcal{N}_\Gamma$ form a partition of $\mathcal{N}$:

$$\mathcal{N} = \left(\bigcup_{i=1}^{N} \mathring{\mathcal{N}}_i \right) \bigcup \mathcal{N}_\Gamma.$$

For a finite element discretization of partial differential equations, the matrix has a sparse structure:

$$\mathring{R}_i A \mathring{R}_j^T = 0 \quad \text{for } i \ne j.$$

If the skeleton unknowns are numbered last, the corresponding block form of the global problem is thus

$$\begin{pmatrix} A_{11} & 0 & \dots & 0 & A_{1\Gamma} \\ 0 & A_{22} & & \vdots & A_{2\Gamma} \\ \vdots & 0 & \ddots & 0 & \vdots \\ 0 & 0 & & A_{NN} & A_{N\Gamma} \\ A_{\Gamma 1} & A_{\Gamma 2} & \dots & A_{\Gamma N} & A_{\Gamma\Gamma} \end{pmatrix} \begin{pmatrix} \mathring{\mathbf{U}}_1 \\ \mathring{\mathbf{U}}_2 \\ \vdots \\ \mathring{\mathbf{U}}_N \\ \mathbf{U}_\Gamma \end{pmatrix} = \begin{pmatrix} \mathring{\mathbf{F}}_1 \\ \mathring{\mathbf{F}}_2 \\ \vdots \\ \mathring{\mathbf{F}}_N \\ \mathbf{F}_\Gamma \end{pmatrix}. \tag{6.39}$$

As in the two-subdomain case for matrix (6.1), it has an arrow shape which allows a multifrontal direct factorization.

In order to define a Neumann–Neumann preconditioner for the substructured problem

$$\boxed{\mathbb{S}\mathbf{U}_\Gamma = \mathbf{F}_\Gamma - \sum_{i=1}^{N} A_{\Gamma i} A_{ii}^{-1} \mathring{\mathbf{F}}_i, \ \text{where } \mathbb{S} := A_{\Gamma\Gamma} - \sum_{i=1}^{N} A_{\Gamma i} A_{ii}^{-1} A_{i\Gamma},} \tag{6.40}$$

we make use of the decomposition induced on the skeleton $\mathcal{N}_\Gamma$ by the geometrical domain decomposition. We thus have the following decomposition of the skeleton:

$$\mathcal{N}_\Gamma = \bigcup_{1 \le i \le N} \mathcal{N}_{\Gamma_i},$$

where the set of indices of interface unknowns of subdomain Ω_i is

$$\boxed{\mathcal{N}_{\Gamma_i} := \{k \in \mathcal{N}_i /\ \mu_k > 1\} = \mathcal{N}_\Gamma \cap \mathcal{N}_i,\ 1 \le i \le N.}$$

Let R_{Γ_i} be the restriction Boolean matrix from the skeleton $\mathcal{N}_\Gamma$ to $\mathcal{N}_{\Gamma_i}$. For a vector of skeleton dof $\mathbf{U}_\Gamma \in \mathbb{R}^{\#\mathcal{N}_\Gamma}$, let $\mathbf{U}_{\Gamma_i} := R_{\Gamma_i}\mathbf{U}_\Gamma = (U_{\Gamma k})_{k\in\mathcal{N}_{\Gamma_i}}$ denote the set of interface degrees of freedom of subdomain Ω_i.

Let D_{Γ_i} be a diagonal matrix of size $\mathbb{R}^{\#\mathcal{N}_{\Gamma_i}} \times \mathbb{R}^{\#\mathcal{N}_{\Gamma_i}}$ whose entries are

$$\boxed{\left(D_{\Gamma_i}\right)_{kk} := \frac{1}{\mu_k} \qquad \forall k \in \mathcal{N}_{\Gamma_i}.}$$

We clearly have a partition of unity on the skeleton:

$$\boxed{\sum_{i=1}^{N} R_{\Gamma_i}^T D_{\Gamma_i} R_{\Gamma_i} = I_{d\Gamma}.}$$

In order to define a Neumann–Neumann preconditioner for (6.40), we first decompose $A_{\Gamma\Gamma}$ into its local contributions coming from the subdomains

$$A_{\Gamma\Gamma} = \sum_{i=1}^{N} A_{\Gamma\Gamma}^{(i)},$$

as was done in the two-subdomain case in (6.32). More precisely, for a Poisson problem, let k, l be the indices of two degrees of freedom on the interface associated with the two basis functions ϕ^k and ϕ^l. The corresponding entry $a_{\Gamma\Gamma}^{kl}$ can be decomposed into a sum $a_{\Gamma\Gamma}^{i,kl}$, where

$$a_{\Gamma\Gamma}^{i,kl} = \int_{\Omega_i} \nabla\phi^k \nabla\phi^l, \quad i = 1, \ldots, N.$$

This yields a decomposition of $\mathbb{S}$:

$$\boxed{\mathbb{S} = \sum_{i=1}^{N} \mathbb{S}_i, \quad \mathbb{S}_i := A_{\Gamma\Gamma}^{(i)} - A_{\Gamma i} A_{ii}^{-1} A_{i\Gamma}.} \tag{6.41}$$

Remark 6.4.1. *Note that in contrast to the two-subdomain case, matrices $\mathbb{S}_i$ have many zero columns and lines and are thus noninvertible. In the original matrix A arising from a finite element discretization of a partial differential operator, all lines and columns related to indices that do not belong to the set $\mathcal{N}_{\Gamma_i}$ are zeros. It is thus not possible to directly define the preconditioner as a weighted sum of the inverses of the local Schur complement $\mathbb{S}_i$ as in the two-subdomain case* (6.34).

Definition 6.13 (Neumann–Neumann for many subdomains). *We define the local Schur complement matrices $S_i \in \mathbb{R}^{\mathcal{N}_{\Gamma_i} \times \mathcal{N}_{\Gamma_i}}$ by*

$$\boxed{S_i := R_{\Gamma_i} \mathbb{S}_i R_{\Gamma_i}^T.}$$

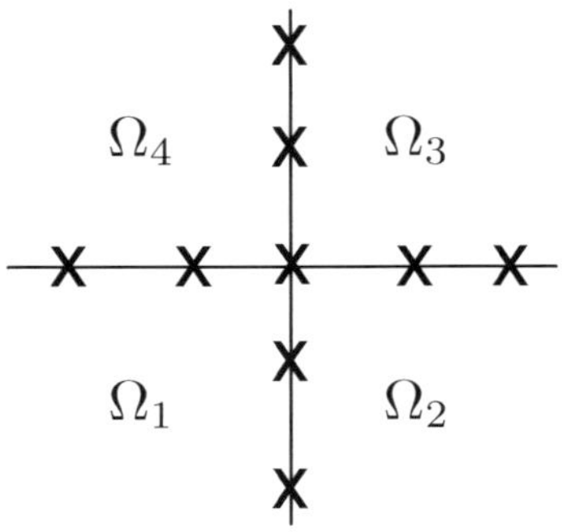

Figure 6.5. *Skeleton and BNN interface unknowns for four subdomains.*

From the above remark, we have

$$\boxed{\mathbb{S}_i = R_{\Gamma_i}^T S_i R_{\Gamma_i}.}$$

From (6.41)*, we have*

$$\mathbb{S} = \sum_{i=1}^{N} R_{\Gamma_i}^T S_i R_{\Gamma_i}. \tag{6.42}$$

The substructuring Neumann–Neumann method consists in solving the interface problem (6.40) *by a conjugate gradient algorithm preconditioned by the interface operator*

$$\boxed{\mathbb{T} := \sum_{i=1}^{N} R_{\Gamma_i}^T D_{\Gamma_i} S_i^{-1} D_{\Gamma_i} R_{\Gamma_i}.} \tag{6.43}$$

Note that it is not necessary to compute the matrix S_i^{-1}. As seen in formula (6.35), it is sufficient to solve a Neumann problem in the subdomains.

6.4.1 ▪ Remarks on FETI

In the two-subdomain case, we have seen at both the continuous and algebraic levels in previous sections that Neumann–Neumann and FETI algorithms are very similar. At the algebraic levels the substructured operators $S(.,0)$ and $\mathbb{T}_{feti}(.,0)$ are both square and have the same size: the number of interface unknowns. When there are cross points (points that belong to three or more subdomains, which is typical in the many-subdomain case), things are more involved and the number of Lagrange multipliers will exceed the number of (nonduplicated) interface unknowns. This is exemplified in Figures 6.5 and 6.6. We have a domain decomposed into four subdomains. The cross point belongs to the four subdomains. Figure 6.5 corresponds to the Neumann–Neumann method where the unknowns related to the substructured operator $S(.,0)$ are the degrees of freedom located on the interfaces without any duplication. Figure 6.6 corresponds to the FETI method where the unknowns related to the FETI operator $\mathbb{T}_{feti}(.,0)$ are the links between the duplicated interface unknowns. They are Lagrange multipliers of the equality constraints on the duplicated interface unknowns. When a dof belongs to two and only two subdomains, there is one associated Lagrange multiplier. The duplication of the cross point of Figure 6.6 yields four unknowns, denoted by u_1^C, u_2^C, u_3^C, and u_4^C. The subscript indicates

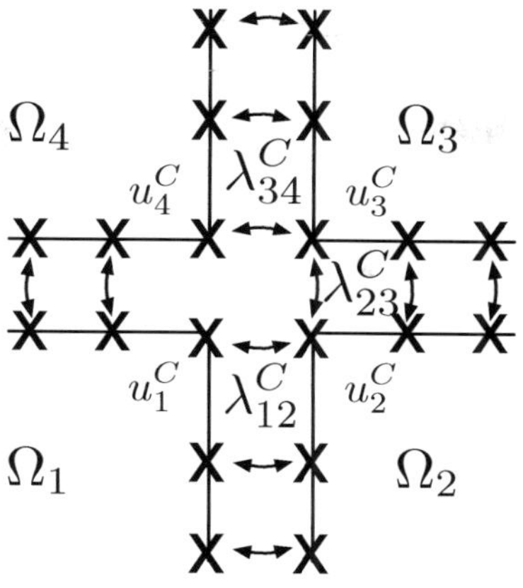

Figure 6.6. *FETI duplicated interfaces and nonredundant Lagrange multipliers for four subdomains.*

the subdomain it comes from. In order to minimize the number of Lagrange multipliers, we choose to impose nonredundant equality constraints such as $u_1^C = u_2^C$, $u_2^C = u_3^C$, and $u_3^C = u_4^C$. They yield three Lagrange multipliers labeled λ_{12}, λ_{23}, and λ_{34}. Thus we have two more unknowns in the FETI method than in the Neumann–Neumann method. An equivalence between a correct generalization of the FETI method and the Neumann–Neumann method, although possible, cannot be simple. We refer the interested reader to, e.g., the books [185, 158]. Papers for the more advanced Schur complement methods are, e.g., those introducing FETI-DP [78], BBDC [47], and the relation between these two algorithms [136].

6.5 ▪ Neumann–Neumann method in FreeFem++

From the point of view of a practical implementation, formulas (6.40) and (6.43) raise two difficulties. The first is the need for a data structure adapted to the interface and the second is the well-posedness of the local Neumann problems, as in the case of floating subdomains for Poisson problems where the constant function **1** is in the kernel of $\mathbb{S}_i$ for $1 \le i \le N$.

We overcome the first difficulty with an implementation based on the global unknowns. It enables us to use the formalism already introduced for the Schwarz method (Chapter 1), based on a global partition of unity (see section 1.3):

$$\mathrm{Id} = \sum_{i=1}^{N} R_i^T D_i R_i \,. \tag{6.44}$$

As for the second difficulty, we regularize the Neumann problems. Since they are involved only in the preconditioning step, the converged solution is not affected. In section 7.8.2, we present a more classical approach to handle a possible kernel in matrix S_i.

We detail here our FreeFem++ implementation. As explained in section 1.7, a Dirichlet boundary condition is penalized with a parameter denoted here by tgv. Denote the matrix of a local "Dirichlet" problem by A_D^i:

$$A_D^i := \begin{pmatrix} A_{ii} & A_{i\Gamma_i} \\ A_{\Gamma_i i} & A_{\Gamma_i\Gamma_i}^{(i)} + \mathrm{tgv}\,\mathrm{Id} \end{pmatrix}. \tag{6.45}$$

For the sake of simplicity, we suppose that tgv is sufficiently large so that

$$(A_D^i)^{-1}\begin{pmatrix} \mathring{F}_i \\ F_\Gamma^i + \text{tgv}\, U_{\Gamma_i} \end{pmatrix} = \begin{pmatrix} A_{ii}^{-1}(\mathring{F}_i - A_{i\Gamma_i} U_{\Gamma_i}) \\ U_{\Gamma_i} \end{pmatrix}. \tag{6.46}$$

Let ε be a small positive parameter and $M^i = (m^{i,kl})_{k,l \in \mathcal{N}_i}$ denote the local mass matrix of the subdomain Ω_i. More precisely, for $k, l \in \mathcal{N}_i$, we define

$$m^{i,kl} := \left(\int_{\Omega_i} \phi_k \phi_l \right).$$

The regularized local Neumann problem A_N^i is defined by

$$A_N^i := \varepsilon M^i + \begin{pmatrix} A_{ii} & A_{i\Gamma_i} \\ A_{\Gamma_i i} & A_{\Gamma_i \Gamma_i}^{(i)} \end{pmatrix}. \tag{6.47}$$

For the sake of simplicity, we assume that the regularizing parameter ε is small enough so that the following formula holds:

$$(A_N^i)^{-1} := \begin{pmatrix} A_{ii}^{-1}(\text{Id} + A_{i\Gamma_i} S_i^{-1} A_{\Gamma_i i} A_{ii}^{-1}) & -A_{ii}^{-1} A_{i\Gamma_i} S_i^{-1} \\ -S_i^{-1} A_{\Gamma_i i} A_{ii}^{-1} & S_i^{-1} \end{pmatrix}. \tag{6.48}$$

Using the algebraic partition of unity, we now define the global counterpart of the interface operator S (see (6.31)).

Definition 6.14. *Let* $\mathbf{U}, \mathbf{F} \in \mathbb{R}^{\#\mathcal{N}}$. *We define* $\widetilde{S}(\mathbf{U}, \mathbf{F})$ *by the following formula:*

$$\boxed{\widetilde{S}(\mathbf{U}, \mathbf{F}) := \mathbf{F} - A \sum_i R_i^T D_i (A_D^i)^{-1} \begin{pmatrix} \mathring{\mathbf{F}}_i \\ \mathbf{F}_\Gamma^i + \text{tgv}\, \mathbf{U}_{\Gamma_i} \end{pmatrix}.} \tag{6.49}$$

This definition deserves some explanations:

- For each $1 \le i \le N$, the application of $(A_D^i)^{-1}$ to the vector $(\mathring{\mathbf{F}}_i, \mathbf{F}_\Gamma^i + \text{tgv}\, \mathbf{U}_{\Gamma_i})^T$ returns a vector local to subdomain Ω_i which is equal to $\mathbf{U}_{\Gamma_i}$ on the interface Γ_i and which verifies (6.39) for the nodes interior to Ω_i as seen from (6.46).
- The local values previously obtained are assembled in a global vector using the partition of unity (6.44).
- The global vector obtained is equal to $\mathbf{U}_\Gamma$ on the skeleton Γ and verifies (6.39) for the interior nodes.
- Finally, $\widetilde{S}(\mathbf{U}, \mathbf{F})$ is the corresponding global residual.

More precisely, we have the following property written in a compact form.

Result 6.5.1. *The link between the global residual and the Schur complement residual is expressed by*

$$\boxed{\forall \mathbf{U}, \mathbf{F} \in \mathbb{R}^{\#\mathcal{N}}, \quad \widetilde{S}(\mathbf{U}, \mathbf{F}) = R_\Gamma^T S(R_\Gamma \mathbf{U}, \mathbf{F}),}$$

where R_Γ *was defined as the restriction operator to the interface unknowns.*

Proof. From (6.46) and (6.49), we have indeed

$$\widetilde{S}(\mathbf{U},\mathbf{F}) = \mathbf{F} - A\sum_i R_i^T D_i \begin{pmatrix} A_{ii}^{-1}(\overset{\circ}{\mathbf{F}}_i - A_{i\Gamma_i}\mathbf{U}_{\Gamma_i}) \\ \mathbf{U}_{\Gamma_i} \end{pmatrix}$$

$$= \mathbf{F} - A\begin{pmatrix} A_{11}^{-1}(\overset{\circ}{\mathbf{F}}_1 - A_{1\Gamma_1}\mathbf{U}_{\Gamma_1}) \\ A_{22}^{-1}(\overset{\circ}{\mathbf{F}}_2 - A_{2\Gamma_2}\mathbf{U}_{\Gamma_2}) \\ \vdots \\ A_{NN}^{-1}(\overset{\circ}{\mathbf{F}}_N - A_{N\Gamma_N}\mathbf{U}_{\Gamma_N}) \\ \mathbf{U}_\Gamma \end{pmatrix} = \begin{pmatrix} 0 \\ 0 \\ \vdots \\ 0 \\ G_\Gamma - \mathbb{S}(\mathbf{U}_\Gamma) \end{pmatrix},$$

where we have used the partition of unity identity (6.44), the fact that the diagonal entries of D_i are 1 for the interior dofs of subdomain Ω_i ($\overset{\circ}{\mathcal{N}}_i$), and the equality

$$A_{i\Gamma_i} R_{\Gamma_i}\mathbf{U} = A_{i\Gamma} R_\Gamma \mathbf{U}. \qquad \square$$

Thus we have seen that solving

$$\boxed{\widetilde{S}(\cdot,0)(\mathbf{U}) = -\widetilde{S}(0,\mathbf{F})} \tag{6.50}$$

amounts to solving (6.40). From Result 6.5.1, it is clear that the following hold:

- System (6.50) is not invertible since the kernel of $\widetilde{S}(\cdot,0)$ is made of vectors $\mathbf{U} \in \mathbb{R}^{\#\mathcal{N}}$ such that $R_\Gamma\mathbf{U} = 0$.
- The range of $\widetilde{S}(\cdot,0)$ is made of vectors whose interior values are such that $-\widetilde{S}(0,\mathbf{F}) \in \text{range}(\widetilde{S}(\cdot,0))$.
- Solutions to (6.50) exist and are unique up to an arbitrary element of $\ker(\widetilde{S}(\cdot,0))$.

In other words, only the interface values of a solution to (6.50) are unique. The correct interior values can be obtained by a parallel postprocessing consisting in setting the interior values to

$$A_{ii}^{-1}(\overset{\circ}{\mathbf{F}}_i - A_{i\Gamma_i}\mathbf{U}_{\Gamma_i}) \quad \forall 1 \le i \le N.$$

A preconditioner for (6.50) is defined by

$$\boxed{\widetilde{\mathbb{T}}(\mathbf{r}) := \left(\sum_{i=1}^N R_i^T D_i (A_{\mathcal{N}}^i)^{-1} D_i R_i\right)\mathbf{r} \quad \text{for } \mathbf{r} \in \mathbb{R}^{\#\mathcal{N}}.} \tag{6.51}$$

Result 6.5.2. *We have the following link between* $\widetilde{\mathbb{T}}$ *defined by* (6.51) *and* $\mathbb{T}$ *defined by* (6.43). *For all* $\mathbf{r}_\Gamma \in \mathbb{R}^{\#\mathcal{N}_\Gamma}$,

$$R_\Gamma(\widetilde{\mathbb{T}} R_\Gamma^T r_\Gamma) = \mathbb{T}(\mathbf{r}_\Gamma).$$

Proof. We first prove the following identity:

$$\widetilde{\mathbb{T}}\begin{pmatrix} 0 \\ 0 \\ \vdots \\ 0 \\ \mathbf{r}_\Gamma \end{pmatrix} = \begin{pmatrix} -A_{11}^{-1}A_{1\Gamma}\mathbb{S}_1^{-1}D_\Gamma\mathbf{r}_\Gamma \\ -A_{22}^{-1}A_{2\Gamma}\mathbb{S}_2^{-1}D_\Gamma\mathbf{r}_\Gamma \\ \vdots \\ -A_{NN}^{-1}A_{N\Gamma}\mathbb{S}_N^{-1}D_\Gamma\mathbf{r}_\Gamma \\ \mathbb{T}(\mathbf{r}_\Gamma) \end{pmatrix}.$$

This shows that the action of $\widetilde{\mathbb{T}}$ on the interface component of the residual amounts to that of $\mathbb{T}$ defined by formula (6.43). Then, for each subdomain $i, 1 \le i \le N$, we have

$$(A_N^i)^{-1} D_i R_i \begin{pmatrix} 0 \\ 0 \\ \vdots \\ 0 \\ \mathbf{r}_\Gamma \end{pmatrix} = (A_N^i)^{-1} \begin{pmatrix} 0 \\ D_{\Gamma_i} \mathbf{r}_{\Gamma_i} \end{pmatrix} = \begin{pmatrix} -A_{ii}^{-1} A_{i\Gamma} \mathbb{S}_i^{-1} D_\Gamma \mathbf{r}_\Gamma \\ R_{\Gamma_i} \mathbb{S}_i^{-1} D_\Gamma \mathbf{r}_\Gamma \end{pmatrix}.$$

For the interior unknowns, D_i is the identity so that

$$D_i (A_N^i)^{-1} D_i R_i \begin{pmatrix} 0 \\ 0 \\ \vdots \\ 0 \\ \mathbf{r}_\Gamma \end{pmatrix} = \begin{pmatrix} -A_{ii}^{-1} A_{i\Gamma} \mathbb{S}_i^{-1} D_\Gamma \mathbf{r}_\Gamma \\ D_{\Gamma_i} R_{\Gamma_i} \mathbb{S}_i^{-1} D_\Gamma \mathbf{r}_\Gamma \end{pmatrix}.$$

Finally, due to the fact that the global partition of unity (6.50) induces an interface partition of unity, we have, by left-multiplying by R_i^T and summing over the subdomains,

$$\widetilde{\mathbb{T}} \begin{pmatrix} 0 \\ 0 \\ \vdots \\ 0 \\ \mathbf{r}_\Gamma \end{pmatrix} = \begin{pmatrix} -A_{11}^{-1} A_{1\Gamma} \mathbb{S}_1^{-1} D_\Gamma \mathbf{r}_\Gamma \\ -A_{22}^{-1} A_{2\Gamma} \mathbb{S}_2^{-1} D_\Gamma \mathbf{r}_\Gamma \\ \vdots \\ -A_{NN}^{-1} A_{N\Gamma} \mathbb{S}_N^{-1} D_\Gamma \mathbf{r}_\Gamma \\ \mathbb{T}(\mathbf{r}_\Gamma) \end{pmatrix}. \qquad \square$$

We summarize as follows:

- We solve the (ill-posed) linear system (6.50) by a conjugate gradient method preconditioned by operator $\widetilde{\mathbb{T}}$ (6.51).
- The solution returned by the Krylov method is correct on the interface, but the interior values need to be corrected (see section 3.5 and the characterization of the kernel and range of $\widetilde{S}(.,0)$ above);
- This correction is achieved by the following postprocessing step:

$$\mathring{\mathbf{U}}_i = A_{ii}^{-1}(\mathring{\mathbf{F}}_i - A_{i\Gamma_i} \mathbf{U}_{\Gamma_i}),$$

  where $\mathbf{U}_{\Gamma_i}$ is the restriction to Γ_i of the interface values $\mathbf{U}_\Gamma$.

6.5.1 ▪ FreeFem++ scripts

We are ready now to provide the FreeFem++ implementation. The beginning is identical to those of the Schwarz routines, except that the decomposition into subdomains is without overlap. Therefore we do not need to add layers to create an overlapping mesh and the different operators are created accordingly. As in the case of the Schwarz algorithm, we still need to create a partition of unity (based on the nonoverlapping decomposition) and the restriction and extension operators. Also, we provide the option of building a coarse space, based on constant local vectors weighted by the partition of unity. Afterwards, we build the local Dirichlet and Neumann problem matrices.

```
mesh Thi=Th; // freefem's trick, formal definition
fespace Vhi(Thi,P1); // freefem's trick, formal definition
Vhi[int] rhsid(npart),rhsin(npart); // local right−hand sides
Vhi[int] auntgv(npart);
plot(part, wait=1,fill=1,ps="Decomp.eps");
for(int i=0;i<npart;++i)
  {
    Thi=aTh[i];
    varf vad(u,v) = int2d(Thi)(eta*u*v+kappa*Grad(u)'*Grad(v))
                                        + on(1,u=g) + int2d(Thi)(f*v) // Dirichlet pb. −Delta (u)=f , u=g on the boundary
                                        + on(10,u=0); // Dirichlet BC on the interface
    varf van(u,v) = int2d(Thi)(etan*u*v+kappa*Grad(u)'*Grad(v))
                                        + on(1,u=g) + int2d(Thi)(f*v) // Dirichlet pb. −Delta (u)=f , u=g on the boundary
                                         ; // Neumann BC on the interface
    aA[i] = vad(Vhi,Vhi,solver=UMFPACK); // local "Dirichlet" matrix
    aN[i] = van(Vhi,Vhi,solver=UMFPACK); // local "Neumann" matrix
    rhsid[i][]= vad(0,Vhi); // local "Dirichlet" rhs
    rhsin[i][]= van(0,Vhi); // local "Neumann" rhs
    varf vaun(u,v) = on(10,u=1);
    auntgv[i][] = vaun(0,Vhi); // array of tgv on Gamma intern, 0 elsewhere
  }
```

Listing 6.1. `./FETI/FreefemProgram/FETI-BNN.edp`

We need also the routine counterparts of the application of the operators $\widetilde{\mathbb{S}}$ and $\widetilde{\mathbb{T}}$. Before that, let us define the local Dirichlet solve for a homogeneous and nonhomogeneous problem defined on the global boundary.

```
func real[int] extendDir(real[int] ud, bool hom)
// Solve local Dirichlet problems with ud at the interface
// homogeneous on the real boundary (hom = 1) or not (hom = 0)
{
   Vh s=0;
   for(int i=0;i<npart;++i)
     {
       real[int] ui = Rih[i]*ud; // local solution
       real[int] bi = ui .* auntgv[i][]; // take into account the interface conditions
       if (hom)
            bi = auntgv[i][] ? bi : 0; // update rhs
       else
            bi = auntgv[i][] ? bi : rhsid[i][];
       ui= aA[i] ^-1 * bi; // solve local Dirichlet problem
       bi = Dih[i]*ui; // use the partition of unity
       s[]+= Rih[i]'*bi; // extend the local solution globally
     }
   return s[];
}
```

Listing 6.2. `./FreefemCommon/matvecDtNtD.idp`

The effect of operator $\widetilde{\mathbb{S}}$ is given by

```
func real[int] A(real[int] &l) // DtN operator
{
  Vh ui, rn, s;
  ui[]=l.*intern[];
  s[] = extendDir(ui[],1);
  rn[]= Aglobal*s[];
  rn[] = rn[] .* intern[];
  return rn[];
}
```

Listing 6.3. `./FreefemCommon/matvecDtNtD.idp`

and that of operator $\widetilde{\mathbb{T}}$ is given by

```
// and the application of the preconditioner
func real[int] Mm1(real[int] &l) // preconditioner = NtD operator
{
  Vh s = 0,rn;
  rn[] = l;
  for(int i=0;i<npart;++i)
     {
       real[int] rb = Rih[i]*rn[];
       real[int] ri = Dih[i]*rb; // local residual
       real[int] ui = aN[i]^-1 * ri; // local solution of the Neumann problem
       real[int] bi = Dih[i]*ui;
       s[]+= Rih[i]'*bi; // extend the local solution globally
     }
  s[] = s[].*intern[];
  return s[];
}
```

Listing 6.4. `./FreefemCommon/matvecDtNtD.idp`

The latter can be used simply as a preconditioner in a conjugate gradient, or one can add a second level by thus obtaining the balancing Neumann–Neumann algorithm.

```
func real[int] BNN(real[int] &u) // precond BNN
{
  real[int] aux2=Mm1(u);
    if(withBNN){
      real[int] aux1 = Q(u);
      aux2 = P(u);
      real[int] aux3 = Mm1(aux2);
      aux2 = PT(aux3);
      aux2 += aux1;
    }
  return aux2;
}
```

Listing 6.5. `./FETI/FreefemProgram/PCG-BNN.idp`

Finally, this is used in the main script in the following way:

```
Vh s, b, un, gd;
include "../../FreefemCommon/matvecDtNtd.idp"
include "../../FreefemCommon/coarsespace.idp"
// compute rhs of the interface problem: b[]
un = 0;
```

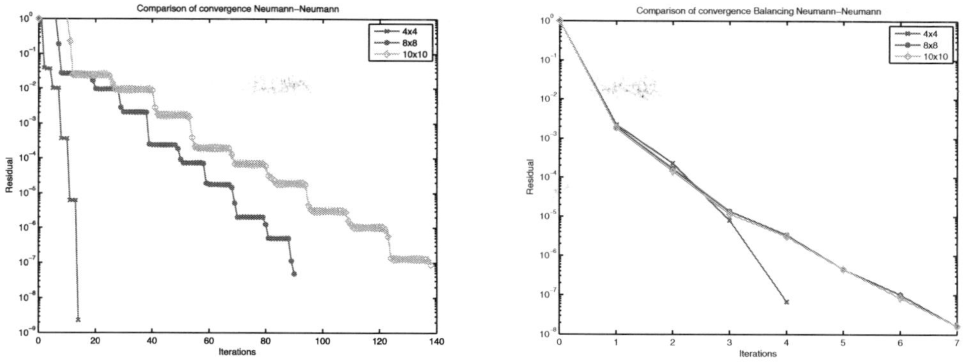

Figure 6.7. *Neumann–Neumann convergence without (left) and with (right) coarse space for a uniform decomposition.*

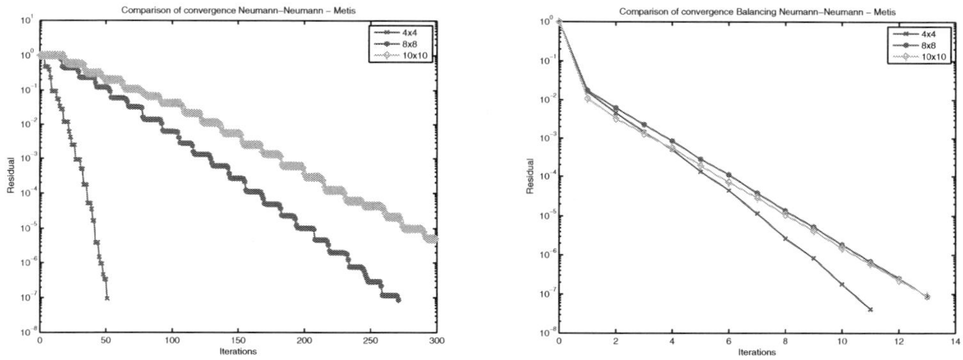

Figure 6.8. *Neumann–Neumann convergence without (left) and with (right) coarse space for a METIS decomposition.*

```
gd[]=g;
un[]+= bord[].*gd[];
s[] = extendDir(un[],0);
b[]= Aglobal*s[];
b[]−= rhsglobal[];
b[] *= −1;
include "PCG−BNN.idp"
Vh lam = 0,sol;
lam[] = lam[].*intern[]; // insures that the first iterate verifies
lam[]+= bord[].*gd[]; // the global BC, from iter=1 on this is true
sol[] = myPCG(lam[],tol,maxit);
sol[] = extendDir(sol[],0);
Vh err = sol−uglob;
plot(sol,cmm=" Final Solution", wait=1,dim=3,fill=1,value=1);
cout << " Final error: " << err[].linfty << endl;
```

Listing 6.6. `./FETI/FreefemProgram/FETI-BNN.edp`

In Figures 6.7 and 6.8, we report the results for uniform and METIS decompositions into 4×4, 8×8, and 10×10 subdomains. We use the Neumann–Neumann method as a preconditioner in a Krylov method. In the first set of tests the method is used without a coarse space, and we see that the iteration number depends linearly on the total number of domains and not only on the number of domains in one direction, as in the case of the Schwarz methods. By adding a coarse space, the behavior becomes insensitive to the number of subdomains as expected. Note that the convergence is better in the case of uniform decompositions than in the case of decompositions using METIS.

6.6 ▪ Nonstandard Neumann–Neumann-type methods

Some algorithmic aspects of systems of PDE-based simulations can be better clarified by means of symbolic computation techniques. This is very important since numerical simulations rely heavily on solving systems of PDEs. Some domain decomposition methods are well understood and efficient for scalar symmetric equations (e.g., Laplacian, biLaplacian) and to some extent for nonsymmetric equations (e.g., convection-diffusion). But they have poor performances and lack robustness when used for symmetric systems of PDEs, and even more so for nonsymmetric complex systems.

In particular, we shall concentrate on Neumann–Neumann and FETI-type algorithms. In some sense, these methods are less optimal when applied to systems of PDEs (e.g., Stokes, Oseen, linear elasticity) than the domain decomposition methods for scalar problems. Indeed, in the case of two subdomains consisting of two half-planes, it is well known that the Neumann–Neumann preconditioner is an exact preconditioner (the preconditioned operator is the identity operator) for the Schur complement equation of scalar equations like the Laplace problem. Unfortunately, this does not hold in the vector case.

In order to achieve this goal, in [32, 33] we have used algebraic methods developed in constructive algebra, D-modules (differential modules), and symbolic computation such as the so-called Smith or Jacobson normal forms and Gröbner basis techniques for transforming a linear system of PDEs into a set of independent scalar PDEs. These algebraic and symbolic methods provide important intrinsic information (e.g., invariants) about the linear system of PDEs to be solved. For instance, we discover that the two-dimensional Stokes system is in some sense equivalent to a biharmonic operator (6.56). In fluid mechanics, it relates to the stream function formulation [11]. These built-in properties need to be taken into account in the design of new numerical methods, which can supersede the usual ones based on a direct extension of the classical scalar methods to linear systems of PDEs. By means of these techniques, it is also possible to transform the linear system of PDEs into a set of decoupled PDEs under certain types of invertible transformations. One of these techniques is the so-called Smith normal form of the matrix of ordinary differential operators associated with the linear system. The Smith normal form has already been successfully applied to open problems in the design of perfectly matched layers (PML). The theory of PML for scalar equations was well developed, and the usage of the Smith normal form allowed these works to be extended to systems of PDEs. A general approach is proposed and applied to the particular case of the compressible Euler equations that model aeroacoustic phenomena in [142] and for shallow-water equations in [10].

For domain decomposition methods, several results have been obtained on compressible Euler equations [58, 59] and Stokes and Oseen systems [60, 61], as well as a new method in the "Smith" spirit which has been derived in [101]. Previously, the computations were performed heuristically; however, in [32, 33] a systematic way to build optimal algorithms for given PDE systems has been shown.

Notation. If R is a ring, then $R^{p\times q}$ is the set of $p \times q$ matrices with entries in R and $\mathrm{GL}_p(R)$ is the group of invertible matrices of $R^{p\times p}$, namely $\mathrm{GL}_p(R) := \{E \in R^{p\times p} \mid \exists\, F \in R^{p\times p} : EF = FE = I_p\}$. An element of $\mathrm{GL}_p(R)$ is called a *unimodular matrix*. A diagonal matrix with elements d_i will be denoted by $\mathrm{diag}(d_1, \dots, d_p)$. If k is a field (e.g.,

$k = \mathbb{Q}, \mathbb{R}, \mathbb{C}$), then $k[x_1,\ldots,x_n]$ is the commutative ring of polynomials in $x_1,\ldots,x_n$ with coefficients in k. In what follows, $k(x_1,\ldots,x_n)$ will denote the field of rational functions in $x_1,\ldots,x_n$ with coefficients in k. Finally, if $r, r' \in R$, then $r' \,|\, r$ means that r' divides r, i.e., there exists $r'' \in R$ such that $r = r''\, r'$.

6.6.1 ▪ Smith normal form of linear systems of PDEs

We first introduce the concept of the *Smith normal form* [176] of a matrix with polynomial entries (see, e.g., [94] or [189, Theorem 1.4]). The Smith normal form is a mathematical technique which is classically used in module theory, linear algebra, symbolic computation, ordinary differential systems, and control theory. It was first developed to study matrices with integer entries.

Theorem 6.15. *Let k be a field, where $R = k[s]$, let p be a positive integer, and let $A \in R^{p\times p}$. Then, there exist two matrices $E \in \mathrm{GL}_p(R)$ and $F \in \mathrm{GL}_p(R)$ such that*

$$A = E\,S\,F,$$

where $S = \mathrm{diag}(d_1,\ldots,d_p)$ and $d_i \in R$ satisfying $d_1 \,|\, d_2 \,|\, \cdots \,|\, d_p$. In particular, we can take $d_i = m_i / m_{i-1}$, where m_i is the greatest common divisor of all the $i \times i$-minors of A (i.e., the determinants of all $i \times i$-submatrices of A), with the convention that $m_0 = 1$. The matrix $S = \mathrm{diag}(d_1,\ldots,d_p) \in R^{p\times p}$ is called a Smith normal form *of A.*

We note that $E \in \mathrm{GL}_p(R)$ is equivalent to $\det(E)$ is an invertible polynomial, i.e., $\det(E) \in k \setminus \{0\}$. Also, in what follows, we shall assume that the d_i's are *monic polynomials*, i.e., their leading coefficients are 1, which will allow us to call the matrix $S = \mathrm{diag}(d_1,\ldots,d_p)$ *the* Smith normal form of A. But the unimodular matrices E and F are not uniquely defined by A. The proof of Theorem 6.15 is constructive and gives an algorithm for computing matrices E, S, and F. The computation of Smith normal forms is available in many computer algebra systems such as Maple, *Mathematica*, Magma, etc.

Consider now the following model problem in $\mathbb{R}^d$ with $d = 2, 3$:

$$\mathscr{L}_d(\mathbf{w}) = \mathbf{g} \quad \text{in } \mathbb{R}^d, \quad |\mathbf{w}(\mathbf{x})| \to 0 \quad \text{for } |\mathbf{x}| \to \infty. \tag{6.52}$$

For instance, $\mathscr{L}_d(\mathbf{w})$ can represent the Stokes/Oseen/linear elasticity operators in dimension d. In the case of the Oseen equation, the diagonal matrix S (6.57) is the product of a Laplace operator and a convection-diffusion operator. We suppose that the inhomogeneous linear system of PDEs (6.52) has constant coefficients; then it can be rewritten as

$$A_d\, \mathbf{w} = \mathbf{g}, \tag{6.53}$$

where $A_d \in R^{p\times p}$, $R = k[\partial_x, \partial_y]$ (resp., $R = k[\partial_x, \partial_y, \partial_z]$) for $d = 2$ (resp., $d = 3$), and k is a field.

In what follows, we shall study the domain decomposition problem in which $\mathbb{R}^d$ is divided into two half-plane subdomains. We assume that the direction normal to the interface of the subdomains is particularized and denoted by ∂_x. If $R_x = k(\partial_y)[\partial_x]$ for $d = 2$ or $R_x = k(\partial_y, \partial_z)[\partial_x]$ for $d = 3$, then, computing the Smith normal form of the matrix $A_d \in R_x^{p\times p}$, we obtain $A_d = E\,S\,F$, where $S \in R_x^{p\times p}$ is a diagonal matrix, $E \in \mathrm{GL}_p(R_x)$, and $F \in \mathrm{GL}_p(R_x)$. The entries of the matrices E, S, F are polynomials in ∂_x, and E and F are unimodular matrices, i.e., $\det(E), \det(F) \in k(\partial_y) \setminus \{0\}$ if $d = 2$, or

$\det(E), \det(F) \in k(\partial_y, \partial_z) \setminus \{0\}$ if $d = 3$. We recall that the matrices E and F are not unique, contrary to S. Using the Smith normal form of A_d, we get

$$A_d\,\mathbf{w} = \mathbf{g} \quad \Leftrightarrow \quad \{\mathbf{w}_s := F\,\mathbf{w},\; S\,\mathbf{w}_s = E^{-1}\mathbf{g}\}. \tag{6.54}$$

In other words, (6.54) is equivalent to the uncoupled linear system

$$S\,\mathbf{w}_s = E^{-1}\,\mathbf{g}. \tag{6.55}$$

Since $E \in \mathrm{GL}_p(R_x)$ and $F \in \mathrm{GL}_p(R_x)$, the entries of their inverses are still polynomial in ∂_x. Thus, applying E^{-1} to the right-hand side $\mathbf{g}$ of $A_d\,\mathbf{w} = \mathbf{g}$ amounts to taking k-linear combinations of derivatives of $\mathbf{g}$ with respect to x. If $\mathbb{R}^d$ is split into two subdomains $\mathbb{R}^- \times \mathbb{R}^{d-1}$ and $\mathbb{R}^+ \times \mathbb{R}^{d-1}$, then the application of E^{-1} and F^{-1} to a vector can be done for each subdomain independently. No communication between the subdomains is necessary.

In conclusion, it is enough to find a domain decomposition algorithm for the uncoupled system (6.55) and then transform it back to the original one (6.53) by means of the invertible matrix F over R_x. This technique can be applied to any linear system of PDEs once it is rewritten in a polynomial form. The uncoupled system acts on the new dependent variables $\mathbf{w}_s$, which we shall further call *Smith variables* since they are issued from the Smith normal form.

Remark 6.6.1. *Since the matrix F is used to transform* (6.55) *to* (6.53) *(see the first equation of the right-hand side of* (6.54)*) and F is not unique, we need to find a matrix F as simple as possible (e.g., F has minimal degree in ∂_x) so as to obtain a final algorithm whose form can be used for practical computations.*

Example 6.16. *Consider the two-dimensional elasticity operator defined by $\mathcal{E}_2(\mathbf{u}) := -\mu\,\Delta\,\mathbf{u} - (\lambda+\mu)\nabla\,\mathrm{div}\,\mathbf{u}$. If we consider the commutative polynomial rings $R = \mathbb{Q}(\lambda,\mu)[\partial_x,\partial_y]$, $R_x = \mathbb{Q}(\lambda,\mu)(\partial_y)[\partial_x] = \mathbb{Q}(\lambda,\mu,\partial_y)[\partial_x]$ and*

$$A_2 = \begin{pmatrix} (\lambda+2\,\mu)\,\partial_x^2 + \mu\,\partial_y^2 & (\lambda+\mu)\,\partial_x\,\partial_y \\ (\lambda+\mu)\,\partial_x\,\partial_y & \mu\,\partial_x^2 + (\lambda+2\,\mu)\,\partial_y^2 \end{pmatrix} \in R^{2\times 2}$$

is the matrix of partial differential operators associated with $\mathcal{E}_2$, i.e., $\mathcal{E}_2(\mathbf{u}) = A_2\,\mathbf{u}$, then the Smith normal form of $A_2 \in R_x^{2\times 2}$ is defined by

$$S_{A_2} = \begin{pmatrix} 1 & 0 \\ 0 & \Delta^2 \end{pmatrix}. \tag{6.56}$$

The particular form of S_{A_2} shows that, over R_x, the system of PDEs for the linear elasticity in $\mathbb{R}^2$ is algebraically equivalent to a biharmonic equation.

Example 6.17. *Consider the two-dimensional Oseen operator $\mathcal{O}_2(\mathbf{w}) = \mathcal{O}_2(\mathbf{v},q) := (c\,\mathbf{v} - \nu\,\Delta\mathbf{v} + \mathbf{b}\cdot\nabla\mathbf{v} + \nabla q, \nabla\cdot\mathbf{v})$, where $\mathbf{b}$ is the convection velocity. If $\mathbf{b} = 0$, then we obtain the Stokes operator $\mathcal{S}_2(\mathbf{w}) = \mathcal{S}_2(\mathbf{v},q) := (c\,\mathbf{v} - \nu\,\Delta\mathbf{v} + \nabla q, \nabla\cdot\mathbf{v})$. If $R = \mathbb{Q}(b_1,b_2,c,\nu)[\partial_x,\partial_y]$, $R_x = \mathbb{Q}(b_1,b_2,c,\nu)(\partial_y)[\partial_x] = \mathbb{Q}(b_1,b_2,c,\nu,\partial_y)[\partial_x]$ and*

$$O_2 = \begin{pmatrix} -\nu(\partial_x^2+\partial_y^2) + b_1\,\partial_x + b_2\,\partial_y + c & 0 & \partial_x \\ 0 & -\nu(\partial_x^2+\partial_y^2) + b_1\,\partial_x + b_2\,\partial_y + c & \partial_y \\ \partial_x & \partial_y & 0 \end{pmatrix}$$

is the matrix of partial differential operators associated with $\mathcal{O}_2$, *i.e.*, $\mathcal{O}_2(\mathbf{w}) = O_2\,\mathbf{w}$, *then the Smith normal form of* $O_2 \in R_x^{3\times 3}$ *is defined by*

$$S_{O_2} = \begin{pmatrix} 1 & 0 & 0 \\ 0 & 1 & 0 \\ 0 & 0 & \Delta L_2 \end{pmatrix}, \quad L_2 = c - \nu\Delta + \mathbf{b}\cdot\nabla. \tag{6.57}$$

From the form of S_{O_2} *we can deduce that the two-dimensional Oseen equations can be mainly characterized by the scalar fourth-order partial differential operator* ΔL_2. *This is not surprising since the stream function formulation of the Oseen equations for* $d = 2$ *gives the same PDE for the stream function.*

Remark 6.6.2. *The above applications of the Smith normal forms suggest that one should design an optimal domain decomposition method for the biharmonic operator* Δ^2 *(resp.,* ΔL_2*) in the case of linear elasticity (resp., the Oseen/Stokes equations) for the two-dimensional problems and then transform it back to the original system.*

6.6.2 ▪ An optimal algorithm for the biharmonic operator

We give here an example of Neumann–Neumann methods in their iterative version for Laplace and biLaplace equations. For simplicity, consider a decomposition of the domain $\Omega = \mathbb{R}^2$ into two half-planes $\Omega_1 = \mathbb{R}^- \times \mathbb{R}$ and $\Omega_2 = \mathbb{R}^+ \times \mathbb{R}$. Let the interface $\{0\} \times \mathbb{R}$ be denoted by Γ and $(\mathbf{n}_i)_{i=1,2}$ be the outward normal of $(\Omega_i)_{i=1,2}$.

We consider the problem

$$\begin{array}{lll} -\Delta u = f & \text{in} & \mathbb{R}^2, \\ |u(\mathbf{x})| \to 0 & \text{for } | & \mathbf{x}| \to \infty, \end{array} \tag{6.58}$$

and the following **Neumann–Neumann algorithm** applied to (6.58):

Let u_Γ^n *be the interface solution at iteration* n. *We obtain* u_Γ^{n+1} *from* u_Γ^n *by the iterative procedure*

$$\left\{ \begin{array}{ll} -\Delta u^{i,n} = f & \textit{in } \Omega_i, \\ u^{i,n} = u_\Gamma^n & \textit{on } \Gamma, \end{array} \right. \qquad \left\{ \begin{array}{ll} -\Delta \tilde{u}^{i,n} = 0 & \textit{in } \Omega_i, \\ \dfrac{\partial \tilde{u}^{i,n}}{\partial \mathbf{n}_i} = -\dfrac{1}{2}\left(\dfrac{\partial u^{1,n}}{\partial \mathbf{n}_1} + \dfrac{\partial u^{2,n}}{\partial \mathbf{n}_2} \right) & \textit{on } \Gamma, \end{array} \right. \tag{6.59}$$

and then $u_\Gamma^{n+1} = u_\Gamma^n + \frac{1}{2}\left(\tilde{u}^{1,n} + \tilde{u}^{2,n}\right)$.

This algorithm is *optimal* in the sense that it converges in two iterations.

Since the biharmonic operator seems to play a key role in the design of a new algorithm for both Stokes and elasticity problems in two dimensions, we need to build an optimal algorithm for it. We consider the following problem:

$$\begin{array}{c} \text{Find } \phi : \mathbb{R}^2 \to \mathbb{R} \text{ such that} \\ \Delta^2 \phi = g \text{ in } \mathbb{R}^2, \quad |\phi(\mathbf{x})| \to 0 \text{ for } |\mathbf{x}| \to \infty. \end{array} \tag{6.60}$$

We also consider the following **"Neumann–Neumann" algorithm** applied to (6.60):

Let $(\phi_\Gamma^n, D\phi_\Gamma^n)$ be the interface solution at iteration n (suppose also that $\phi_\Gamma^0 = \phi^0|_\Gamma$, $D\phi_\Gamma^0 = (\Delta\phi^0)_\Gamma$). We obtain $(\phi_\Gamma^{n+1}, D\phi_\Gamma^n)$ from $(\phi_\Gamma^n, D\phi_\Gamma^n)$ by the iterative procedure

$$\left\{\begin{array}{rll} -\Delta^2\phi^{i,n} = f & \text{in } \Omega_i, \\ \phi^{i,n} = \phi_\Gamma^n & \text{on } \Gamma, \\ \Delta\phi^{i,n} = D\phi_\Gamma^n & \text{on } \Gamma, \end{array}\right. \quad \left\{\begin{array}{rll} -\Delta^2\tilde{\phi}^{i,n} = 0 & & \text{in } \Omega_i, \\ \dfrac{\partial\tilde{\phi}^{i,n}}{\partial \mathbf{n}_i} = -\dfrac{1}{2}\left(\dfrac{\partial\phi^{1,n}}{\partial \mathbf{n}_1} + \dfrac{\partial\phi^{2,n}}{\partial \mathbf{n}_2}\right) & & \text{on } \Gamma, \\ \dfrac{\partial\Delta\tilde{\phi}^{i,n}}{\partial \mathbf{n}_i} = -\dfrac{1}{2}\left(\dfrac{\partial\Delta\phi^{1,n}}{\partial \mathbf{n}_1} + \dfrac{\partial\Delta\phi^{2,n}}{\partial \mathbf{n}_2}\right) & & \text{on } \Gamma, \end{array}\right. \tag{6.61}$$

and then $\phi_\Gamma^{n+1} = \phi_\Gamma^n + \frac{1}{2}\left(\tilde{\phi}^{1,n} + \tilde{\phi}^{2,n}\right)$, $D\phi_\Gamma^{n+1} = D\phi_\Gamma^n + \frac{1}{2}\left(\tilde{\Delta}\phi^{1,n} + \tilde{\Delta}\phi^{2,n}\right)$.

This is a generalization of the Neumann–Neumann algorithm for the Δ operator and is also *optimal* (the proof can be found in [60]).

Now, in the case of the two-dimensional linear elasticity, ϕ represents the second component of the vector of Smith variables, that is, $\phi = (\mathbf{w}_s)_2 = (F\mathbf{u})_2$, where $\mathbf{u} = (u, v)$ is the displacement field. Hence, we need to replace ϕ with $(F\mathbf{u})_2$ in the algorithm for the biLaplacian and then simplify it using algebraically admissible operations. Thus, one can obtain an optimal algorithm for the Stokes equations or linear elasticity depending on the form of F. From here comes the necessity of properly choosing the matrix F (which is not unique), used to define the Smith normal form, in order to obtain a "good" algorithm for the systems of PDEs from the optimal one applied to the biharmonic operator. In [58] and [60], the computation of the Smith normal forms for the Euler equations and the Stokes equations was done by hand or by using the Maple command *Smith*. Surprisingly, the corresponding matrices F have provided good algorithms for the Euler equations and the Stokes equations even if the approach was entirely heuristic.

The efficiency of our algorithms relies heavily on the simplicity of the Smith variables, that is, on the entries of the unimodular matrix F used to compute the Smith normal form of the matrix A. Within a constructive *algebraic analysis* approach, we developed a method for constructing many possible Smith variables (*completion problem*). Taking into account the physical aspects, the user can then choose the simplest one among them. Also, in the algorithms presented in the previous sections, we have equations in the domains Ω_i and interface conditions on Γ obtained heuristically. We have designed an automatic way to reduce the interface conditions with respect to the equations in the domains (*reduction problem*). For more details and explicit computations, we refer the reader to [32, 33].

6.6.3 ▪ Some optimal algorithms

After performing the completion and the reduction of the interface conditions, we can give examples of optimal algorithms.

Example 6.18. *Consider the elasticity operator*

$$\mathcal{E}_d\,\mathbf{u} = -\mathrm{div}\,\sigma(\mathbf{u}), \quad \sigma(\mathbf{u}) = \mu(\nabla\mathbf{u} + (\nabla\mathbf{u})^T) + \lambda\,\mathrm{div}\,\mathbf{u}\,I_d.$$

If $d = 2$, then the completion algorithm gives two possible choices for F:

$$F = \begin{pmatrix} -\frac{\partial_x(\mu\partial_x^2 - \lambda\partial_y^2)}{(\lambda+\mu)\partial_y^3} & 1 \\ 1 & 0 \end{pmatrix}, \quad F = \begin{pmatrix} 1 & -\frac{(\lambda+\mu)\partial_x((3\mu+2\lambda)\partial_y^2 + (2\mu+\lambda)\partial_x^2)}{\partial_y^3} \\ 0 & 1 \end{pmatrix}. \tag{6.62}$$

By replacing ϕ in the Neumann–Neumann algorithm for the biLaplacian by $(F\mathbf{u})$ and rewriting the interface conditions, using the equations inside the domain like in [60], we get two different algorithms for the elasticity system. Note that in the first case of (6.62), $\phi = u$, and in the second one, $\phi = v$ (where $\mathbf{u} = (u, v)$). Below, we shall write in detail the algorithm in the second case. To simplify the writing, we denote $u_\tau = \mathbf{u}\cdot\tau$, $u_\mathbf{n} = \mathbf{u}\cdot\mathbf{n}$, $\sigma_{\mathbf{nn}}(\mathbf{u}) = (\sigma(\mathbf{u})\cdot\mathbf{n})\cdot\mathbf{n}$, $\sigma_{\mathbf{n}\tau}(\mathbf{u}) = (\sigma(\mathbf{u})\cdot\mathbf{n})\cdot\tau$ with τ the tangential vector and n the normal vector.

Remark 6.6.3. *We found an algorithm with a mechanical meaning: Find the tangential part of the normal stress and the normal displacement at the interface so that the normal part of the normal stress and the tangential displacement on the interface match. This is very similar to the original Neumann–Neumann algorithm, which means that the implementation effort of the new algorithm from an existing Neumann–Neumann algorithm is negligible (the same type of quantities — displacement fields and efforts — are imposed at the interfaces), except that the new algorithm requires knowledge of some geometric quantities, such as normal and tangential vectors. Note also that, with the adjustment of the definition of tangential quantities for $d = 3$, the algorithm is the same and is also similar to the results in [60].*

Let $(u_\Gamma^n, \sigma_\Gamma^n)$ be the interface solution at iteration n (suppose also that $u_\Gamma^0 = (u_\tau^0)|_\Gamma$, $\sigma_\Gamma^0 = (\sigma_{s\mathbf{nn}}(u^0))|_\Gamma$). We obtain $(u_\Gamma^{n+1}, \sigma_\Gamma^n)$ from $(u_\Gamma^n, \sigma_\Gamma^n)$ by the following iterative procedure

$$
\left\{
\begin{aligned}
\mathscr{E}_2(\mathbf{u}^{i,n}) &= f && \text{in } \Omega_i,\\
u_{\tau_i}^{1,n} &= u_\Gamma^n && \text{on } \Gamma,\\
\sigma_{\mathbf{n}_i\mathbf{n}_i}(\mathbf{u}^{i,n}) &= \sigma_\Gamma^n && \text{on } \Gamma,
\end{aligned}
\right.
\qquad
\left\{
\begin{aligned}
\mathscr{E}_2(\tilde{\mathbf{u}}^{i,n}) &= 0 && \text{in } \Omega_i,\\
\tilde{\mathbf{u}}_{\tau_i}^{i,n} &= -\frac{1}{2}\left(\mathbf{u}_{\mathbf{n}_1}^{1,n} + \mathbf{u}_{\mathbf{n}_2}^{2,n}\right) && \text{on } \Gamma,\\
\sigma_{\mathbf{n}_i\tau_i}(\tilde{\mathbf{u}}^{i,n}) &= -\frac{1}{2}\left(\sigma_{\mathbf{n}_1\tau_1}(\mathbf{u}^{1,n}) + \sigma_{\mathbf{n}_2\tau_2}(\mathbf{u}^{2,n})\right) && \\
& && \text{on } \Gamma,
\end{aligned}
\right.
\tag{6.63}
$$

and $u_\Gamma^{n+1} = u_\Gamma^n + \frac{1}{2}\left(\tilde{u}_{\tau_1}^{1,n} + \tilde{u}_{\tau_2}^{2,n}\right)$, $\sigma_\Gamma^{n+1} = \sigma_\Gamma^n + \frac{1}{2}\left(\sigma_{\mathbf{n}_1\mathbf{n}_1}(\tilde{\mathbf{u}}^{1,n}) + \sigma_{\mathbf{n}_2\mathbf{n}_2}(\tilde{\mathbf{u}}^{2,n})\right)$.

All algorithms and interface conditions are derived for problems posed on the whole space, since for the time being, this is the only way to treat these problems from the algebraic point of view. The effect of the boundary condition on bounded domains cannot be quantified with the same tools. All the algorithms are designed at the PDE level, and it is very important to choose the right discrete framework in order to preserve the optimal properties. For example, in the case of linear elasticity a good candidate would be the TDNNS finite elements that can be found in [157], and the algorithms obtained by these algebraic techniques have been used to design a FETI-TDNNS method [156].

Chapter 7

Robust coarse spaces via generalized eigenproblems: The GenEO method

We have already analyzed in Chapter 5 two-level additive Schwarz methods for a Poisson problem with a coarse space motivated by heuristic ideas. The condition number estimates were based on a tool specific to domain decomposition methods and multigrid methods, namely the notion of stable decomposition (Definition 5.10), as well as on functional analysis results such as Poincaré inequalities or trace theorems and on a numerical analysis interpolation inequality (5.8). In this chapter, we present the GenEO (Generalized Eigenvalue in the Overlap) method to build coarse spaces for which any targeted convergence rate can be achieved. The construction is as algebraic as possible. We shall make heavy use of a generalization of the stable decomposition notion, namely the fictitious space lemma, see section 7.2.1. This abstract result is based on writing domain decomposition methods as a product of three linear operators.

As also mentioned previously, the main motivation for building a robust coarse space for a two-level additive Schwarz method is to achieve scalability when solving highly heterogeneous problems, i.e., problems for which the convergence properties do not depend on the variation of the coefficient. Recently, for scalar elliptic problems, operator-dependent coarse spaces have been built in the case where coefficients are not resolved by the subdomain partition; see, e.g., [62, 70, 84, 83, 102, 148, 160, 172, 173]. We have seen in Chapter 5 that a very useful tool for building coarse spaces for which the corresponding two-level method is robust, regardless of the partition into subdomains or the coefficient distribution, is the solution of local generalized eigenvalue problems. In this spirit, for the Darcy problem, [84] proposes solving local generalized eigenvalue problems on overlapping coarse patches; local contributions are then "glued together" via a partition of unity to obtain a global coarse space. More recently, [83, 70] and [148, 62] have built on these ideas and proposed different coarse spaces based on different generalized eigenvalue problems. The theoretical results in [62] rely on uniform (in the coefficients) weighted Poincaré inequalities and have two drawbacks: (i) at the theoretical level some assumptions are needed on the coefficient distribution in the overlaps, and (ii) the arguments cannot be generalized easily to the case of systems of PDEs because the maximum principle no longer holds.

To remedy these drawbacks, in [178] we proposed a coarse space construction based on Generalized Eigenproblems in the Overlap (which we will refer to as the *GenEO coarse space*); the method was previously introduced in [177] by the same authors. This particular construction has been applied successfully to positive definite systems of PDEs discretized by finite elements with only a few extra assumptions. The resulting generalized eigenvalue problems are closely related but different from those proposed in [70]. Which of the two approaches is better in practice in terms of stability versus coarse space size is still the object of ongoing research; see, for example, [187].

In this chapter, the GenEO method as well as most classical two-level methods are presented in a different light, under a common framework. Moreover, their convergence can be proven in an abstract setting, provided that the assumptions of the fictitious space lemma are satisfied. Before stating this lemma, we first reformulate the two-level additive Schwarz method in this framework.

7.1 ▪ Reformulation of the additive Schwarz method

In this section we take the first steps in the use of the abstract theory of the "fictitious space lemma" (Lemma 7.4) by better formalizing the additive Schwarz method (ASM). In order to simplify the presentation, we will place ourselves directly in an algebraic setting where the set of degrees of freedom $\mathcal{N}$ is decomposed into N subsets $(\mathcal{N}_i)_{1\le i\le N}$. In this framework we also have a partition of unity,

$$\sum_{i=1}^{N} R_i^T D_i R_i = \mathrm{Id},$$

as defined in (1.25) in section 1.3. The coarse space is of size $\#\mathcal{N}_0$ and is spanned by the columns of a rectangular matrix R_0^T of size $\#\mathcal{N}\times\#\mathcal{N}_0$. Note that most of the assumptions of this lemma are verified without the precise definition of a coarse space. That is why the latter will be specified later on in Definition 7.16 only when the stable decomposition property requires it.

In the fictitious space lemma (see Lemma 7.4) a certain number of abstract ingredients are needed. These ingredients can be easily identified in the case of the ASM. This intuitive introduction will give the general flavor of the methods exposed later on. We will come to the abstract framework after this short presentation.

Definition 7.1 (ASM components in the fictitious space lemma). *Two Hilbert spaces H, H_D, two other associated bilinear forms and induced scalar products, and the $\mathcal{R}_{ASM,2}$ operator between them are defined as follows:*

- *Space $H := \mathbb{R}^{\#\mathcal{N}}$ is endowed with the standard Euclidean scalar product. We consider another bilinear form a defined by*

$$a : H\times H\to\mathbb{R},\quad (\mathbf{U},\mathbf{V})\longmapsto a(\mathbf{U},\mathbf{V}) := \mathbf{V}^T A\mathbf{U} \tag{7.1}$$

where A is the matrix of the problem we want to solve. Recall that matrix A is symmetric positive definite.

- *Space H_D, defined as the product space*

$$H_D := \mathbb{R}^{\#\mathcal{N}_0}\times\prod_{i=1}^{N}\mathbb{R}^{\#\mathcal{N}_i}, \tag{7.2}$$

is endowed with a standard scalar Euclidean product. For $\mathcal{U} = (\mathbf{U}_i)_{0\le i\le N}$, $\mathcal{V} = (\mathbf{V}_i)_{0\le i\le N}$ with $\mathbf{U}_i,\mathbf{V}_i\in\mathbb{R}^{\#\mathcal{N}_i}$. The bilinear form b is defined by

$$\begin{aligned} b : H_D\times H_D &\longrightarrow \mathbb{R},\\ (\mathcal{U},\mathcal{V}) &\longmapsto b(\mathcal{U},\mathcal{V}) := \sum_{i=0}^{N} a(R_i^T\mathbf{U}_i, R_i^T\mathbf{V}_i) = \sum_{i=0}^{N}\mathbf{V}_i^T(R_i A R_i^T)\mathbf{U}_i\\ &:= \mathcal{V}^T B\,\mathcal{U}, \end{aligned} \tag{7.3}$$

where $B : H_D \to H_D$ is the operator defined by

$$\forall\, \mathcal{U} = (\mathbf{U}_i)_{0\le i\le N} \in H_D \quad B(\mathcal{U}) := (R_i A R_i^T \mathbf{U}_i)_{0\le i\le N}\,. \tag{7.4}$$

Note that B is actually a block diagonal operator whose inverse is

$$\forall\, \mathcal{U} = (\mathbf{U}_i)_{0\le i\le N} \in H_D \quad B^{-1}(\mathcal{U}) := ((R_i A R_i^T)^{-1} \mathbf{U}_i)_{0\le i\le N}\,.$$

- *The linear operator $\mathcal{R}_{ASM,2}$ is defined as*

$$\mathcal{R}_{ASM,2} : H_D \longrightarrow H, \qquad (\mathbf{U}_i)_{0\le i\le N} \mapsto \sum_{i=0}^{N} R_i^T \mathbf{U}_i\,. \tag{7.5}$$

After having settled these ingredients, we can proceed to the reformulation of the additive Schwarz method.

Lemma 7.2 (ASM in terms of the fictitious space lemma). *The two-level ASM preconditioner*

$$\boxed{M_{ASM,2}^{-1} = \sum_{i=0}^{N} R_i^T (R_i A R_i^T)^{-1} R_i} \tag{7.6}$$

can be rewritten as

$$\boxed{M_{ASM,2}^{-1} = \mathcal{R}_{ASM,2} B^{-1} \mathcal{R}_{ASM,2}^*\,,} \tag{7.7}$$

where the operator $\mathcal{R}_{ASM,2}^* : H \to H_D$ is the adjoint of the operator $\mathcal{R}_{ASM,2}$ with respect to the standard Euclidean scalar products and operator B is defined in (7.4).

Proof. First of all, note that by definition $\mathcal{R}_{ASM,2}^*$ can be written as

$$(\mathcal{R}_{ASM,2}^*(\mathbf{U}), \mathcal{V})_2 := (\mathbf{U}, \mathcal{R}_{ASM,2}(\mathcal{V}))_2 \quad \forall \mathbf{U} \in H,\ \mathcal{V} := (\mathbf{V}_i)_{0\le i\le N} \in H_D$$

or in other words

$$\sum_{i=0}^{N} \mathbf{V}_i^T (\mathcal{R}_{ASM,2}^*(\mathbf{U}))_i := \mathbf{U}^T \mathcal{R}_{ASM,2}(\mathcal{V})\,,$$

that is,

$$\sum_{i=0}^{N} \mathbf{V}_i^T (\mathcal{R}_{ASM,2}^*(\mathbf{U}))_i := \mathbf{U}^T \sum_{i=0}^{N} R_i^T \mathbf{V}_i = \sum_{i=0}^{N} \mathbf{V}_i^T R_i \mathbf{U}\,.$$

Since this equality is valid for arbitrary $\mathbf{V}_i$, we have the identification

$$\mathcal{R}_{ASM,2}^*(\mathbf{U}) = (R_i \mathbf{U})_{0\le i\le N}\,, \tag{7.8}$$

which leads to the rewriting (7.7) of the additive Schwarz method. □

The application of the preconditioner in terms of these operators can be explained as follows:

- According to (7.8), the rightmost operator $\mathcal{R}_{ASM,2}^*$ decomposes a global vector in H into local components in H_D.
- The middle operator B^{-1} corresponds to solving a coarse problem and N local Dirichlet problems.
- $\mathcal{R}_{ASM,2}$ interpolates the modified local components into a global vector in H.

As we shall see in the rest of this chapter, this abstract form is also valid for many domain decomposition methods, such as the balancing Neumann–Neumann preconditioner (BNN) and a modified version of the optimized Schwarz method (OSM). It will enable us to both analyze their condition number and propose new coarse space constructions.

7.2 ▪ Mathematical foundation

In this section we present a few abstract lemmas that are needed in what follows. Lemmas 7.4 and 7.9 are at the core of the study of many multilevel methods. Lemma 7.7 is specific to the GenEO method that we will introduce.

7.2.1 ▪ Fictitious space lemma

We first state the lemma proved in [150, 149] as it is written in [105]. In its proof we need the following auxiliary result.

Lemma 7.3 (Auxiliary result). *Let m be an integer, and let $A_1, A_2 \in \mathbb{R}^{m\times m}$ be two symmetric positive definite matrices. Suppose there exists a constant c such that*

$$(A_1 u, u) \le c\,(A_2 u, u) \ \ \forall u \in \mathbb{R}^m .$$

Then, $A_2^{-1}A_1$ has real eigenvalues that are bounded by constant c.

Lemma 7.4 (Fictitious space lemma). *Let H and H_D be two Hilbert spaces, with the scalar products denoted by $(\cdot,\cdot)$ and $(\cdot,\cdot)_D$. Let the symmetric positive bilinear forms $a : H\times H \to \mathbb{R}$ and $b : H_D \times H_D \to \mathbb{R}$, generated by the SPD operators $A : H \to H$ and $B : H_D \to H_D$, respectively (i.e., $(Au, v) = a(u, v)$ for all $u, v \in H$ and $(B u_D, v_D)_D = b(u_D, v_D)$ for all $u_D, v_D \in H_D$). Suppose that there exists a linear operator $\mathcal{R} : H_D \to H$ such that the following hold:*

- *$\mathcal{R}$ is surjective.*
- *There exists a positive constant c_R such that*

$$a(\mathcal{R} u_D, \mathcal{R} u_D) \le c_R \cdot b(u_D, u_D) \ \ \forall u_D \in H_D . \tag{7.9}$$

- *There exists a positive constant c_T such that for all $u \in H$ there exists $u_D \in H_D$ with $\mathcal{R} u_D = u$ and*

$$c_T \cdot b(u_D, u_D) \le a(\mathcal{R} u_D, \mathcal{R} u_D) = a(u, u). \tag{7.10}$$

We introduce the adjoint operator $\mathcal{R}^* : H \to H_D$ by $(\mathcal{R} u_D,\, u) = (u_D,\, \mathcal{R}^* u)_D$ for all $u_D \in H_D$ and $u \in H$.

Then, we have the spectral estimate

$$\boxed{c_T \cdot a(u, u) \le a\left(\mathcal{R} B^{-1} \mathcal{R}^* A u,\ u\right) \le c_R \cdot a(u, u) \ \ \forall u \in H,} \tag{7.11}$$

which proves that the eigenvalues of operator $\mathcal{R} B^{-1} \mathcal{R}^* A$ are bounded from below by c_T and from above by c_R with sharp bounds for the spectrum of $\mathcal{R} B^{-1} \mathcal{R}^* A$ given by the best possible constants c_T and c_R in the above inequalities.

Proof. We will give a proof of the spectral estimates only in the finite dimensional case. First note that operator $\mathcal{R}B^{-1}\mathcal{R}^* : H \mapsto H$ is symmetric by definition. Its positive definiteness is easy to check. For any $u \in H$, we have

$$(\mathcal{R}B^{-1}\mathcal{R}^* u, u) = (B^{-1}\mathcal{R}^* u, \mathcal{R}^* u)_D \geq 0.$$

Since B is SPD the above term is zero if and only if $\mathcal{R}^* u = 0$. Since $\mathcal{R}$ is surjective, it follows that $\mathcal{R}^*$ is one-to-one. Thus, $\mathcal{R}^* u = 0$ implies that $u = 0$.

We first prove the upper bound of the spectral estimate (7.11). First note that (7.9) is equivalent to

$$(\mathcal{R}^* A \mathcal{R} u_D, u_D)_D \leq c_R (B u_D, u_D)_D \quad \forall u_D \in H_D.$$

Using Lemma 7.3, the eigenvalues of $B^{-1}\mathcal{R}^* A \mathcal{R} : H_D \mapsto H_D$ are real and bounded from above by c_R. This bound carries over to operator $\mathcal{R}B^{-1}\mathcal{R}^* A : H \mapsto H$. Indeed, for any positive integer n we have

$$(\mathcal{R}B^{-1}\mathcal{R}^* A)^n = \mathcal{R}\,(B^{-1}\mathcal{R}^* A \mathcal{R})^{n-1} B^{-1}\mathcal{R}^* A.$$

Thus, for any $u \in H \setminus \{0\}$

$$\begin{aligned}
&\|(\mathcal{R}B^{-1}\mathcal{R}^* A)^n u\|^{1/n} = \|\mathcal{R}\,(B^{-1}\mathcal{R}^* A \mathcal{R})^{n-1} B^{-1}\mathcal{R}^* A u\|^{1/n} \\
&\leq \|\mathcal{R}\|^{1/n}_{H_D \mapsto H} \|(B^{-1}\mathcal{R}^* A \mathcal{R})^{n-1}\|^{1/n} \|B^{-1}\mathcal{R}^* A u\|^{1/n}_{H \mapsto H_D}.
\end{aligned}$$

By taking the limit of the previous inequality as n tends to infinity, we obtain that the spectral radius of $\mathcal{R}B^{-1}\mathcal{R}^* A$ is bounded from above by that of $B^{-1}\mathcal{R}^* A \mathcal{R}$, which is itself less than or equal to c_R.

We now prove the lower bound c_T. For all $u \in H$, let $u_D \in H_D$ such that $\mathcal{R}(u_D) = u$ satisfies estimate (7.10). We have

$$\begin{aligned}
a(u,u) &= a(\mathcal{R}(u_D), u) = (\mathcal{R}(u_D), A u) = (u_D, \mathcal{R}^* A u)_D \\
&= (u_D, B B^{-1}\mathcal{R}^* A u)_D = b(u_D, B^{-1}\mathcal{R}^* A u) \\
&\leq b(u_D, u_D)^{1/2}\, b(B^{-1}\mathcal{R}^* A u, B^{-1}\mathcal{R}^* A u)^{1/2} \\
&\leq \frac{1}{\sqrt{c_T}}\, a(u,u)^{1/2}\, b(B^{-1}\mathcal{R}^* A u, B^{-1}\mathcal{R}^* A u)^{1/2}.
\end{aligned} \tag{7.12}$$

Dividing by $a(u,u)^{1/2}/\sqrt{c_T}$ and squaring we get

$$\begin{aligned}
c_T\, a(u,u) &\leq (\mathcal{R}^* A u, B^{-1}\mathcal{R}^* A u)_D \\
&= (A u, \mathcal{R}B^{-1}\mathcal{R}^* A u) = a(u, \mathcal{R}B^{-1}\mathcal{R}^* A u).
\end{aligned}$$

For a proof valid in the infinite dimensional case as well and for a proof of the optimality of the spectral estimate see [150, 149] or [105]. □

Remark 7.2.1. *Lemma 7.4 is crucial in the definition of domain decomposition algorithms, and it requires a few elements of explanation:*

- *If $\mathcal{R}$ were invertible, constant c_T^{-1} in estimate (7.10) would be a continuity constant of $\mathcal{R}^{-1}$.*

- *In order to apply this lemma to the ASM, we have defined $\mathcal{R}$ as a sum of vectors local to subdomains; see (7.5). Then, estimate (7.10) is often known as the "stable decomposition property"; see [190] or [185]. In the following we use this name even when the operator $\mathcal{R}$ also involves projections.*

Note that the bilinear operators from Lemma 7.4 can also be related to an optimization problem.

Lemma 7.5 (Related optimization problem). *For all $u \in H$, we have*

$$a((\mathcal{R}B^{-1}\mathcal{R}^*A)^{-1}u,\, u) = \min_{u_D | \mathcal{R}(u_D)=u} b(u_D,\, u_D).$$

Proof. By definition, this is equivalent to proving

$$((\mathcal{R}B^{-1}\mathcal{R}^*)^{-1}u,\, u) = \min_{u_D | \mathcal{R}(u_D)=u} (Bu_D,\, u_D)_D\,.$$

We solve this constrained quadratic minimization problem using a Lagrangian:

$$\begin{array}{rccl} \mathcal{L}: & H_D \times H & \to & \mathbb{R}, \\ & (u_D, \lambda) & \mapsto & \frac{1}{2}(Bu_D, u_D)_D - (\lambda, \mathcal{R}(u_D) - u). \end{array}$$

By differentiating the Lagrangian $\mathcal{L}$, we get the optimality system

$$Bu_D = \mathcal{R}^*(\lambda) \text{ and } \mathcal{R}(u_D) = u\,,$$

so that $\lambda = (\mathcal{R}B^{-1}\mathcal{R}^*)^{-1}u$ and $u_D = B^{-1}\mathcal{R}^*(\mathcal{R}B^{-1}\mathcal{R}^*)^{-1}u$. This leads to $(Bu_D, u_D)_D = ((\mathcal{R}B^{-1}\mathcal{R}^*)^{-1}u,\, u)$. ☐

7.2.2 ▪ Symmetric generalized eigenvalue problem

Important remark. In order to avoid further confusion, we warn the reader that in this section, the notation A or B does not refer to the matrices of a linear system to be solved but to abstract linear operators.

Let V be a Hilbert space of dimension n with the scalar product denoted by $(\cdot,\cdot)$. Let A and B be symmetric positive linear semidefinite operators from V on V. We first assume that operator B is also definite. We introduce the following generalized eigenvalue problem:

$$\boxed{\begin{array}{c}\text{Find } (\mathbf{y}_k, \mu_k) \in V \times \mathbb{R} \text{ such that} \\ A\mathbf{y}_k = \mu_k\, B\mathbf{y}_k\,. \end{array}} \tag{7.13}$$

Since operator B is symmetric positive definite, the eigenvalues of (7.13) can be chosen so that they form a basis of V that is both A-orthogonal and B-orthonormal,

$$(B\mathbf{y}_k, \mathbf{y}_l) = \delta_{kl},\ \ (A\mathbf{y}_k, \mathbf{y}_l) = 0 \text{ for } k \neq l\,,$$

where δ_{kl} is the classical Kronecker symbol. We have the following result verified by the solutions of this problem.

Lemma 7.6. *Let $\tau > 0$ and define the space related to the eigenpairs of the problem* (7.13) *as*

$$Y_\tau := \mathrm{Span}\left\{\mathbf{y}_k \mid \mu_k < \frac{1}{\tau}\right\}. \tag{7.14}$$

Let ξ_τ denote the projection on Y_τ parallel to $\mathrm{Span}\{\mathbf{y}_k \mid \mu_k \geq \frac{1}{\tau}\}$.

Then, for all $\mathbf{y} \in V$ the following inequality holds:

$$\boxed{(\mathbf{y}-\xi_\tau(\mathbf{y}), B(\mathbf{y}-\xi_\tau(\mathbf{y})) \leq \tau(A\mathbf{y},\mathbf{y}).} \tag{7.15}$$

Proof. Let $\mathbf{y} \in V$. We have

$$\mathbf{y} = \sum_{k=1}^{n}(B\mathbf{y},\mathbf{y}_k)\mathbf{y}_k = \underbrace{\sum_{\mu_k<\frac{1}{\tau}}(B\mathbf{y},\mathbf{y}_k)\mathbf{y}_k}_{\in Y_\tau} + \underbrace{\sum_{\mu_k\geq\frac{1}{\tau}}(B\mathbf{y},\mathbf{y}_k)\mathbf{y}_k}_{\in \mathrm{Span}\{\mathbf{y}_k \mid \mu_k \geq \frac{1}{\tau}\}}.$$

Thus, we have

$$\mathbf{y}-\xi_\tau(\mathbf{y}) = \sum_{\mu_k\geq\frac{1}{\tau}}(B\mathbf{y},\mathbf{y}_k)\mathbf{y}_k,$$

so that using the B-orthonormality of the eigenvector basis,

$$(B(\mathbf{y}-\xi_\tau(\mathbf{y})), \mathbf{y}-\xi_\tau(\mathbf{y})) = \sum_{\mu_k\geq\frac{1}{\tau}}(B\mathbf{y},\mathbf{y}_k)^2 \leq \tau \sum_{\mu_k\geq\frac{1}{\tau}}(B\mathbf{y},\mathbf{y}_k)^2\mu_k. \tag{7.16}$$

On the other hand, using the A-orthogonality of the eigenvector basis, we have

$$\begin{aligned}
(A\mathbf{y},\mathbf{y}) &= \left(\sum_{k=1}^{n}(B\mathbf{y},\mathbf{y}_k)A\mathbf{y}_k, \sum_{l=1}^{n}(B\mathbf{y},\mathbf{y}_l)\mathbf{y}_l\right)\\
&= \left(\sum_{k=1}^{n}(B\mathbf{y},\mathbf{y}_k)A\mathbf{y}_k, \mathbf{y}\right)\\
&= \sum_{k=1}^{n}(B\mathbf{y},\mathbf{y}_k)(A\mathbf{y},\mathbf{y}_k)\\
&= \sum_{k=1}^{n}(B\mathbf{y},\mathbf{y}_k)^2\mu_k \geq \sum_{\mu_k\geq\frac{1}{\tau}}(B\mathbf{y},\mathbf{y}_k)^2\mu_k.
\end{aligned} \tag{7.17}$$

Combining (7.16) and (7.17) ends the proof. □

We need to consider the case where both operators A and B may be indefinite. Let P be the orthogonal projection on range(A). Since A is symmetric, P is actually a projection parallel to ker(A). We introduce the following generalized eigenvalue problem:

$$\boxed{\begin{array}{c}\text{Find } (\mathbf{x}_k, \lambda_k) \in \mathrm{range}(A) \times \mathbb{R} \text{ such that}\\ PBP\,\mathbf{x}_k = \lambda_k\, A\mathbf{x}_k.\end{array}} \tag{7.18}$$

Note that operator PBP is symmetric positive from range(A) into itself and that matrix A, seen as a linear operator from range(A) into itself, is symmetric positive definite. Thus the eigenvalues of (7.18) can be chosen so that they form both an A- and PBP-orthogonal basis of range(A). We have the following result verified by the solutions of this problem.

Lemma 7.7. *Let $\tau > 0$ and define the space related to the eigenpairs of the problem (7.18):*

$$Z_\tau := \ker(A) \bigoplus \mathrm{Span}\{\mathbf{x}_k \,|\, \lambda_k > \tau\}. \tag{7.19}$$

Let π_τ denote the projection on Z_τ parallel to $\mathrm{Span}\{\mathbf{x}_k \,|\, \lambda_k \le \tau\}$.

Then, for all $\mathbf{x} \in V$ the following inequality holds:

$$\boxed{(\mathbf{x} - \pi_\tau(\mathbf{x}), B(\mathbf{x} - \pi_\tau(\mathbf{x})) \le \tau (A\mathbf{x}, \mathbf{x}).} \tag{7.20}$$

Proof. Let C be the symmetric linear operator defined as follows:

$$\begin{aligned} C \ : \ \mathrm{range}(A) &\to \mathrm{range}(A), \\ \mathbf{x} &\longmapsto PBP\mathbf{x}. \end{aligned}$$

Let $m := \dim \mathrm{range}(A) = n - \dim \ker(A)$. By normalizing the family of eigenvectors $(\mathbf{x}_k)_{1 \le k \le m}$ from (7.18) so that they form an A-orthonormal basis of range(A), we get

$$P\mathbf{x} = \sum_{k=1}^{m} (AP\mathbf{x}, \mathbf{x}_k)\mathbf{x}_k.$$

Note that the eigenvectors are C-orthogonal as well,

$$(C\mathbf{x}_k, \mathbf{x}_l) = (\lambda_k A\mathbf{x}_k, \mathbf{x}_l) = \lambda_k \delta_{kl}.$$

For any $\mathbf{x} \in V$, we have the following decomposition into components of ker(A) and range(A):

$$\mathbf{x} = \underbrace{(\mathbf{x} - P\mathbf{x})}_{\in \ker(A)} + \underbrace{\sum_{k=1}^{m} (AP\mathbf{x}, \mathbf{x}_k)\mathbf{x}_k}_{\in \mathrm{range}(A)}.$$

In this spirit, the orthogonal projection $\pi_\tau(\mathbf{x})$ of $\mathbf{x}$ on Z_τ is

$$\pi_\tau(\mathbf{x}) := \underbrace{(\mathbf{x} - P\mathbf{x})}_{\in \ker(A)} + \underbrace{\sum_{\lambda_k > \tau} (AP\mathbf{x}, \mathbf{x}_k)\mathbf{x}_k}_{\in \mathrm{Span}\{\mathbf{x}_k |\, \lambda_k > \tau\}}.$$

Note that π_τ is a projection parallel to $\mathrm{Span}\{\mathbf{x}_k | \lambda_k \le \tau\}$. By estimating now the first term of (7.20) we see that

$$\mathbf{x} - \pi_\tau(\mathbf{x}) = \sum_{\lambda_k \le \tau} (AP\mathbf{x}, \mathbf{x}_k)\mathbf{x}_k.$$

By plugging this formula into $(B(\mathbf{x} - \pi_\tau(\mathbf{x})), \mathbf{x} - \pi_\tau(\mathbf{x}))$ and by using the fact that the eigenvectors $\mathbf{x}_k$ and $\mathbf{x}_l$ belong to range(A) ($\mathbf{x}_k = P\mathbf{x}_k$) and their C-orthogonality, we get

$$
\begin{aligned}
(B(\mathbf{x}-\pi_\tau(\mathbf{x})),\mathbf{x}-\pi_\tau(\mathbf{x})) &= \left(\sum_{k|\lambda_k\le\tau}(AP\mathbf{x},\mathbf{x}_k)B\mathbf{x}_k,\ \sum_{l|\lambda_l\le\tau}(AP\mathbf{x},\mathbf{x}_l)\mathbf{x}_l\right)\\
&= \sum_{l|\lambda_l\le\tau}\sum_{k|\lambda_k\le\tau}(AP\mathbf{x},\mathbf{x}_k)(AP\mathbf{x},\mathbf{x}_l)(B\mathbf{x}_k,\mathbf{x}_l)\\
&= \sum_{l|\lambda_l\le\tau}\sum_{k|\lambda_k\le\tau}(AP\mathbf{x},\mathbf{x}_k)(AP\mathbf{x},\mathbf{x}_l)(BP\mathbf{x}_k,P\mathbf{x}_l) \qquad (7.21)\\
&= \sum_{l|\lambda_l\le\tau}\sum_{k|\lambda_k\le\tau}(AP\mathbf{x},\mathbf{x}_k)(AP\mathbf{x},\mathbf{x}_l)(C\mathbf{x}_k,\mathbf{x}_l)\\
&= \sum_{\lambda_k\le\tau}(AP\mathbf{x},\mathbf{x}_k)^2\lambda_k.
\end{aligned}
$$

We can easily find the upper bounds of this expression:

$$
\begin{aligned}
\sum_{\lambda_k\le\tau}(AP\mathbf{x},\mathbf{x}_k)^2\lambda_k &\le \tau\sum_{\lambda_k\le\tau}(AP\mathbf{x},\mathbf{x}_k)^2 \le \tau\sum_{k=1}^{m}(AP\mathbf{x},\mathbf{x}_k)^2\\
&= \tau\left(AP\mathbf{x},\sum_{k=1}^{m}(AP\mathbf{x},\mathbf{x}_k)\mathbf{x}_k\right) = \tau(AP\mathbf{x},P\mathbf{x}) = \tau(A\mathbf{x},\mathbf{x}).
\end{aligned} \qquad (7.22)
$$

From (7.21) and (7.22), the conclusion follows. □

We have defined two generalized eigenvalue problems with no apparent connection, (7.13) and (7.18). They allow the definition of two different spaces, (7.14) and (7.19). In the following we would like to analyze the relationship between these two spaces.

Lemma 7.8 (Relationship between Y_τ and Z_τ). *We suppose that B is a positive definite operator. We can distinguish between these cases:*

- *If A is positive definite, then $Y_\tau = Z_\tau$.*
- *If A is not positive definite, then $Y_\tau = Z_\tau$ if $PBP = PB$.*

Proof. Consider first the case when A is positive definite. In this case the projection on the range of A is the identity $P = I$ and $\ker(A) = \emptyset$. Thus problem (7.18) reduces to

$$B\mathbf{x}_k = \lambda_k A\mathbf{x}_k \Leftrightarrow A^{-1}B\mathbf{x}_k = \lambda_k\mathbf{x}_k,$$

while (7.13) will be equivalent to

$$A^{-1}B\mathbf{y}_k = \frac{1}{\mu_k}\mathbf{y}_k.$$

We can thus conclude that

$$Y_\tau = \operatorname{Span}\left\{\mathbf{y}_k \mid \mu_k < \frac{1}{\tau}\right\} = \operatorname{Span}\{\mathbf{x}_k \mid \lambda_k > \tau\} = Z_\tau.$$

If A is not definite but B is, we can now left-multiply (7.13) by $B^{-1/2}$, which results in

$$\underbrace{B^{-1/2}AB^{-1/2}}_{\bar{A}}\underbrace{B^{1/2}\mathbf{y}_k}_{\bar{\mathbf{y}}_k} = \mu_k\underbrace{B^{1/2}\mathbf{y}_k}_{\bar{\mathbf{y}}_k} \Leftrightarrow \bar{A}\bar{\mathbf{y}}_k = \mu_k\bar{\mathbf{y}}_k, \qquad (7.23)$$

which is a standard eigenvalue problem. First note that

$$\mathrm{range}(\bar{A}) = B^{-1/2}\mathrm{range}(A),\ \ker(\bar{A}) = B^{1/2}\ker(A).$$

Let us analyze problem (7.23). If $\mu_k = 0$, then $\bar{\mathbf{y}}_k \in \ker(\bar{A})$. Otherwise

$$\mu_k \neq 0 \Rightarrow \bar{\mathbf{y}}_k = \frac{1}{\mu_k}\bar{A}\bar{\mathbf{y}}_k \in \mathrm{range}(\bar{A}) \Leftrightarrow \bar{P}\bar{\mathbf{y}}_k = \frac{1}{\mu_k}\bar{A}\bar{\mathbf{y}}_k, \tag{7.24}$$

with $\bar{P}$ being the projection onto the range$(\bar{A})$ parallel to $\ker(\bar{A})$.

Consider now the similar problem

$$\begin{aligned}
&\bar{P}\bar{\mathbf{x}}_k = \lambda_k \bar{A}\bar{\mathbf{x}}_k \\
\Leftrightarrow\ &\bar{P}\bar{\mathbf{x}}_k = \lambda_k B^{-1/2} A \underbrace{B^{-1/2}\bar{\mathbf{x}}_k}_{\mathbf{x}_k} \\
\Leftrightarrow\ &\underbrace{B^{1/2}\bar{P}B^{-1/2}}_{P} B\mathbf{x}_k = \lambda_k A\mathbf{x}_k \\
\Leftrightarrow\ &PB\mathbf{x}_k = \lambda_k A\mathbf{x}_k.
\end{aligned} \tag{7.25}$$

We see that

$$\begin{aligned}
B^{1/2}Y_\tau &= \mathrm{Span}\left\{B^{1/2}\mathbf{y}_k \mid \mu_k < \frac{1}{\tau}\right\} = \mathrm{Span}\left\{\bar{\mathbf{y}}_k \mid \mu_k < \frac{1}{\tau}\right\} \\
&= \ker(\bar{A}) \oplus \mathrm{Span}\{\bar{\mathbf{x}}_k : \lambda_k > \tau\} \\
&= B^{1/2}\ker(A) + B^{1/2}\mathrm{Span}\{\mathbf{x}_k : \lambda_k > \tau\} \\
\Rightarrow Y_\tau &= \ker(A) + \mathrm{Span}\{\mathbf{x}_k : \lambda_k > \tau\}
\end{aligned}$$

with $\mathbf{x}_k$ being the solution of the generalized eigenvalue problem (7.25). This leads to the conclusion that $Y_\tau = Z_\tau$ if $PBP = PB$. □

Remark 7.2.2 (Counterexample). *Problems (7.18) and (7.25) are not equivalent in general. Indeed,*

$$PB\mathbf{x}_k = PB\left(\mathbf{x}_k^{\ker(A)} + \mathbf{x}_k^{\mathrm{range}(A)}\right) = PBP\mathbf{x}_k + PB\mathbf{x}_k^{\ker(A)}$$

and the quantity $PB\mathbf{x}_k^{\ker(A)}$ *is not necessarily null. Consider now the following example:*

$$A = \begin{pmatrix} 1 & 0 \\ 0 & 0 \end{pmatrix},\ B = \begin{pmatrix} 2 & 1 \\ 1 & 1 \end{pmatrix} \Rightarrow P = \begin{pmatrix} 1 & 0 \\ 0 & 0 \end{pmatrix}.$$

The solution of eigenproblem (7.18) is

$$\begin{aligned}
\begin{pmatrix} 1 & 0 \\ 0 & 0 \end{pmatrix}\mathbf{y} = \mu\begin{pmatrix} 2 & 1 \\ 1 & 1 \end{pmatrix}\mathbf{y} \quad &\Leftrightarrow \quad \begin{pmatrix} y_1 \\ 0 \end{pmatrix} = \mu\begin{pmatrix} 2y_1 + y_2 \\ y_1 + y_2 \end{pmatrix} \\
&\Leftrightarrow \quad (\mathbf{y}, \mu) \in \left\{\left[\begin{pmatrix} 0 \\ 1 \end{pmatrix}, 0\right], \left[\begin{pmatrix} 1 \\ -1 \end{pmatrix}, 1\right]\right\}.
\end{aligned}$$

The solution of eigenproblem (7.18) *is*

$$\begin{pmatrix} 2 & 0 \\ 0 & 0 \end{pmatrix}\mathbf{x} = \lambda \begin{pmatrix} 1 & 0 \\ 0 & 0 \end{pmatrix}\mathbf{x} \quad \Leftrightarrow \quad \begin{pmatrix} 2x_1 \\ 0 \end{pmatrix} = \lambda \begin{pmatrix} x_1 \\ 0 \end{pmatrix}$$

$$\Leftrightarrow \quad (\mathbf{x},\lambda) \in \left\{ \left[\begin{pmatrix} 0 \\ 1 \end{pmatrix}, 0 \right], \left[\begin{pmatrix} 1 \\ 0 \end{pmatrix}, 2 \right] \right\}.$$

Consider now $\tau = 1.5$. *Then we have*

$$Y_\tau = \mathrm{Span}\left\{ \begin{pmatrix} 0 \\ 1 \end{pmatrix} \right\}, \; Z_\tau = \mathrm{Span}\left\{ \begin{pmatrix} 0 \\ 1 \end{pmatrix}, \begin{pmatrix} 1 \\ 0 \end{pmatrix} \right\} = \mathbb{R}^2.$$

Therefore $Y_\tau \neq Z_\tau$.

7.2.3 ▪ Auxiliary lemma

In the following, we make frequent use of the following lemma.

Lemma 7.9. *Let* $M, n, (n_i)_{1\le i\le M}$ *be positive integers, let* $Q_i \in \mathbb{R}^{n\times n_i}$, $1 \le i \le M$, *be rectangular matrices, and let* $A \in \mathbb{R}^{n\times n}$ *be a symmetric positive definite matrix. Let* $\widetilde{k}_0$ *be an integer defined by*

$$\widetilde{k}_0 := \max_{1\le j\le M} \#\{i \,|\, Q_i^T A Q_j \neq 0\}.$$

Then for all $\mathbf{U}_i \in \mathbb{R}^{n_i}$, $1 \le i \le M$, *we have the following estimate:*

$$\boxed{\left(\sum_{i=1}^M Q_i \mathbf{U}_i\right)^T A \left(\sum_{i=1}^M Q_i \mathbf{U}_i\right) \le \widetilde{k}_0 \sum_{i=1}^M \mathbf{U}_i^T (Q_i^T A Q_i) \mathbf{U}_i.} \tag{7.26}$$

Proof. By the Cauchy–Schwarz inequality, we have

$$\begin{aligned} \left(\sum_{i=1}^M Q_i \mathbf{U}_i\right)^T A \left(\sum_{i=1}^M Q_i \mathbf{U}_i\right) &= \sum_{i,j | Q_i^T A Q_j \neq 0} (\mathbf{U}_i^T Q_i^T) A \left(Q_j \mathbf{U}_j\right) \\ &\le \sum_{i,j | Q_i^T A Q_j \neq 0} \left(\mathbf{U}_i^T Q_i^T A Q_i \mathbf{U}_i\right)^{1/2} \left(\mathbf{U}_j^T Q_j^T A Q_j \mathbf{U}_j\right)^{1/2}. \end{aligned} \tag{7.27}$$

Let us introduce the connectivity matrix $C \in \mathbb{R}^M \times \mathbb{R}^M$ defined as follows:

$$C_{ij} = \begin{cases} 1 \text{ if } Q_i^T A Q_j \neq 0, \; 1 \le i, j \le M, \\ 0 \text{ otherwise.} \end{cases} \tag{7.28}$$

$\mathbf{v} \in \mathbb{R}^M$, $\mathbf{v} = (v_i)_{1\le i\le M}$, is the vector of norms defined as

$$\mathbf{v} := \left((\mathbf{U}_1^T Q_1^T A Q_1 \mathbf{U}_1)^{1/2}, \ldots, (\mathbf{U}_M^T Q_M^T A Q_M \mathbf{U}_M)^{1/2} \right)^T. \tag{7.29}$$

Note that we have

$$\|\mathbf{v}\|_2^2 = \sum_{i=1}^M v_i^2 = \sum_{i=1}^M \mathbf{U}_i^T (Q_i^T A Q_i) \mathbf{U}_i \tag{7.30}$$

and matrix C is symmetric as a consequence of the symmetry of A. With notations (7.28) and (7.29), estimate (7.27) becomes

$$\left(\sum_{i=1}^{M} Q_i \mathbf{U}_i\right)^T A \left(\sum_{i=1}^{M} Q_i \mathbf{U}_i\right) \leq \mathbf{v}^T C \mathbf{v}. \tag{7.31}$$

Note that we also have by definition of an operator norm

$$\mathbf{v}^T C \mathbf{v} \leq \|C\|_2 \cdot \|v\|_2^2. \tag{7.32}$$

Since matrix C is symmetric, its 2-norm is given by the largest eigenvalue in the modulus, which is less than the infinity norm of C. It can be easily checked that $\|C\|_\infty = \widetilde{k}_0$, and by consequence we will have $\|C\|_2 \leq \widetilde{k}_0$. Finally, from estimates (7.30), (7.31), and (7.32), we have

$$\left(\sum_{i=1}^{M} Q_i \mathbf{U}_i\right)^T A \left(\sum_{i=1}^{M} Q_i \mathbf{U}_i\right) \leq \widetilde{k}_0 \sum_{i=1}^{M} v_i^2 = \widetilde{k}_0 \sum_{i=1}^{M} \mathbf{U}_i^T (Q_i^T A Q_i) \mathbf{U}_i,$$

which ends the proof. □

7.3 ▪ Finite element setting

We recall here the finite element setting that is used in order to build the coarse space. Consider first the variational formulation of a problem for a given open domain $\Omega \subset \mathbb{R}^d$ ($d = 2$ or 3),

$$a_\Omega(u, v) = l(v) \quad \forall v \in V, \tag{7.33}$$

where $l(v)$ denotes a linear form over a Hilbert space V. The problem we consider is given through a symmetric positive definite bilinear form that is defined in terms of an integral over any open set $\omega \subset \mathbb{R}^d$ for some integer d. Typical examples are the Darcy equation (**K** is a diffusion tensor),

$$a_\omega(u, v) := \int_\omega \mathbf{K} \nabla u \cdot \nabla v \, dx, \tag{7.34}$$

or the elasticity system (C is the fourth-order stiffness tensor and $\varepsilon(\boldsymbol{u})$ is the strain tensor of a displacement field $\boldsymbol{u}$)

$$a_\omega(\boldsymbol{u}, \boldsymbol{v}) := \int_\omega C \, \varepsilon(\boldsymbol{u}) : \varepsilon(\boldsymbol{v}) \, dx.$$

We consider a discretization of the variational problem (7.33) by the finite element (FE) method on a mesh $\mathcal{T}_h$ of Ω: $\overline{\Omega} = \bigcup_{c \in \mathcal{T}_h} c$. The approximation space is then denoted by V_h, $\{\phi_k\}_{k \in \mathcal{N}}$ denotes its basis, and $\mathcal{N}$ is the related index set. Let us write the discrete FE problem that we want to solve:

$$\begin{array}{c} \text{Find } u_h \in V_h \text{ such that} \\ a_\Omega(u_h, v_h) = l(v_h) \quad \forall v_h \in V_h, \end{array} \tag{7.35}$$

which gives the matrix form

$$A\mathbf{U} = b, A_{ij} = a_\Omega(\phi_j, \phi_i), b_i = l(\phi_i) \quad \forall i, j \in \mathcal{N},$$

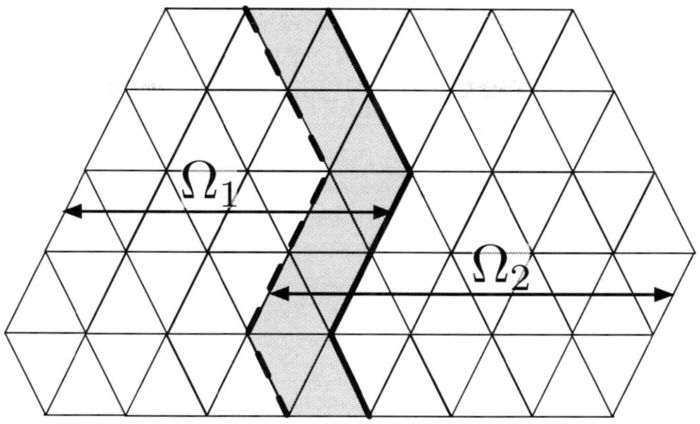

Figure 7.1. *Initial geometric decomposition of the domain.*

so that

$$u_h = \sum_{k\in\mathcal{N}} u_k\,\phi_k,\ \mathbf{U} = (u_k)_{k\in\mathcal{N}}.$$

Domain Ω is decomposed into N subdomains $(\Omega_i)_{1\le i\le N}$ so that all subdomains are a union of cells of the mesh $\mathcal{T}_h$, see Figure 7.1. This decomposition induces a natural decomposition of the set of indices $\mathcal{N}$ into N subsets of indices $(\mathcal{N}_j)_{1\le j\le N}$ as was done in (1.28):

$$\mathcal{N}_j := \{k \in \mathcal{N} : \text{meas}(\text{Supp}(\phi_k)\cap\Omega_j) \neq 0\},\ 1\le j\le N. \tag{7.36}$$

Let $\widetilde{A}^j$ be the $\#\mathcal{N}_j \times \#\mathcal{N}_j$ matrix defined by

$$\boxed{\mathbf{V}_j^T \widetilde{A}^j \mathbf{U}_j := a_{\Omega_j}\left(\sum_{k\in\mathcal{N}_j} \mathbf{U}_{jk}\phi_k, \sum_{k\in\mathcal{N}_j} \mathbf{V}_{jk}\phi_k\right),\ \ \mathbf{U}_j, \mathbf{V}_j \in \mathbb{R}^{\mathcal{N}_j}.} \tag{7.37}$$

When the bilinear form a results from a Laplace problem, the previous matrix corresponds to the discretization of local Neumann boundary value problems. For this reason we will call it the "Neumann" matrix, even in a more general setting.

7.4 ▪ GenEO coarse space for additive Schwarz

In order to apply the fictitious space lemma to the ASM, we use the framework defined in section 7.1. Note that most of the assumptions of this lemma are verified without the precise definition of a coarse space. That is why it will be specified later on in Definition 7.16 only when needed.

Let us start with the surjectivity and continuity of $\mathcal{R}_{ASM,2}$, which do not depend on the choice of the coarse space.

Lemma 7.10 (Surjectivity of $\mathcal{R}_{ASM,2}$). *The operator $\mathcal{R}_{ASM,2}$ defined by* (7.5) *is surjective.*

Proof. From the partition of unity, as defined in (1.25) in section 1.3, we have for all $\mathbf{U} \in H$

$$\mathbf{U} = \mathcal{R}_{ASM,2}(\mathcal{U}),\ \text{with}\ \mathcal{U} = (0, (D_i R_i \mathbf{U})_{1\le i\le N}),$$

which proves the surjectivity of $\mathcal{R}_{ASM,2}$. $\square$

Lemma 7.11 (Continuity of $\mathcal{R}_{ASM,2}$). *Let*

$$k_0 := \max_{1\le i\le N} \#\left\{j \mid R_j A R_i^T \neq 0\right\} \tag{7.38}$$

be the maximum multiplicity of the interaction between subdomains plus one. Then for all $\mathcal{U} \in H_D$, we have

$$a(\mathcal{R}_{ASM,2}(\mathcal{U}), \mathcal{R}_{ASM,2}(\mathcal{U})) \le 2k_0\, b(\mathcal{U}, \mathcal{U}),$$

which proves the continuity of operator $\mathcal{R}_{ASM,2}$ (i.e., hypothesis (7.9) from Lemma 7.4), with

$$c_R = 2k_0$$

as a continuity constant.

Proof. Let $\mathcal{U} := (\mathbf{U}_i)_{0\le i\le N} \in H_D$. Then we have by definition

$$a(\mathcal{R}_{ASM,2}(\mathcal{U}), \mathcal{R}_{ASM,2}(\mathcal{U})) = \left(\sum_{i=0}^N R_i^T \mathbf{U}_i\right)^T A \left(\sum_{i=0}^N R_i^T \mathbf{U}_i\right) \tag{7.39}$$

and

$$b(\mathcal{U}, \mathcal{U}) = \sum_{i=0}^N \left(R_i^T \mathbf{U}_i\right)^T A \left(R_i^T \mathbf{U}_i\right). \tag{7.40}$$

Applying Lemma 7.9, with $M = N+1$, $Q_i = R_i^T$ $(0 \le i \le N)$, would not yield a sharp estimate. Indeed, we note that $R_0 A R_i^T \neq 0$, $0 \le i \le N$, which means the constant $\tilde{k}_0$ in Lemma 7.9 would be equal to $N+1$. Thus, we proceed in two steps. Since A is symmetric positive, we have

$$\begin{aligned}\left(\sum_{i=0}^N R_i^T \mathbf{U}_i\right)^T A \left(\sum_{i=0}^N R_i^T \mathbf{U}_i\right) \le 2\Bigg(& (R_0^T \mathbf{U}_0)^T A (R_0^T \mathbf{U}_0) \\ & + \left(\sum_{i=1}^N R_i^T \mathbf{U}_i\right)^T A \left(\sum_{i=1}^N R_i^T \mathbf{U}_i\right)\Bigg).\end{aligned} \tag{7.41}$$

We use Lemma 7.9 to estimate the last term, and we have

$$\begin{aligned}\left(\sum_{i=0}^N R_i^T \mathbf{U}_i\right)^T A \left(\sum_{i=0}^N R_i^T \mathbf{U}_i\right) &\le 2\Bigg((R_0^T \mathbf{U}_0)^T A (R_0^T \mathbf{U}_0) \\ &\quad + k_0 \sum_{i=1}^N \left(R_i^T \mathbf{U}_i\right)^T A \left(R_i^T \mathbf{U}_i\right)\Bigg) \\ &\le 2k_0 \sum_{i=0}^N \left(R_i^T \mathbf{U}_i\right)^T A \left(R_i^T \mathbf{U}_i\right).\end{aligned} \tag{7.42}$$

□

7.4.1 ▪ Some estimates for a stable decomposition with $\mathcal{R}_{ASM,2}$

We now focus on the stable decomposition, estimate (7.10), which requires further analysis and is dependent on the choice of the coarse space given here by its tall and skinny column-matrix form $Z_0 = R_0^T$. Nevertheless, we note that some intermediate results, valid for arbitrary coarse space, simplify the requirement of (7.10). For example, the following lemma is valid whatever the choice of $Z_0 = R_0^T$.

Lemma 7.12. *Let* $\mathbf{U} \in H$ *and* $\mathcal{U} := (\mathbf{U}_i)_{0\le i\le N} \in H_D$ *such that* $\mathbf{U} = \mathcal{R}_{ASM,2}(\mathcal{U})$. *Then, we have*

$$b(\mathcal{U},\mathcal{U}) \le 2a(\mathbf{U},\mathbf{U}) + (2k_0+1)\sum_{i=1}^{N} a(R_i^T\mathbf{U}_i, R_i^T\mathbf{U}_i). \tag{7.43}$$

Proof. By definition of $\mathcal{R}_{ASM,2}$ (see (7.5)) and the Cauchy–Schwarz inequality, we have

$$\begin{aligned}
b(\mathcal{U},\mathcal{U}) &= a(R_0^T\mathbf{U}_0, R_0^T\mathbf{U}_0) + \sum_{i=1}^{N} a(R_i^T\mathbf{U}_i, R_i^T\mathbf{U}_i)\\
&= a\left(\mathbf{U} - \sum_{i=1}^{N} R_i^T\mathbf{U}_i, \mathbf{U} - \sum_{i=1}^{N} R_i^T\mathbf{U}_i\right) + \sum_{i=1}^{N} a(R_i^T\mathbf{U}_i, R_i^T\mathbf{U}_i)\\
&\le 2\left[a(\mathbf{U},\mathbf{U}) + a\left(\sum_{i=1}^{N} R_i^T\mathbf{U}_i, \sum_{i=1}^{N} R_i^T\mathbf{U}_i\right)\right] + \sum_{i=1}^{N} a(R_i^T\mathbf{U}_i, R_i^T\mathbf{U}_i)\\
&\le 2\left[a(\mathbf{U},\mathbf{U}) + \sum_{1\le i,j\le N \backslash R_j A R_i^T \neq 0} a(R_i^T\mathbf{U}_i, R_j^T\mathbf{U}_j)\right] + \sum_{i=1}^{N} a(R_i^T\mathbf{U}_i, R_i^T\mathbf{U}_i).
\end{aligned}$$

Applying Lemma 7.9, the middle term can be bounded by

$$k_0 \sum_{i=1}^{N} a(R_i^T\mathbf{U}_i, R_i^T\mathbf{U}_i)$$

and the conclusion follows. □

Note that the result (7.43) is insufficient to yield a spectral estimate of the ASM preconditioner. We still have to bound from above the subdomain energy terms $a(R_i^T\mathbf{U}_i, R_i^T\mathbf{U}_i)$, $1 \le i \le N$, by the global energy term $a(\mathbf{U}, \mathbf{U})$. In order to do this, we first introduce an estimate of $a(\mathbf{U}, \mathbf{U})$ from below in terms of a sum of some local energy terms (see (7.44)) and then infer a construction of the coarse space in Definition 7.16.

Lemma 7.13. *Let k_1 be the maximal multiplicity of the subdomain intersection, i.e., the largest integer m such that there exist m different subdomains whose intersection has a nonzero measure.*

Then, for all $\mathbf{U} \in \mathbb{R}^{\mathcal{N}}$, *we have*

$$\sum_{j=1}^{N} (R_j\mathbf{U})^T \widetilde{A}^j R_j\mathbf{U} \le k_1 \mathbf{U}^T A\mathbf{U} = k_1 a(\mathbf{U},\mathbf{U}), \tag{7.44}$$

where matrices $\widetilde{A}^j$ *are defined by* (7.37).

Proof. This property makes use of the finite element setting of section 7.3. Let $u_h := \sum_{k\in\mathcal{N}} U_k\phi_k$; then, by definition,

$$a_\Omega(u_h,u_h) = \mathbf{U}^T A\mathbf{U} \text{ and } a_{\Omega_j}(u_h,u_h) = (R_j\mathbf{U})^T \widetilde{A}^j R_j\mathbf{U}.$$

Since at most k_1 of the subdomains have a nonzero measure intersection, the sum $\sum_{j=1}^{N} a_{\Omega_j}(u_h, u_h)$ cannot exceed k_1 times $a_\Omega(u_h, u_h)$. Let us be more explicit in the case of a Darcy equation (7.34). Inequality (7.44) reads as

$$\sum_{j=1}^{N} \int_{\Omega_j} \mathbf{K} |\nabla u_h|^2 \, dx \leq k_1 \int_\Omega \mathbf{K} |\nabla u_h|^2 \, dx . \qquad \square$$

7.4.2 ▪ Definition of the GenEO coarse space

All the previous results are valid independently of the choice of the $(\mathbf{U}_j)_{0\leq j\leq N}$ in the decomposition of U and of the coarse space. They cannot give access to an estimate of the constant c_T in the stable decomposition estimate required by condition (7.10) of the fictitious space lemma, Lemma 7.4. But, they are important steps since they enable a *local* construction of a *global* coarse space so that the constant c_T can be chosen a priori less than 1/2; see Lemma 7.12. Actually, we see from (7.43) and (7.44) that it is sufficient, for some given parameter $\tau > 0$, to define $(\mathbf{U}_j)_{1\leq j\leq N}$ such that

$$\sum_{j=1}^{N} a(R_j^T \mathbf{U}_j, R_j^T \mathbf{U}_j) \leq \tau \sum_{j=1}^{N} (R_j \mathbf{U})^T \widetilde{A}^j R_j \mathbf{U}, \tag{7.45}$$

which can be satisfied by demanding that for all subdomains j we have

$$a(R_j^T \mathbf{U}_j, R_j^T \mathbf{U}_j) \leq \tau \; (R_j \mathbf{U})^T \widetilde{A}^j R_j \mathbf{U}. \tag{7.46}$$

This estimate will have as a consequence the stability of the decomposition, and a short computation shows that we can take $c_T := (2+(2k_0+1)k_1\,\tau)^{-1}$ in (7.10). The definition of the $\mathbf{U}_j$ which satisfy (7.46) for a given threshold τ will be a consequence of the definition of the coarse space.

We now detail the construction of the GenEO coarse space.

To start with, we apply the abstract lemma (Lemma 7.7) to the following generalized eigenvalue problem in each subdomain.

Definition 7.14 (Generalized eigenvalue problem). *For all subdomains* $1 \leq j \leq N$, *let*

$$\widetilde{B}_j := D_j (R_j A R_j^T) D_j .$$

Let $\widetilde{P}_j$ *be the projection on* range$(\widetilde{A}^j)$ *parallel to* ker$(\widetilde{A}^j)$. *Consider the following generalized eigenvalue problem:*

$$\boxed{\begin{array}{c} \text{Find } (\widetilde{\mathbf{U}}_{jk}, \lambda_{jk}) \in \operatorname{range}(\widetilde{A}^j) \times \mathbb{R} \\ \text{such that } \widetilde{P}_j \widetilde{B}_j \widetilde{P}_j \, \widetilde{\mathbf{U}}_{jk} = \lambda_{jk} \widetilde{A}^j \widetilde{\mathbf{U}}_{jk} . \end{array}}$$

Define also

$$\tilde{Z}_{j\tau} := \ker(\widetilde{A}^j) \bigoplus \operatorname{Span}\{\widetilde{\mathbf{U}}_{jk} \,|\, \lambda_{jk} > \tau\}$$

and the local projection $\widetilde{\pi}_j$ *on* $\widetilde{Z}_{j\tau}$ *parallel to* Span$\{\widetilde{\mathbf{U}}_{jk} \,|\, \lambda_{jk} \leq \tau\}$.

With this notation, Lemma 7.7 translates into the following.

Lemma 7.15 (Intermediate lemma). *For all subdomains $1 \le j \le N$ and $\widetilde{\mathbf{U}}_j \in \mathbb{R}^{\mathcal{N}_j}$, we have*

$$\left(R_j^T D_j (I_d - \tilde{\pi}_j)\widetilde{\mathbf{U}}_j\right)^T A \left(R_j^T D_j (I_d - \tilde{\pi}_j)\widetilde{\mathbf{U}}_j\right) \le \tau\, \widetilde{\mathbf{U}}_j^T \widetilde{A}^j \widetilde{\mathbf{U}}_j\,. \tag{7.47}$$

We see now that estimate (7.46) can be obtained directly from (7.47) provided that the $\mathbf{U}_j$ are such that the left-hand sides of (7.46) and (7.47) are the same, that is,

$$\boxed{\mathbf{U}_j := D_j (I_d - \tilde{\pi}_j) R_j \mathbf{U}.} \tag{7.48}$$

It remains now to define the coarse space component $\mathbf{U}_0$ and the coarse space interpolation operator, such that $\sum_{j=0}^N R_j^T \mathbf{U}_j = \mathbf{U}$. From the previous results we can infer that this decomposition is stable.

Definition 7.16 (GenEO coarse space). *The GenEO coarse space is defined as a sum of local contributions weighted with the partition of unity:*

$$V_0 := \bigoplus_{j=1}^N R_j^T D_j \widetilde{Z}_{j\tau}\,. \tag{7.49}$$

Let $Z_0 \in \mathbb{R}^{\#\mathcal{N} \times dim(V_0)}$ be a column matrix so that V_0 is spanned by its columns. We denote its transpose by $R_0 := Z_0^T$.

Note that for all $W_0 \in V_0$, we have the following equality:

$$W_0 = R_0^T ((R_0 R_0^T)^{-1}) R_0 W_0\,.$$

Theorem 7.17 (GenEO stable decomposition). *Let $\mathbf{U} \in \mathbb{R}^{\mathcal{N}}$ for all subdomains $1 \le j \le N$; we define $\mathbf{U}_j$ like in (7.48) and $\mathbf{U}_0$ by*

$$\boxed{\mathbf{U}_0 := (R_0 R_0^T)^{-1} R_0 \left(\sum_{j=1}^N R_j^T D_j \tilde{\pi}_j R_j \mathbf{U} \right)}$$

so that

$$R_0^T \mathbf{U}_0 := \sum_{j=1}^N R_j^T D_j \tilde{\pi}_j R_j \mathbf{U}.$$

Let $c_T := (2 + (2k_0 + 1)k_1 \tau)^{-1}$ and let $\mathcal{U}$ denote $(\mathbf{U}_i)_{0 \le i \le N} \in H_D$.
Then, $\mathcal{U}$ is a c_T-stable decomposition of $\mathbf{U}$ since we have

$$\mathcal{R}_{ASM,2}(\mathcal{U}) = \sum_{j=0}^N R_j^T \mathbf{U}_j = \mathbf{U} \quad \text{and} \quad c_T b(\mathcal{U}, \mathcal{U}) \le \mathbf{U}^T A \mathbf{U} = a(\mathbf{U}, \mathbf{U}).$$

Proof. We first check that $\mathcal{R}(\mathcal{U}) = U$. By definition, we have

$$\begin{aligned}\mathbf{U} &= \sum_{j=1}^N R_j^T D_j R_j \mathbf{U} \\ &= \underbrace{\sum_{j=1}^N R_j^T D_j \tilde{\pi}_j R_j \mathbf{U}}_{R_0^T \mathbf{U}_0} + \sum_{j=1}^N R_j^T \underbrace{D_j (I_d - \tilde{\pi}_j) R_j \mathbf{U}}_{\mathbf{U}_j} = \sum_{j=0}^N R_j^T \mathbf{U}_j.\end{aligned}$$

We now prove the second part of the theorem, the stability of the decomposition. By Lemma 7.12, we have

$$b(\mathcal{U},\mathcal{U}) \le 2\,a(\mathbf{U},\mathbf{U}) + (2k_0+1)\sum_{i=1}^{N} a(R_i^T\mathbf{U}_i, R_i^T\mathbf{U}_i). \tag{7.50}$$

Then, by definition of $\mathbf{U}_i$, $1 \le i \le N$, in (7.48)

$$a(R_i^T\mathbf{U}_i, R_i^T\mathbf{U}_i) = \left(R_i^T D_i (I_d - \tilde{\pi}_i) R_i \mathbf{U}\right)^T A \left(R_i^T D_i (I_d - \tilde{\pi}_i) R_i \mathbf{U}\right) \tag{7.51}$$

and by Lemma 7.15 we thus have

$$a(R_i^T\mathbf{U}_i, R_i^T\mathbf{U}_i) \le \tau\,(R_i\mathbf{U})^T \widetilde{A}^i\,(R_i\mathbf{U})\,.$$

Summing over all subdomains and using Lemma 7.13, we have

$$\sum_{i=1}^{N} a(R_i^T\mathbf{U}_i, R_i^T\mathbf{U}_i) \le k_1\,\tau\,\mathbf{U}^T A\mathbf{U}.$$

Finally, using (7.50) and (7.51), we have

$$b(\mathcal{U},\mathcal{U}) \le (2 + (2k_0+1)k_1\,\tau)\,\mathbf{U}^T A\mathbf{U}. \qquad \square \tag{7.52}$$

Combining Lemma 7.11 and (7.52), we have thus proved the following.

Theorem 7.18. *The eigenvalues of the two-level Schwarz preconditioned system satisfy the following estimate:*

$$\boxed{\frac{1}{2+(2k_0+1)k_1\,\tau} \le \lambda(M_{ASM,2}^{-1}A) \le 2k_0\,.}$$

Due to the definition of the coarse space, we see that the condition number of the preconditioned problem will not depend on the number of the subdomains but only on the parameters k_0, k_1, and τ. Parameter τ can be chosen arbitrarily small at the expense of a large coarse space.

7.5 ▪ Hybrid Schwarz with GenEO

Another version of the Schwarz preconditioner, which we will call the hybrid Schwarz method (HSM), can also be formalized in the framework of the fictitious space lemma. The coarse space is as before based on the same generalized eigensolves. The difference lies in the way the coarse space correction is applied; see [132] and [184]. Let P_0 (see (7.58)) denote the a-orthogonal projection on the GenEO coarse space V_0 defined above (Definition 7.16). Let us denote by M_{ASM}^{-1} the one-level Schwarz method

$$M_{ASM}^{-1} := \sum_{i=1}^{N} R_i^T (R_i A R_i^T)^{-1} R_i\,.$$

Following [132], the hybrid Schwarz preconditioner is defined as follows:

$$\boxed{M_{HSM}^{-1} := R_0^T(R_0AR_0^T)^{-1}R_0 + (I_d - P_0)M_{ASM}^{-1}(I_d - P_0^T).} \tag{7.53}$$

We shall see at the end how to implement this preconditioner efficiently in a preconditioned conjugate gradient (PCG) method.

We first study the spectral properties of this preconditioner. In order to apply the fictitious space lemma (Lemma 7.4), it suffices in the previous setup to modify the linear operator $\mathcal{R}$ (7.5) and replace it by the following definition.

Definition 7.19. *For $\mathcal{U} = (\mathbf{U}_i)_{0\le i\le N}$, $\mathcal{R}_{HSM} : H_D \longrightarrow H$ is defined by*

$$\mathcal{R}_{HSM}(\mathcal{U}) := R_0^T\,\mathbf{U}_0 + \sum_{i=1}^{N}(I_d - P_0)R_i^T\,\mathbf{U}_i\,. \tag{7.54}$$

We check the assumptions of the fictitious space lemma when $\mathcal{R}_{HSM}$ replaces $\mathcal{R}_{ASM,2}$ in Definition 7.1. This will give us the condition number estimate (7.57).

Lemma 7.20 (Surjectivity of $\mathcal{R}_{HSM}$). *The operator $\mathcal{R}_{HSM}$ is surjective.*

Proof. For all $\mathbf{U} \in H$, we have

$$\mathbf{U} = P_0\,\mathbf{U} + (I_d - P_0)\mathbf{U} = P_0\,\mathbf{U} + \sum_{i=1}^{N}(I_d - P_0)R_i^T D_i R_i \mathbf{U}\,.$$

Since $P_0\,\mathbf{U} \in V_0$, there exists $\mathbf{U}_0 \in R^{\#\mathcal{N}_0}$ such that $P_0\,\mathbf{U} = R_0^T\,\mathbf{U}_0$. Thus, we have

$$\mathbf{U} = R_0^T\,\mathbf{U}_0 + \sum_{i=1}^{N}(I_d - P_0)R_i^T(D_i R_i \mathbf{U}),$$

or, in other words

$$\mathcal{R}_{HSM}(\mathbf{U}_0, (D_i R_i \mathbf{U})_{1\le i\le N}) = \mathbf{U},$$

which proves the surjectivity. □

Lemma 7.21 (Continuity of $\mathcal{R}_{HSM}$). *Let k_0 be defined as in (7.38).*
Then, for all $\mathcal{U} = (\mathbf{U}_i)_{0\le i\le N} \in H_D$, the following estimate holds:

$$a(\mathcal{R}_{HSM}(\mathcal{U}), \mathcal{R}_{HSM}(\mathcal{U})) \le k_0\, b(\mathcal{U}, \mathcal{U}). \tag{7.55}$$

Proof. Since P_0 and $I_d - P_0$ are a-orthogonal projections, we also have, making use of Lemma 7.9,

$$
\begin{aligned}
a(\mathcal{R}_{HSM}(\mathcal{U}),\mathcal{R}_{HSM}(\mathcal{U})) &= a(R_0^T\mathbf{U}_0,R_0^T\mathbf{U}_0)\\
&\quad + a\left((I_d-P_0)\sum_{i=1}^N R_i^T\mathbf{U}_i,(I_d-P_0)\sum_{i=1}^N R_i^T\mathbf{U}_i\right)\\
&\le a(R_0^T\mathbf{U}_0,R_0^T\mathbf{U}_0)+a\left(\sum_{i=1}^N R_i^T\mathbf{U}_i,\sum_{i=1}^N R_i^T\mathbf{U}_i\right)\\
&\le a(R_0^T\mathbf{U}_0,R_0^T\mathbf{U}_0)+k_0\sum_{i=1}^N a(R_i^T\mathbf{U}_i,R_i^T\mathbf{U}_i)\\
&\le k_0\sum_{i=0}^N a(R_i^T\mathbf{U}_i,R_i^T\mathbf{U}_i)=k_0\, b(\mathcal{U},\mathcal{U}). \qquad \square
\end{aligned}
$$

Lemma 7.22 (Stable decomposition with $\mathcal{R}_{HSM}$). *Let $\mathbf{U}\in H$. For $1\le j\le N$, we define*

$$\mathbf{U}_j := D_j(I_d-\tilde{\pi}_j)R_j\mathbf{U}$$

and $\mathbf{U}_0\in\mathbb{R}^{\mathcal{N}_0}$ such that

$$R_0^T\mathbf{U}_0=P_0\mathbf{U}.$$

We define $\mathcal{U}:=(\mathbf{U}_i)_{1\le i\le N}\in H_D$.

Then, we have

$$\mathcal{R}_{HSM}(\mathcal{U})=\mathbf{U}$$

and the stable decomposition property is verified:

$$b(\mathcal{U},\mathcal{U})\le(1+k_1\tau)a(\mathbf{U},\mathbf{U}). \tag{7.56}$$

Proof. We first check that we indeed have a decomposition, i.e., that the equality $\mathcal{R}_{HSM}(\mathcal{U})=\mathbf{U}$ holds. Note that for all $1\le j\le N$ we have

$$R_j^T D_j\tilde{\pi}_j R_j\mathbf{U}\in V_0\Rightarrow(I_d-P_0)R_j^T D_j\tilde{\pi}_j R_j\mathbf{U}=0.$$

We have

$$
\begin{aligned}
\mathbf{U} &= P_0\mathbf{U}+(I_d-P_0)\mathbf{U}\\
&= P_0\mathbf{U}+(I_d-P_0)\sum_{j=1}^N R_j^T D_j R_j\mathbf{U}\\
&= R_0^T\mathbf{U}_0+(I_d-P_0)\sum_{j=1}^N R_j^T D_j(I_d-\tilde{\pi}_j)R_j\mathbf{U}=\mathcal{R}_{HSM}(\mathcal{U}).
\end{aligned}
$$

The last thing to do is to check the stability of this decomposition. Using Lemma 7.15, Lemma 7.13 and the fact that P_0 is an a-orthogonal projection, we have

$$
\begin{aligned}
b(\mathcal{U},\mathcal{U}) &= a(R_0^T\mathbf{U}_0,R_0^T\mathbf{U}_0)\\
&\quad+\sum_{j=1}^N a(R_j^T\underbrace{D_j(I_d-\tilde{\pi}_j)R_j\mathbf{U}}_{\mathbf{U}_j},R_j^T\underbrace{D_j(I_d-\tilde{\pi}_j)R_j\mathbf{U}}_{\mathbf{U}_j})\\
&\le a(P_0\mathbf{U},P_0\mathbf{U})+\tau\sum_{j=1}^N\left(R_j\mathbf{U}\right)^T\tilde{A}^j\left(R_j\mathbf{U}\right)\\
&\le a(\mathbf{U},\mathbf{U})+k_1\tau a(\mathbf{U},\mathbf{U})\le(1+k_1\tau)a(\mathbf{U},\mathbf{U}). \qquad \square
\end{aligned}
$$

Previous lemmas lead to the condition number estimate of the following algorithm.

Theorem 7.23 (Hybrid Schwarz algorithm). *Let τ be a user-defined parameter for building the GenEO coarse space as in Definition 7.14.*

The eigenvalues of the hybrid Schwarz preconditioned system satisfy the following estimate:

$$\boxed{\frac{1}{1+k_1\,\tau} \leq \lambda(M_{HSM}^{-1}A) \leq k_0.} \tag{7.57}$$

Note that a hybrid preconditioner will lead to a better condition number than for the additive one (Theorem 7.18) and consequently to a faster convergence.

7.5.1 ▪ Efficient implementation of the hybrid Schwarz method

As written in (7.53), the application of the hybrid preconditioner involves several applications of the projection P_0. We see in this section that a clever choice of the initial guess in the PCG algorithm reduces the cost of the algorithm; see [132].

Note first that the matrix form of P_0 is

$$P_0 = R_0^T(R_0AR_0^T)^{-1}R_0A. \tag{7.58}$$

Indeed, this formula clearly defines a projection since

$$\begin{aligned}(R_0^T(R_0AR_0^T)^{-1}R_0A)^2 &= R_0^T(R_0AR_0^T)^{-1}R_0AR_0^T(R_0AR_0^T)^{-1}R_0A\\ &= R_0^T(R_0AR_0^T)^{-1}R_0A.\end{aligned}$$

This projection is A-orthogonal since for all vectors $\boldsymbol{U}$, $\boldsymbol{V}$, we have

$$\begin{aligned}(R_0^T(R_0AR_0^T)^{-1}R_0A\,\boldsymbol{U},\,\boldsymbol{V})_A &= (R_0^T(R_0AR_0^T)^{-1}R_0A\,\boldsymbol{U},A\boldsymbol{V})\\ &= (\boldsymbol{U},R_0^T(R_0AR_0^T)^{-1}R_0A\,\boldsymbol{V})_A.\end{aligned}$$

Finally, the range of $R_0^T(R_0AR_0^T)^{-1}R_0A$ is V_0 since for all $\boldsymbol{U} \in V_0$, there exists $\boldsymbol{W}$ such that $\boldsymbol{U} = R_0^T\boldsymbol{W}$ and we have

$$R_0^T(R_0AR_0^T)^{-1}R_0A\boldsymbol{U} = R_0^T(R_0AR_0^T)^{-1}R_0AR_0^T\boldsymbol{W} = R_0^T\boldsymbol{W} = \boldsymbol{U}.$$

From (7.58), we can rewrite definition (7.53) of M_{HSM}^{-1} as

$$M_{HSM}^{-1} := P_0A^{-1} + (I_d - P_0)M_{ASM}^{-1}(I_d - P_0^T).$$

Note also that

$$P_0^TA = AR_0^T(R_0AR_0^T)^{-1}R_0A = AP_0.$$

These relations yield the following expression for the preconditioned operator:

$$\boxed{M_{HSM}^{-1}A = P_0 + (I_d - P_0)M_{ASM}^{-1}A(I_d - P_0).} \tag{7.59}$$

When solving the linear system $A\boldsymbol{x} = \boldsymbol{b}$ with the PCG preconditioned by M_{HSM}^{-1} (7.53) the method seeks an approximation to the solution in the Krylov space

$$\mathcal{K}^n(M_{HSM}^{-1}A,\tilde{\boldsymbol{r}}_0) := \{\tilde{\boldsymbol{r}}_0, M_{HSM}^{-1}A\,\tilde{\boldsymbol{r}}_0, \ldots, (M_{HSM}^{-1}A)^{n-1}\tilde{\boldsymbol{r}}_0\},$$

where

$$\tilde{r}_0 := M_{HSM}^{-1}(b - Ax_0)$$

is the initial preconditioned residual. If $\tilde{r}_0$ is chosen so that

$$P_0 \tilde{r}_0 = 0,$$

we have a simplification in the expression of the Krylov subspace:

$$\mathcal{K}^n(M_{HSM}^{-1}A, \tilde{r}_0) = \{\tilde{r}_0, (I_d - P_0)M_{ASM}^{-1}A\tilde{r}_0, \ldots, ((I_d - P_0)M_{ASM}^{-1})^{n-1}\tilde{r}_0\}.$$

This can be easily proved using formula (7.59) and the fact that $P_0^2 = P_0$:

$$\begin{aligned} M_{HSM}^{-1}A\tilde{r}_0 &= (P_0 + (I_d - P_0)M_{ASM}^{-1}A(I_d - P_0))\tilde{r}_0 \\ &= (I_d - P_0)M_{ASM}^{-1}A\tilde{r}_0, \end{aligned}$$

$$\begin{aligned} (M_{HSM}^{-1}A)^2\tilde{r}_0 &= (P_0 + (I_d - P_0)M_{ASM}^{-1}A(I_d - P_0))(I_d - P_0)M_{ASM}^{-1}A\tilde{r}_0 \\ &= (I_d - P_0)M_{ASM}^{-1}A(I_d - P_0)M_{ASM}^{-1}A\tilde{r}_0 \\ &= ((I_d - P_0)M_{ASM}^{-1}A)^2\tilde{r}_0, \\ &\ \ \vdots \end{aligned}$$

This means that in the PCG method, it is sufficient to consider that the preconditioner is

$$(I_d - P_0)M_{ASM}^{-1}.$$

In order to have $P_0\tilde{r}_0 = 0$, we can choose, for example,

$$x_0 = R_0^T(R_0AR_0^T)^{-1}R_0 b.$$

By using (7.53) and (7.58) we see that

$$P_0M_{HSM}^{-1} = P_0R_0^T(R_0AR_0^T)^{-1}R_0 = R_0^T(R_0AR_0^T)^{-1}R_0,$$

which leads to

$$\begin{aligned} P_0\tilde{r}_0 &= P_0M_{HSM}^{-1}(b - AR_0^T(R_0AR_0^T)^{-1}R_0 b) \\ &= R_0^T(R_0AR_0^T)^{-1}R_0(b - AR_0^T(R_0AR_0^T)^{-1}R_0 b) \\ &= 0. \end{aligned}$$

To sum up, the PCG algorithm (see Algorithm 3.2 in section 3.3.1) for the hybrid Schwarz method takes the form given in Algorithm 7.1.

7.6 ▪ FreeFem++ implementation

We illustrate the GenEO coarse space on a Darcy problem with two layers of very high and different heterogeneities; see Figure 7.2:

$$-\operatorname{div}(\alpha\nabla u) = f \ \text{ in } \Omega, \tag{7.60}$$

Algorithm 7.1 PCG algorithm for the hybrid Schwarz method

Compute $x_0 := R_0^T(R_0AR_0^T)^{-1}R_0\, b$,
$\mathbf{r}_0 := b - A\mathbf{x}_0$,
$\mathbf{z}_0 = (I_d - P_0)M_{ASM}^{-1}\mathbf{r}_0$
$\mathbf{p}_0 = \mathbf{z}_0$.
for $i = 0, 1, \ldots$ **do**
  $\rho_i = (\mathbf{r}_i, \mathbf{z}_i)_2$
  $\mathbf{q}_i = A\mathbf{p}_i$
  $\alpha_i = \dfrac{\rho_i}{(\mathbf{p}_i, \mathbf{q}_i)_2}$
  $\mathbf{x}_{i+1} = \mathbf{x}_i + \alpha_i \mathbf{p}_i$
  $\mathbf{r}_{i+1} = \mathbf{r}_i - \alpha_i \mathbf{q}_i$
  $\mathbf{z}_{i+1} = (I_d - P_0)M_{ASM}^{-1}\mathbf{r}_{i+1}$
  $\rho_{i+1} = (\mathbf{r}_{i+1}, \mathbf{z}_{i+1})_2$
  $\beta_{i+1} = \dfrac{\rho_{i+1}}{\rho_i}$
  $\mathbf{p}_{i+1} = \mathbf{z}_{i+1} + \beta_{i+1}\mathbf{p}_i$
  check convergence; continue if necessary
end for

where $\alpha : \Omega \mapsto \mathbb{R}$ is defined by

$$\alpha = \begin{cases} 1.e6 & \text{if} \quad .2 < y < .4, \\ 1.e5 & \text{if} \quad .6 < y < .8, \\ 1 & \text{else.} \end{cases}$$

The implementation of the hybrid Schwarz (7.53) method is based on three FreeFem++ scripts given here. The coarse space (7.49) is built in `GENEO.idp`. We see that the main differences with respect to the standard coarse grid are the definition of the local data under the form of weighted matrices and the construction of the coarse space. The other parts are identical to the standard coarse spaces; only the size of the coarse space is bigger.

```
for(int i=0;i<npart;++i)
{
  mesh Thi = aTh[i];
  fespace Vhi(Thi,P1);
  Vhi[int] eV(abs(nev));
  real[int] ev(abs(nev));
  if (nev > 0){//GENEO coarse space
  int k = EigenValue(aN[i],aAweighted[i],sym=true,sigma=0,value=ev,vector=eV);
  cout << "Eigenvalues in the subdomain "<< i <<endl;
  k=min(k,nev);   //sometimes the no of converged eigenvalues is bigger than nev.
  cout << ev <<endl;
  }
  else// Nicolaides Coarse space
  {
    eV[0] = 1.;
  }

  for(int j=0;j<abs(nev);++j){
    real[int] zitemp = Dih[i]*eV[j][];
    int k = i*abs(nev)+j;
```

```
    Z[k][]=Rih[i]'*zitemp;
  }
}
```

Listing 7.1. `../THEORIECOARSEGRID/FreefemProgram/GENEO.idp`

In `PCG-CS.idp`, we define the preconditioner M_{HSM}^{-1} (7.53) and the preconditioned conjugate gradient method.

```
func real[int] AS2(real[int] & r){
    real[int] z = Q(r);
    real[int] aux = AS(r);
    z += aux;
     return z;
}
func real[int] BNN(real[int] &u) // precond BNN
{
    real[int] aux1 = Q(u);
    real[int] aux2 = P(u);
    real[int] aux3 = AS(aux2);//aux3 = AS(P(u))
    aux2 = PT(aux3);// aux2 = PT(AS(P(u)))
    aux2 += aux1;// aux2 = Q(u) + PT(AS(P(u)))
  return aux2;
}
/*# fin BNNPrecond #*/
/*# debutCGSolve #*/
func real[int] myPCG(real[int] xi,real eps, int nitermax)
{
  ofstream filei("Convprec.m"); // MATLAB convergence history file
  ofstream fileignu("Convprec.gnu"); // Gnuplot convergence history file
    Vh r, un, p, zr, zrel, rn, w, er;
    un[] = xi;
    r[] = A(un[]);
    r[] -= rhsglobal[];
    r[] *= -1.0;
    zr[] = BNN(r[]);
    real resinit=sqrt(zr[]'*zr[]);
    p = zr;
    for(int it=0;it<nitermax;++it)
      {
        plot(un,value=1,wait=1,fill=1,dim=3,cmm="Approximate solution at iteration "+it);
        real relres = sqrt(zr[]'*zr[])/resinit;
        cout << "It: "<< it << " Relative residual = " << relres << endl;
        int j = it+1;
        filei << "relres("+j+")=" << relres << ";" << endl;
        fileignu << relres << endl;
        if(relres < eps)
          {
           cout << "CG has converged in " + it + " iterations " << endl;
           cout << "The relative residual is " + relres << endl;
           break;
          }
        w[] = A(p[]);
        real alpha = r[]'*zr[];
        real aux2 = alpha;
        real aux3 = w[]'*p[];
        alpha /= aux3; // alpha = (rj,zj)/(Apj,pj);
        un[] += alpha*p[]; // xj+1 = xj + alpha*p;
```

```
        r[] -= alpha*w[]; // rj+1 = rj - alpha*Apj;
        zr[] = BNN(r[]); // zj+1 = M^-1*rj+1;
        real beta = r[]'*zr[];
        beta /= aux2; // beta = (rj+1,zj+1)/(rj,zj);
        p[] *= beta;
        p[] += zr[];
    }
    return un[];
}
```

Listing 7.2. ../THEORIECOARSEGRID/FreefemProgram/PCG-CS.idp

The script of the "main" program is given by AS2-PCG-GENEO.edp.

```
/*# debutPartition #*/
include "../../FreefemCommon/dataGENEO.edp"
include "../../FreefemCommon/decomp.idp"
include "../../FreefemCommon/createPartition.idp"
SubdomainsPartitionUnity(Th,part[],sizeovr,aTh,Rih,Dih,Ndeg,AreaThi);
plot(part,wait=1,fill=1,ps="partition.eps");
/*# endPartition #*/
/*# debutGlobalData #*/
Aglobal = vaglobal(Vh,Vh,solver = UMFPACK); // global matrix
rhsglobal[] = vaglobal(0,Vh); // global rhs
uglob[] = Aglobal^-1*rhsglobal[];
/*# finGlobalData #*/
/*# debutLocalData #*/
for(int i = 0;i<npart;++i)
  {
    cout << " Domain :" << i << "/" << npart << endl;
    matrix aT = Aglobal*Rih[i]';
    aA[i] = Rih[i]*aT;
    set(aA[i],solver = UMFPACK); // direct solvers using UMFPACK
    varf valocal(u,v) = int2d(aTh[i])(eta*u*v+ka*(Grad(u)'*Grad(v)))
                        +on(1,u=g);
    fespace Vhi(aTh[i],P1);
    aN[i]= valocal(Vhi,Vhi);
    set(aN[i], solver = UMFPACK);
    matrix atimesxi = aA[i] * Dih[i];
    aAweighted[i] = Dih[i] * atimesxi;
    set(aAweighted[i], solver = UMFPACK);
  }
/*# finLocalData #*/
/*# debutPCGSolve #*/
include "../../FreefemCommon/matvecAS.idp"
include "GENEO.idp"
include "PCG-CS.idp"
Vh un = 0, sol; // initial guess un and final solution
cout << "Schwarz Dirichlet algorithm" << endl;
sol[] = myPCG(un[], tol, maxit); // PCG with initial guess un
plot(sol,cmm=" Final solution", wait=1,dim=3,fill=1,value=1);
Vh er = sol-uglob;
cout << " Final relative error: " << er[].linfty/sol[].linfty << endl;
/*# finPCGSolve #*/
```

Listing 7.3. ../THEORIECOARSEGRID/FreefemProgram/AS2-PCG-GENEO.edp

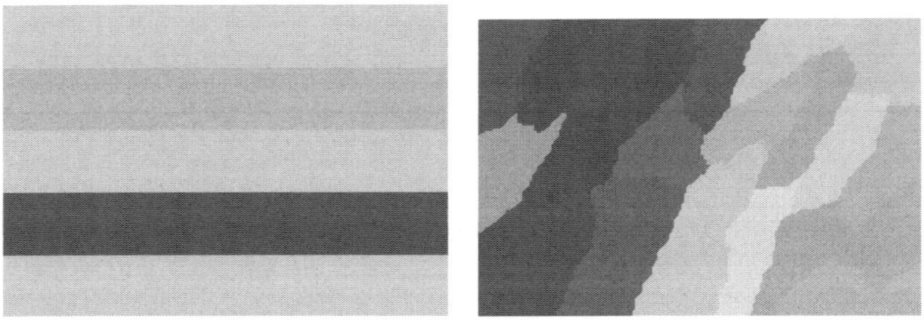

Figure 7.2. *Diffusion coefficient (left) and subdomain partition (right).*

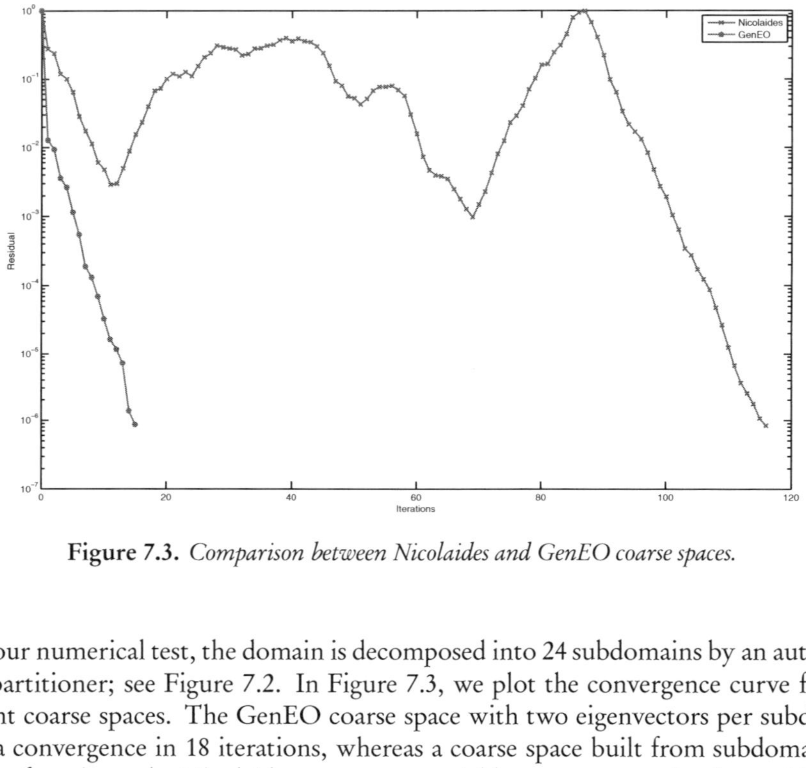

Figure 7.3. *Comparison between Nicolaides and GenEO coarse spaces.*

In our numerical test, the domain is decomposed into 24 subdomains by an automatic mesh partitioner; see Figure 7.2. In Figure 7.3, we plot the convergence curve for two different coarse spaces. The GenEO coarse space with two eigenvectors per subdomain yields a convergence in 18 iterations, whereas a coarse space built from subdomainwise constant functions, the Nicolaides coarse space, yields a convergence in almost 120 iterations.

7.7 ▪ SORAS-GenEO-2 coarse space for the optimized Schwarz method

7.7.1 ▪ Symmetrized ORAS method

The purpose of this section is to define a general framework for building an adaptive coarse space for optimized Schwarz methods (see Chapter 2) for decomposition into overlapping subdomains. We show that we can achieve the same robustness as we have in this chapter for Schwarz [178] and FETI-BDD [179] domain decomposition methods with GenEO coarse spaces. For this, a nonstandard symmetric variant of the ORAS method as well as two generalized eigenvalue problems are introduced in [106].

For all subdomains $1 \leq i \leq N$, let B_i be an SPD matrix of size $\#\mathcal{N}_i \times \#\mathcal{N}_i$, which comes typically from the discretization of boundary value local problems using optimized transmission conditions. The ORAS preconditioner (see Chapter 2), is defined as [181]

$$M_{ORAS,1}^{-1} := \sum_{i=1}^{N} R_i^T D_i B_i^{-1} R_i \,. \tag{7.61}$$

Due to matrices D_i, this preconditioner is not symmetric. Note that in the special case $B_i = R_i A R_i^T$, the ORAS algorithm reduces to the RAS algorithm [21]. The symmetrized variant of RAS is the additive Schwarz method (ASM), $M_{ASM,1}^{-1} := \sum_{i=1}^{N} R_i^T A_i^{-1} R_i$, which has been extensively studied and for which various coarse spaces have been analyzed. For the ORAS method, it seems at first glance that we should accordingly study the following symmetrized variant:

$$M_{OAS,1}^{-1} := \sum_{i=1}^{N} R_i^T B_i^{-1} R_i \,. \tag{7.62}$$

For reasons explained in Remark 7.7.1, we introduce another nonstandard variant of the ORAS preconditioner (7.61), the symmetrized ORAS (SORAS) algorithm:

$$M_{SORAS,1}^{-1} := \sum_{i=1}^{N} R_i^T D_i B_i^{-1} D_i R_i \,. \tag{7.63}$$

7.7.2 ▪ Two-level SORAS algorithm

In order to define the two-level SORAS algorithm, we introduce two generalized eigenvalue problems.

First, for all subdomains $1 \leq i \leq N$, we consider the following problem.

Definition 7.24.

$$\begin{aligned} &\textit{Find } (\mathbf{U}_{ik}, \mu_{ik}) \in \mathbb{R}^{\#\mathcal{N}_i} \setminus \{0\} \times \mathbb{R} \textit{ such that} \\ &D_i R_i A R_i^T D_i \mathbf{U}_{ik} = \mu_{ik} B_i \mathbf{U}_{ik} \;. \end{aligned} \tag{7.64}$$

Let $\gamma > 0$ be a user-defined threshold. We define $Z_{geneo}^{\gamma} \subset \mathbb{R}^{\#\mathcal{N}}$ as the vector space spanned by the family of vectors $(R_i^T D_i \mathbf{U}_{ik})_{\mu_{ik} > \gamma, 1 \leq i \leq N}$ corresponding to eigenvalues larger than γ.

In order to define the second generalized eigenvalue problem, we make use of "Neumann" subdomain matrices $(\widetilde{A}^j)_{1 \leq j \leq N}$ (see (7.37)).

Definition 7.25. *We introduce the following generalized eigenvalue problem:*

$$\begin{aligned} &\textit{Find } (\mathbf{V}_{jk}, \lambda_{jk}) \in \mathbb{R}^{\#\mathcal{N}_i} \setminus \{0\} \times \mathbb{R} \textit{ such that} \\ &\widetilde{A}^i \mathbf{V}_{ik} = \lambda_{ik} B_i \mathbf{V}_{ik} \;. \end{aligned} \tag{7.65}$$

Let $\tau > 0$ be a user-defined threshold. We define $Z_{geneo}^{\tau} \subset \mathbb{R}^{\#\mathcal{N}}$ as the vector space spanned by the family of vectors $(R_i^T D_i \mathbf{V}_{ik})_{\lambda_{ik} < \tau, 1 \leq i \leq N}$ corresponding to eigenvalues smaller than τ.

We are now ready to define the SORAS two-level preconditioner.

Definition 7.26 (The SORAS-GenEO-2 preconditioner). *Let P_0 denote the A-orthogonal projection on the coarse space*

$$Z_{GenEO\text{-}2} := Z^{\gamma}_{geneo} \bigoplus Z^{\tau}_{geneo}.$$

The two-level SORAS-GenEO 2 preconditioner is defined as follows:

$$M^{-1}_{SORAS,2} := P_0 A^{-1} + (I_d - P_0) \sum_{i=1}^{N} R_i^T D_i B_i^{-1} D_i R_i (I_d - P_0^T). \tag{7.66}$$

This definition is reminiscent of the balancing domain decomposition preconditioner [132] introduced for Schur complement-based methods. Note that the coarse space is now defined by two generalized eigenvalue problems instead of one in [178, 179] for ASM and FETI-BDD methods. We have the following theorem.

Theorem 7.27 (Spectral estimate for the SORAS-GenEO-2 preconditioner). *Let k_0 be the maximal multiplicity of the subdomain interactions (see (7.38)), k_1 be the maximal multiplicity of subdomains intersection (see Lemma 7.13) and let $\gamma, \tau > 0$ be arbitrary constants used in Definitions 7.25 and 7.26.*

Then, the eigenvalues of the two-level preconditioned operator satisfy the spectral estimate

$$\frac{1}{1+\dfrac{k_1}{\tau}} \le \lambda(M^{-1}_{SORAS,2} A) \le \max(1, k_0\, \gamma),$$

where $\lambda(M^{-1}_{SORAS,2} A)$ is an eigenvalue of the preconditioned operator.

The proof is based on the fictitious space lemma [150].

Remark 7.7.1. *An analysis of a two-level version of the preconditioner M^{-1}_{OAS} (7.62) following the same path yields the following two generalized eigenvalue problems:*

$$\begin{array}{c} \textit{Find } (\mathbf{U}_{jk}, \mu_{jk}) \in \mathbb{R}^{\#\mathcal{N}_i} \setminus \{0\} \times \mathbb{R} \textit{ such that} \\ A^i \mathbf{U}_{ik} = \mu_{ik} B_i \mathbf{U}_{ik}\,, \end{array}$$

and

$$\begin{array}{c} \textit{Find } (\mathbf{V}_{jk}, \lambda_{jk}) \in \mathbb{R}^{\#\mathcal{N}_i} \setminus \{0\} \times \mathbb{R} \textit{ such that} \\ \widetilde{A}^i \mathbf{V}_{ik} = \lambda_{ik} D_i B_i D_i \mathbf{V}_{ik}\,. \end{array}$$

In the general case for $1 \le i \le N$, matrices D_i may have zero entries for boundary degrees of freedom since they are related to a partition of unity. Moreover, very often matrices B_i and A_i differ only by the interface conditions, that is, for entries corresponding to boundary degrees of freedom. Therefore, matrix $D_i B_i D_i$ on the right-hand side of the last generalized eigenvalue problem is not impacted by the choice of the interface conditions of the one-level optimized Schwarz method. This cannot lead to efficient adaptive coarse spaces.

7.7.3 ▪ Nearly incompressible elasticity

Although our theory does not apply in a straightforward manner to saddle point problems, we use it for those difficult problems for which it is not possible to preserve both symmetry and positivity. Note that generalized eigenvalue problems (7.88) and (7.65)

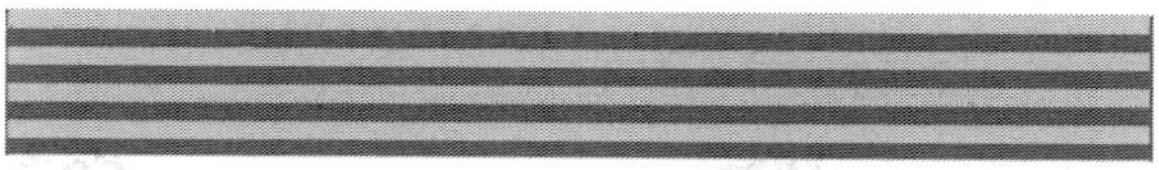

Figure 7.4. *2D Elasticity: coefficient distribution of steel and rubber.*

Table 7.1. *2D Elasticity GMRES iteration counts for a solid made of steel and rubber. Simulations made with FreeFem++* [108].

		AS	SORAS	AS+ZEM		SORAS +ZEM		AS-GenEO		SORAS geneo2	
dof	N	iter.	iter.	iter.	*dim*	iter.	*dim*	iter.	*dim*	iter.	*dim*
35841	8	150	184	117	*24*	79	*24*	110	*184*	13	*145*
70590	16	276	337	170	*48*	144	*48*	153	*400*	17	*303*
141375	32	497	>1000	261	*96*	200	*96*	171	*800*	22	*561*
279561	64	>1000	>1000	333	*192*	335	*192*	496	*1600*	24	*855*
561531	128	>1000	>1000	329	*384*	400	*384*	>1000	*2304*	29	*1220*
1077141	256	>1000	>1000	369	*768*	>1000	*768*	>1000	*3840*	36	*1971*

still make sense if A is the matrix of a saddle point problem and matrices B_i and $\widetilde{A}_i$ are properly defined for each subdomain $1 \le i \le N$. The new coarse space was tested quite successfully on Stokes and nearly incompressible elasticity problems with a discretization based on saddle point formulations in order to avoid locking phenomena. We first report 2D results for the latter case where comparisons are made with other methods in terms of iteration counts only. The mechanical properties of a solid can be characterized by its Young modulus E and Poisson ratio ν or alternatively by its Lamé coefficients λ and μ. These coefficients relate to each other by the following formulas:

$$\lambda = \frac{E\nu}{(1+\nu)(1-2\nu)} \quad \text{and} \quad \mu = \frac{E}{2(1+\nu)}. \tag{7.67}$$

The variational problem consists in finding $(\boldsymbol{u}_h, p_h) \in \mathcal{V}_h := \mathbb{P}_2^d \cap H_0^1(\Omega) \times \mathbb{P}_1$ such that for all $(\boldsymbol{v}_h, q_h) \in \mathcal{V}_h$

$$\begin{cases} \int_\Omega 2\mu \underline{\underline{\varepsilon}}(\boldsymbol{u}_h) : \underline{\underline{\varepsilon}}(\boldsymbol{v}_h) dx & -\int_\Omega p_h \operatorname{div}(\boldsymbol{v}_h) dx = \int_\Omega \boldsymbol{f}\, \boldsymbol{v}_h dx, \\ -\int_\Omega \operatorname{div}(\boldsymbol{u}_h) q_h dx & -\int_\Omega \frac{1}{\lambda} p_h q_h = 0. \end{cases} \tag{7.68}$$

Matrix $\widetilde{A}_i$ arises from the variational formulation (7.68) where the integration over domain Ω is replaced by the integration over subdomain Ω_i and the finite element space $\mathcal{V}_h$ is restricted to subdomain Ω_i. Matrix B_i corresponds to a Robin problem and is the sum of matrix $\widetilde{A}_i$ and the matrix of the following variational formulation restricted to the same finite element space:

$$\int_{\partial\Omega_i \setminus \partial\Omega} \frac{2\alpha\mu(2\mu+\lambda)}{\lambda + 3\mu} \boldsymbol{u}_h \cdot \boldsymbol{v}_h \text{ with } \alpha = 10 \text{ in our test.}$$

We test various domain decomposition methods for a heterogeneous beam of eight layers of steel $(E_1, \nu_1) = (210 \cdot 10^9, 0.3)$ and rubber $(E_2, \nu_2) = (0.1 \cdot 10^9, 0.4999)$; see Figure 7.4. The beam is clamped on its left and right sides. Iteration counts for a relative tolerance of 10^{-6} are given in Table 7.1. We compare the one-level additive Schwarz (AS) and SORAS methods, the two-level AS and SORAS methods with a coarse space consisting of rigid body motions which are zero-energy modes (ZEM), and finally AS with a GenEO coarse space and SORAS with the GenEO-2 coarse space defined in Definition 7.26 with $\tau = 0.4$

and $\gamma = 10^3$. Columns *dim* refer to the total size of the coarse space of a two-level method. Eigenvalue problem (7.65) accounts for roughly 90% of the GenEO-2 coarse space size.

We performed large 3D simulations on 8192 cores of an IBM Blue Gene/Q machine with 1.6 GHz Power A2 processors for both elasticity (200 million dofs in 200 sec.) and Stokes (200 million dofs in 150 sec.) equations. Computing facilities were provided by an IDRIS-GENCI project.

7.8 ▪ Balancing Neumann–Neumann

In the same spirit as in the previous sections of this chapter, the purpose is to reformulate the classical Neumann–Neumann algorithms in the framework of the fictitious space lemma. Note that GenEO versions of the Neumann–Neumann and FETI algorithms were first introduced in [179] and then analyzed in [180].

We first recall the Neumann–Neumann algorithm introduced in Chapter 6. The aim is to solve the substructured system (6.40),

$$\mathbb{S}\,U_\Gamma = G_\Gamma \in \mathbb{R}^{\#\mathcal{N}_\Gamma}, \tag{7.69}$$

where $\mathbb{S}$ is the Schur complement of the global matrix (6.39). This system is defined on the interface degrees of freedom indexed by $\mathcal{N}_\Gamma$. The set $\mathcal{N}_\Gamma$ is decomposed into subdomain interface degrees of freedom:

$$\mathcal{N}_\Gamma = \bigcup_{i=1}^{N} \mathcal{N}_{\Gamma_i}.$$

For each subdomain $1 \le i \le N$, let R_{Γ_i} be the restriction operator from the skeleton $\mathbb{R}^{\#\mathcal{N}_\Gamma}$ onto the subdomain interface $\mathbb{R}^{\#\mathcal{N}_{\Gamma_i}}$, and let $D_{\Gamma_i} \in \mathbb{R}^{\#\mathcal{N}_{\Gamma_i} \times \#\mathcal{N}_{\Gamma_i}}$ be invertible diagonal matrices so that we have the following partition of unity on the skeleton:

$$\sum_{i=1}^{N} R_{\Gamma_i}^T D_{\Gamma_i} R_{\Gamma_i} = I_{\mathbb{R}^{\#\mathcal{N}_\Gamma}}. \tag{7.70}$$

In this context, a Schwarz preconditioner for problem (7.69),

$$M_{ASM,2}^{-1} := \sum_{i=1}^{N} R_{\Gamma_i}^T (R_{\Gamma_i} \mathbb{S} R_{\Gamma_i}^T)^{-1} R_{\Gamma_i},$$

has been considered in [97, 98]. The Neumann–Neumann (NN) preconditioner works differently; see Definition 6.13. It is based on a decomposition of the global Schur complement operator $\mathbb{S}$ into a sum of symmetric positive (but usually not definite) operators. Let $(\mathbb{S}_i)_{1\le i\le N}$ be given by (6.41); recall that we have

$$\mathbb{S} = \sum_{i=1}^{N} \mathbb{S}_i.$$

In section 6.4, we have defined local operators for $1 \le i \le N$:

$$S_i := R_{\Gamma_i} \mathbb{S}_i R_{\Gamma_i}^T,$$

where $S_i \in \mathbb{R}^{\#\mathcal{N}_{\Gamma_i} \times \#\mathcal{N}_{\Gamma_i}}$ are subdomain contributions. Thus, we have

$$\mathbb{S} = \sum_{i=1}^{N} R_{\Gamma_i}^T S_i R_{\Gamma_i} \tag{7.71}$$

and (7.69) can be rewritten as

$$\sum_{i=1}^{N} R_{\Gamma_i}^T S_i R_{\Gamma_i} U_\Gamma = G_\Gamma . \tag{7.72}$$

When the operators S_i are invertible, the Neumann-Neumann preconditioner [14] is defined as

$$M_{NN}^{-1} := \sum_{i=1}^{N} R_{\Gamma_i}^T D_{\Gamma_i} S_i^{-1} D_{\Gamma_i} R_{\Gamma_i} . \tag{7.73}$$

7.8.1 ▪ Easy Neumann–Neumann

We start with the analysis of the one-level Neumann-Neumann (NN) method assuming that the local matrices S_i are invertible.

Definition 7.28 (NN components in the fictitious space lemma). *In the framework of the fictitious space lemma (Lemma 7.4) we define*

- *the space of the skeleton*

$$H := \mathbb{R}^{\#\mathcal{N}_\Gamma}$$

endowed with the standard Euclidean scalar product and the bilinear form $a : H \times H \mapsto \mathbb{R}$

$$a(U,V) := (\mathbb{S}\,U,\, V) \quad \forall U, V \in H,$$

- *the product space of the subdomain interfaces*

$$H_D := \prod_{i=1}^{N} \mathbb{R}^{\mathcal{N}_{\Gamma_i}}$$

endowed with the standard Euclidean scalar product and the bilinear form b

$$\begin{array}{rcl} b : H_D \times H_D & \longrightarrow & \mathbb{R}, \\ ((U_i)_{1\le i\le N}, (V_i)_{1\le i\le N}) & \longmapsto & \displaystyle\sum_{i=1}^{N} (S_i U_i, V_i), \end{array}$$

- *the linear operator* $\mathcal{R}_{NN}$ *as*

$$\begin{aligned} \mathcal{R}_{NN} : H_D &\longrightarrow H, \\ (U_i)_{1\le i\le N} &\mapsto \sum_{i=1}^{N} R_{\Gamma_i}^T D_{\Gamma_i} U_i \end{aligned} \tag{7.74}$$

With this notation, the preconditioner M_{NN}^{-1} *is given by* $\mathcal{R}_{NN} B^{-1} \mathcal{R}_{NN}^*$.

Note that the operator $\mathcal{R}_{NN}$ is surjective since from (7.70) we have for all $U \in H$

$$U = \sum_{i=1}^{N} R_{\Gamma_i}^T D_{\Gamma_i} R_{\Gamma_i} U = \mathcal{R}_{NN}((R_{\Gamma_i} U)_{1\le i\le N}) .$$

Contrarily to the Schwarz method, the stable decomposition is easily checked and is satisfied with $c_T = 1$. Indeed, letting $U \in H$, we have the natural decomposition $U = \sum_{i=1}^{N} R_{\Gamma_i}^T D_{\Gamma_i} R_{\Gamma_i} U$. In other words, letting $\mathcal{U} := (R_{\Gamma_i} U)_{1\le i\le N}$, we have $U = \mathcal{R}_{NN}(\mathcal{U})$. By definition of the bilinear form b we have

$$b(\mathcal{U},\mathcal{U})=\sum_{i=1}^{N}(S_i R_{\Gamma_i} U, R_{\Gamma_i} U)=\left(\sum_{i=1}^{N} R_{\Gamma_i}^T S_i R_{\Gamma_i} U, U\right)=(\mathbb{S}U,U)=a(U,U).$$

We now study the stability of $\mathcal{R}_{NN}$, which is more intricate.

Lemma 7.29 (Continuity of the operator $\mathcal{R}_{NN}$). *Let*

$$\boxed{k_2 := \max_{1\le i\le N} \#\left\{ j \mid R_{\Gamma_j} \mathbb{S} R_{\Gamma_i}^T \neq 0 \right\}} \tag{7.75}$$

be the maximum multiplicity of the interaction between subinterfaces via the Schur complement operator. Let also

$$\tau_{\max} := \max_{1\le i\le N} \max_{U_i\in\mathbb{R}^{\#\mathcal{N}_i}\setminus\{0\}} \frac{(\mathbb{S} R_{\Gamma_i}^T D_{\Gamma_i} U_i, R_{\Gamma_i}^T D_{\Gamma_i} U_i)}{(S_i U_i, U_i)}. \tag{7.76}$$

Then, for all $\mathcal{U}\in H_D$ we have

$$a(\mathcal{R}_{NN}(\mathcal{U}), \mathcal{R}_{NN}(\mathcal{U})) \le c_R\, b(\mathcal{U},\mathcal{U})$$

with $c_R := k_2\, \tau_{\max}$.

Proof. Let $\mathcal{U}=(U_i)_{1\le i\le N}\in H_D$; then

$$a(\mathcal{R}_{NN}(\mathcal{U}),\mathcal{R}_{NN}(\mathcal{U}))=\left(\mathbb{S}\left(\sum_{i=1}^{N} R_{\Gamma_i}^T D_{\Gamma_i} U_i\right), \sum_{j=1}^{N} R_{\Gamma_j}^T D_{\Gamma_j} U_j\right).$$

It is sufficient to apply Lemma 7.9, where matrix A is $\mathbb{S}$, to get

$$a(\mathcal{R}_{NN}(\mathcal{U}),\mathcal{R}_{NN}(\mathcal{U}))\le k_2 \sum_{i=1}^{N}\left(\mathbb{S} R_{\Gamma_i}^T D_{\Gamma_i} U_i, R_{\Gamma_i}^T D_{\Gamma_i} U_i\right). \tag{7.77}$$

We now estimate the continuity constant of the linear operator $\mathcal{R}_{NN}$. From (7.77) and the definition of $\tau_{\max}$, we have

$$\begin{aligned} a(\mathcal{R}_{NN}(\mathcal{U}),\mathcal{R}_{NN}(\mathcal{U})) &\le k_2 \sum_{i=1}^{N}(\mathbb{S} R_{\Gamma_i}^T D_{\Gamma_i} U_i, R_{\Gamma_i}^T D_{\Gamma_i} U_i)\\ &\le k_2\, \tau_{\max} \sum_{i=1}^{N}(S_i U_i, U_i) = k_2\, \tau_{\max}\, b(\mathcal{U},\mathcal{U}). \qquad \square \end{aligned}$$

Note that k_2 is typically larger than k_0 defined in (7.38) as the number of neighbors of a subdomain. This is due to the sparsity pattern of $\mathbb{S}$ as illustrated in Figure 7.5 for a $P1$ finite element discretization. The entries of $\mathbb{S} R_{\Gamma_i}^T$ are not zero on the interfaces drawn with a continuous line. On the dashed interfaces, the entries of $\mathbb{S} R_{\Gamma_i}^T$ are zero. Note that k_2 is typically the number of neighbors j of neighbors k of a subdomain i as can be seen from the equality

$$k_2 = \max_{1\le i\le N} \#\left\{ j \mid \sum_{k=1}^{N} R_{\Gamma_j} R_{\Gamma_k}^T S_k R_{\Gamma_k} R_{\Gamma_i}^T \neq 0 \right\} \tag{7.78}$$

obtained using formula (7.71) in the definition of k_2 (7.75).

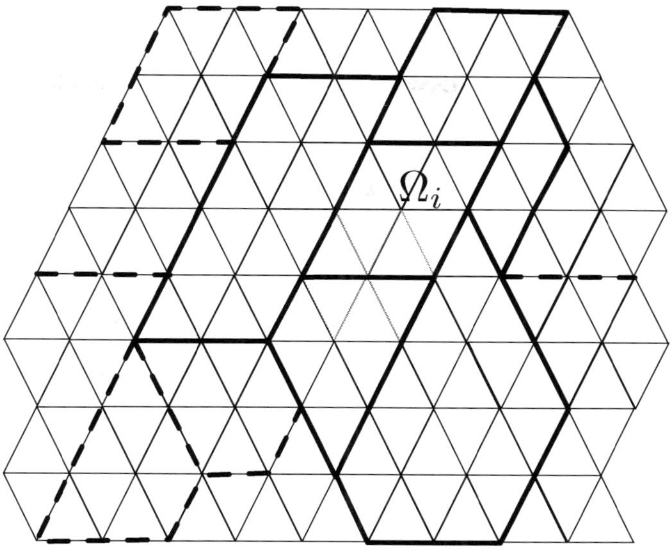

Figure 7.5. *Interface skeleton and support of* $\mathbb{S} R_{\Gamma_i}^T$.

By applying the fictitious space lemma (Lemma 7.4) we have just proved the following.

Theorem 7.30 (Spectral estimate for easy Neumann-Neumann). *Let* $\tau_{\max}$ *be defined by* (7.76) *and* k_2 *by* (7.75). *Then, the eigenvalues of the Neumann-Neumann preconditioned* (7.73) *system satisfy the following estimate:*

$$\boxed{1 \le \lambda(M_{NN}^{-1}\,\mathbb{S}) \le k_2\,\tau_{\max}.}$$

This result is of little practical use since it assumes that the local Neumann subproblems are well-posed, which is not always the case. Moreover, we have studied only the one-level method.

We address the former issue in the next section and a GenEO coarse space construction is given in section 7.8.3.

7.8.2 ▪ Neumann–Neumann with ill-posed subproblems

In the case of a Poisson problem and for floating subdomains (subdomains which do not touch the boundary of the global domain), constant functions are in the kernel of the Neumann subproblems, and thus the local Schur complements have a nontrivial kernel. For elasticity problems, one has to consider rigid body motions. Thus, we have to take into account the possibility of having a nonzero kernel for the local Schur complement S_i, $1 \le i \le N$, since for stationary problems this will often be the case. For each subdomain $1 \le i \le N$, the operator S_i is symmetric and we have the orthogonal direct sum (denoted by the symbol $\overset{\perp}{\oplus}$)

$$\mathbb{R}^{\mathcal{N}_{\Gamma_i}} = \ker S_i \overset{\perp}{\oplus} \operatorname{range} S_i$$

and we denote by

$$\boxed{P_i \text{ the orthogonal projection from } \mathbb{R}^{\mathcal{N}_{\Gamma_i}} \text{ on } \ker S_i \text{ and parallel to } \operatorname{range} S_i\,.}$$

Note that S_i induces an isomorphism from range S_i into itself whose inverse will be denoted by $S_i^\dagger$.

$$S_i^\dagger : \text{range}\, S_i \to \text{range}\, S_i \,.$$

In order to capture the part of the solution that will come from the local kernels S_i ($1 \le i \le N$), let Z_i be a rectangular matrix of size $\#\mathcal{N}_{\Gamma_i} \times \dim(\ker S_i)$ whose columns are a basis of $\ker S_i$. We form a rectangular matrix Z_0 of size $\#\mathcal{N}_\Gamma \times \sum_{i=1}^N \dim(\ker S_i)$ by concatenation of the Z_i's:

$$Z_0 := (R_{\Gamma_i}^T D_{\Gamma_i} Z_i)_{1\le i\le N} \,.$$

Let W_0 be the vector space spanned by the columns of Z_0. We introduce the

projection P_0 from $\mathbb{R}^{\#\mathcal{N}_\Gamma}$ on W_0 and which is $\mathbb{S}$ orthogonal.

Note that if Z_0 is full rank, the matrix form of the linear operator P_0 is

$$P_0 = Z_0(Z_0^T \mathbb{S} Z_0)^{-1} Z_0^T \mathbb{S} \,. \tag{7.79}$$

Definition 7.31 (Balancing NN and the fictitious space lemma). *In the framework of the fictitious space lemma (Lemma 7.4) we define the following:*

- *The Hilbert space*

$$H := \mathbb{R}^{\mathcal{N}_\Gamma}$$

with its standard Euclidean scalar product denoted by $(\cdot,\cdot)$ is also endowed with the bilinear form a defined as

$$a(\mathbf{U},\mathbf{V}) := \mathbf{V}^T \mathbb{S} \mathbf{U} \,.$$

- *The product space defined by*

$$H_D := W_0 \times \prod_{i=1}^N \text{range}\, S_i$$

with its standard Euclidean scalar product denoted by $(\cdot,\cdot)$ is also endowed with the bilinear form b defined as

$$\begin{aligned} b : H_D \times H_D &\longrightarrow \mathbb{R}, \\ ((\mathbf{U}_i)_{0\le i\le N}, (\mathbf{V}_i)_{0\le i\le N}) &\longmapsto \mathbf{V}_0^T \mathbb{S} \mathbf{U}_0 + \sum_{i=1}^N \mathbf{V}_i^T S_i \mathbf{U}_i \,. \end{aligned}$$

- *The linear operator $\mathcal{R}_{BNN}$ is defined as*

$$\begin{aligned} \mathcal{R}_{BNN} : H_D &\longrightarrow H, \\ (\mathbf{U}_i)_{0\le i\le N} &\longmapsto \mathbf{U}_0 + (I_d - P_0) \sum_{i=1}^N R_{\Gamma_i}^T D_{\Gamma_i} \mathbf{U}_i \,. \end{aligned} \tag{7.80}$$

Lemma 7.32 (Balancing Neumann–Neumann preconditioner). *Using the notation in Definition 7.31, the balancing Neumann–Neumann preconditioner is*

$$\boxed{\mathcal{R}_{BNN} B^{-1} \mathcal{R}_{BNN}^* = P_0 \mathbb{S}^{-1} + (I_d - P_0) \sum_{i=1}^N R_{\Gamma_i}^T D_{\Gamma_i} S_i^\dagger (I_d - P_i) D_{\Gamma_i} R_{\Gamma_i} (I_d - P_0)^T .} \tag{7.81}$$

Proof. We first need to find $\mathcal{R}^*_{BNN}$. By definition, we have

$$\mathcal{R}^*_{BNN} : H \longrightarrow H_D,\\ \mathbf{U} \longmapsto \mathcal{R}^*_{BNN}(\mathbf{U})$$

such that

$$\forall \mathcal{V} \in H_D,\ \mathcal{V}^T \mathcal{R}^*_{BNN}(\mathbf{U}) = \mathcal{R}_{BNN}(\mathcal{V})^T \mathbf{U}.$$

For all $\mathcal{V} := (\mathbf{V}_i)_{0\le i\le N} \in H_D$, the above equation is

$$\mathbf{V}_0^T \mathcal{R}^*_{BNN}(\mathbf{U})_0 + \sum_{i=1}^N \mathbf{V}_i^T \mathcal{R}^*_{BNN}(\mathbf{U})_i = \left(\mathbf{V}_0 + (I_d - P_0)\sum_{i=1}^N R_{\Gamma_i}^T D_{\Gamma_i} \mathbf{V}_i\right)^T \mathbf{U}$$
$$= \mathbf{V}_0^T \mathbf{U} + \sum_{i=1}^N \mathbf{V}_i^T D_{\Gamma_i} R_{\Gamma_i} (I_d - P_0)^T \mathbf{U}.$$

Since H_D is a product space, for arbitrary $V_0 \in W_0$ and $V_i \in \mathbb{R}^{\#\mathcal{N}_{\Gamma_i}}$ $(1 \le i \le N)$ we can identify each term in the sum. Thus we have

$$\mathcal{R}^*_{BNN}(\mathbf{U})_0 = P_0 \mathbf{U}$$

and since, for all $1 \le i \le N$, $I_d - P_i$ is the orthogonal projection from $\mathbb{R}^{\mathcal{N}_{\Gamma_i}}$ on range S_i, we have

$$\mathcal{R}^*_{BNN}(\mathbf{U})_i = (I_d - P_i) D_{\Gamma_i} R_{\Gamma_i} (I_d - P_0)^T \mathbf{U}.$$

We now identify operator $B : H_D \to H_D$ related to the bilinear form b and then B^{-1}. Operator B is a block diagonal operator whose diagonal entries are denoted by $(B_i)_{0\le i\le N}$. It is clear that for all subdomains $1 \le i \le N$, B_i is the restriction of S_i from range S_i into itself whose inverse is $S_i^\dagger$. As for B_0, we actually identify its inverse directly. For all U_0, $V_0 \in W_0$, we have

$$(V_0, U_0)_{\mathbb{S}} = (V_0, \mathbb{S}U_0) = (V_0, B_0 U_0)$$
$$= (V_0, \mathbb{S}\mathbb{S}^{-1} B_0 U_0) = (V_0, \mathbb{S}^{-1} B_0 U_0)_{\mathbb{S}} = (V_0, P_0 \mathbb{S}^{-1} B_0 U_0)_{\mathbb{S}}.$$

We can infer that $P_0 \mathbb{S}^{-1} B_0 : W_0 \to W_0$ is the identity operator so that $B_0^{-1} = P_0 \mathbb{S}^{-1}$. Note that when formula (7.79) is valid, we have

$$P_0 \mathbb{S}^{-1} = Z_0 (Z_0^T \mathbb{S} Z_0)^{-1} Z_0^T.$$

This yields the final form of the preconditioner $\mathcal{R}_{BNN} B^{-1} \mathcal{R}^*_{BNN}$ which is called the balancing Neumann–Neumann preconditioner; see [133, 125] and [65]:

$$\boxed{M_{BNN}^{-1} := P_0 \mathbb{S}^{-1} + (I_d - P_0) \sum_{i=1}^N R_{\Gamma_i}^T D_{\Gamma_i} S_i^\dagger (I_d - P_i) D_{\Gamma_i} R_{\Gamma_i} (I_d - P_0)^T.} \qquad \square \qquad (7.82)$$

We now check the assumptions of the fictitious space lemma (Lemma 7.4).

Lemma 7.33 (Surjectivity of $\mathcal{R}_{BNN}$). *The operator $\mathcal{R}_{BNN}$ is surjective.*

Proof. From the partition of unity (7.70), we have for all $\mathbf{U} \in H$

$$\begin{aligned}
\mathbf{U} &= P_0\mathbf{U} + (I_d - P_0)\mathbf{U} = P_0\mathbf{U} + (I_d - P_0)(I_d - P_0)\mathbf{U} \\
&= P_0\mathbf{U} + (I_d - P_0)\sum_{i=1}^{N} R_{\Gamma_i}^T D_{\Gamma_i} R_{\Gamma_i}(I_d - P_0)\mathbf{U} \\
&= P_0\mathbf{U} + (I_d - P_0)\sum_{i=1}^{N} R_{\Gamma_i}^T D_{\Gamma_i}(I_d - P_i) R_{\Gamma_i}(I_d - P_0)\mathbf{U} \\
&\quad + (I_d - P_0)\sum_{i=1}^{N} R_{\Gamma_i}^T D_{\Gamma_i} P_i\, R_{\Gamma_i}(I_d - P_0)\mathbf{U}.
\end{aligned} \tag{7.83}$$

The last term is zero since $P_i\, R_{\Gamma_i}(I_d - P_0)\mathbf{U} \in \ker S_i$, and thus

$$\sum_{i=1}^{N} R_{\Gamma_i}^T D_{\Gamma_i} P_i\, R_{\Gamma_i}(I_d - P_0)\mathbf{U} \in W_0$$

and $I_d - P_0$ is a projection parallel to W_0. Finally we have proved that for a given $\mathbf{U} \in H$, if we define $\mathcal{U} \in H_D$ by

$$\mathcal{U} := (P_0\mathbf{U}, ((I_d - P_i)R_{\Gamma_i}(I_d - P_0)\mathbf{U})_{1\le i\le N}), \tag{7.84}$$

we have $\mathcal{R}_{BNN}(\mathcal{U}) = \mathbf{U}$. □

Lemma 7.34 (Stable decomposition). *The stable decomposition property* (7.10) *is verified with an optimal constant* $c_T = 1$.

Proof. Let $\mathbf{U} \in H$. We use the decomposition defined in (7.84). By using $S_i\, P_i \equiv 0$ (P_i projects on $\ker S_i$), the symmetry of S_i, (7.71), and the $\mathbb{S}$ orthogonality of projection P_0, we have

$$\begin{aligned}
b(\mathcal{U}, \mathcal{U}) &= (\mathbb{S}P_0\mathbf{U}, P_0\mathbf{U}) \\
&\quad + \sum_{i=1}^{N}(S_i(I_d - P_i)R_{\Gamma_i}(I_d - P_0)\mathbf{U}, (I_d - P_i)R_{\Gamma_i}(I_d - P_0)\mathbf{U}) \\
&= (\mathbb{S}P_0\mathbf{U}, P_0\mathbf{U}) \\
&\quad + \sum_{i=1}^{N}(S_i R_{\Gamma_i}(I_d - P_0)\mathbf{U}, (I_d - P_i)R_{\Gamma_i}(I_d - P_0)\mathbf{U}) \\
&= (\mathbb{S}P_0\mathbf{U}, P_0\mathbf{U}) + \sum_{i=1}^{N}(S_i R_{\Gamma_i}(I_d - P_0)\mathbf{U}, R_{\Gamma_i}(I_d - P_0)\mathbf{U}) \\
&= (\mathbb{S}P_0\mathbf{U}, P_0\mathbf{U}) + (\mathbb{S}(I_d - P_0)\mathbf{U}, (I_d - P_0)\mathbf{U}) \\
&= (\mathbb{S}\mathbf{U}, \mathbf{U}) = a(\mathbf{U}, \mathbf{U}). \quad \square
\end{aligned} \tag{7.85}$$

We now consider the stability of the linear operator $\mathcal{R}_{BNN}$.

Lemma 7.35 (Continuity of $\mathcal{R}_{BNN}$). *Let* k_2 *be defined as in* (7.75). *Let* $\tau_{\max}$ *be defined by*

$$\tau_{\max} := \max_{1\le i\le N} \max_{\mathbf{U}_i \in \mathrm{range} S_i \setminus\{0\}} \frac{(\mathbb{S} R_{\Gamma_i}^T D_{\Gamma_i}\mathbf{U}_i, R_{\Gamma_i}^T D_{\Gamma_i}\mathbf{U}_i)}{(S_i\mathbf{U}_i, \mathbf{U}_i)}. \tag{7.86}$$

Then, the stability property (7.9) *is verified with the constant* $c_R = \max(1, k_2\tau_{\max})$.

Proof. Let $\mathcal{U} \in H_D$; by the $\mathbb{S}$-orthogonality of the projection P_0 and Lemma 7.9 with A substituted by $\mathbb{S}$, we have

$$\begin{aligned}
a(\mathcal{R}_{BNN}(\mathcal{U}), \mathcal{R}_{BNN}(\mathcal{U})) &= (\mathbb{S}\mathcal{R}_{BNN}(\mathcal{U}), \mathcal{R}_{BNN}(\mathcal{U}))\\
&= \left(\mathbb{S}(P_0\mathbf{U}_0 + (I_d - P_0)\sum_{i=1}^{N} R_{\Gamma_i}^T D_{\Gamma_i}\mathbf{U}_i), P_0\mathbf{U}_0 + (I_d - P_0)\sum_{i=1}^{N} R_{\Gamma_i}^T D_{\Gamma_i}\mathbf{U}_i\right)\\
&= (\mathbb{S}P_0\mathbf{U}_0, P_0\mathbf{U}_0) + \left(\mathbb{S}(I_d - P_0)\sum_{i=1}^{N} R_{\Gamma_i}^T D_{\Gamma_i}\mathbf{U}_i, (I_d - P_0)\sum_{i=1}^{N} R_{\Gamma_i}^T D_{\Gamma_i}\mathbf{U}_i\right)\\
&\leq (\mathbb{S}\mathbf{U}_0, \mathbf{U}_0) + \left(\mathbb{S}\sum_{i=1}^{N} R_{\Gamma_i}^T D_{\Gamma_i}\mathbf{U}_i, \sum_{i=1}^{N} R_{\Gamma_i}^T D_{\Gamma_i}\mathbf{U}_i\right) \qquad (7.87)\\
&\leq (\mathbb{S}\mathbf{U}_0, \mathbf{U}_0) + k_2\sum_{i=1}^{N}\left(\mathbb{S}R_{\Gamma_i}^T D_{\Gamma_i}\mathbf{U}_i, R_{\Gamma_i}^T D_{\Gamma_i}\mathbf{U}_i\right)\\
&\leq \max(1, k_2\tau_{\max})\left((\mathbb{S}\mathbf{U}_0, \mathbf{U}_0) + \sum_{i=1}^{N}(S_i\mathbf{U}_i, \mathbf{U}_i)\right)\\
&= \max(1, k_2\tau_{\max})\, b(\mathcal{U}, \mathcal{U}),
\end{aligned}$$

where $\tau_{\max}$ is defined by (7.86). □

By applying the fictitious space lemma (Lemma 7.4) we have just proved the following.

Theorem 7.36 (Spectral estimate for BNN). *Let* $\tau_{\max}$ *be defined by* (7.86) *and* k_2 *by* (7.75). *Then, the eigenvalues of the Neumann–Neumann preconditioned* (7.82) *system satisfy the following estimate:*

$$\boxed{1 \leq \lambda(M_{BNN}^{-1}\,\mathbb{S}) \leq \max(1, k_2\,\tau_{\max}).}$$

Constant $\tau_{\max}$ in (7.86) can be large, and thus the balancing Neumann–Neumann preconditioner (7.82) can be inefficient. For this reason, in the next section, we define a coarse space which allows us to guarantee any targeted convergence rate.

7.8.3 ▪ GenEO BNN

We introduce an adaptive coarse space based on a generalized eigenvalue problem on the interface of each subdomain $1 \leq i \leq N$ which allows us to guarantee a targeted convergence rate. We introduce the following eigenvalue problem:

$$\begin{aligned}
&\text{Find } (\mathbf{U}_{ik}, \mu_{ik}) \in \mathbb{R}^{\#\mathcal{N}_{\Gamma_i}} \setminus \{0\} \times \mathbb{R} \text{ such that}\\
&S_i\,\mathbf{U}_{ik} = \mu_{ik}\, D_{\Gamma_i} R_{\Gamma_i} \mathbb{S} R_{\Gamma_i}^T D_{\Gamma_i} \mathbf{U}_{ik}\,.
\end{aligned} \qquad (7.88)$$

Matrix $D_{\Gamma_i} R_{\Gamma_i} \mathbb{S} R_{\Gamma_i}^T D_{\Gamma_i}$ is symmetric positive definite (D_{Γ_i} is invertible) so that we can apply Lemma 7.6 to it. Let $\tau > 0$ be a user-defined threshold; for each subdomain $1 \leq i \leq N$, we introduce a subspace of $W_i \subset \mathbb{R}^{\mathcal{N}_{\Gamma_i}}$:

$$W_i := \operatorname{Span}\left\{\mathbf{U}_{ik} \mid \mu_{ik} < \frac{1}{\tau}\right\}. \qquad (7.89)$$

Note that the zero eigenvalue ($\mu_{ik} = 0$) corresponds to the kernel of S_i so that we have $\ker(S_i) \subset W_i$. Now, let

$$\boxed{\xi_i \text{ denote projection from } \mathbb{R}^{\mathcal{N}_{\Gamma_i}} \text{ on } W_i \text{ parallel to Span}\left\{ \mathbf{U}_{ik} \mid \mu_{ik} \geq \frac{1}{\tau} \right\}.}$$

From these definitions, we can easily check the following.

Lemma 7.37. *For all subdomains $1 \leq i \leq N$ and $\mathbf{U}_i \in \mathbb{R}^{\#\mathcal{N}_{\Gamma_i}}$, we have*

$$\left(R_{\Gamma_i}^T D_{\Gamma_i}(I_d - \xi_i)\mathbf{U}_i\right)^T \mathbb{S} R_{\Gamma_i}^T D_{\Gamma_i}(I_d - \xi_i)\mathbf{U}_i) \leq \tau\, \mathbf{U}_i^T S_i \mathbf{U}_i\,. \tag{7.90}$$

Proof. It is sufficient to apply Lemma 7.6 with $V := \mathbb{R}^{\#\mathcal{N}_{\Gamma_i}}$, operator $A := S_i$, and operator $B := D_{\Gamma_i} R_{\Gamma_i} \mathbb{S} R_{\Gamma_i}^T D_{\Gamma_i}$. □

Definition 7.38 (GenEO coarse space). *Let W_i, $1 \leq i \leq N$, be the ith local component of the GenEO coarse space as defined in (7.89). Let Z_i be a rectangular matrix of size $\#\mathcal{N}_{\Gamma_i} \times \dim W_i$ whose columns are a basis of W_i. We form a rectangular matrix Z_{GenEO} of size $\#\mathcal{N}_\Gamma \times \sum_{i=1}^N \dim W_i$ by a weighted concatenation of the Z_i's:*

$$Z_{GenEO} := (R_{\Gamma_i}^T D_{\Gamma_i} Z_i)_{1 \leq i \leq N}\,.$$

Let W_{GenEO} be the vector space spanned by the columns of Z_{GenEO} and

$$\boxed{\text{projection } P_g \text{ from } \mathbb{R}^{\#\mathcal{N}_\Gamma} \text{ on } W_{GenEO} \text{ which is } \mathbb{S} \text{ orthogonal.}} \tag{7.91}$$

If the columns of Z_{GenEO} are independent, the matrix form of P_g is

$$P_g = Z_{GenEO}(Z_{GenEO}^T \mathbb{S} Z_{GenEO})^{-1} Z_{GenEO}^T \mathbb{S}\,.$$

The proof is similar to that of formula (7.58) that was done in the context of the hybrid Schwarz method in section 7.5. Note that $\ker(S_j) \subset W_j$ for all $1 \leq j \leq N$ so that we have $W_0 \subset W_{GenEO}$.

Definition 7.39 (BNN GenEO). *The balancing Neumann–Neumann GenEO preconditioner is defined by*

$$\boxed{M_{BNNG}^{-1} := P_g \mathbb{S}^{-1} + (I_d - P_g) \sum_{i=1}^N R_{\Gamma_i}^T D_{\Gamma_i} S_i^\dagger (I_d - P_i) D_{\Gamma_i} R_{\Gamma_i} (I_d - P_g)^T\,.} \tag{7.92}$$

In order to study this new preconditioner we use the same framework as for the balancing Neumann–Neumann method, except that the natural coarse space W_0 is replaced by the GenEO coarse space W_{GenEO} in the definition of the product space H_D,

$$H_D := W_{GenEO} \times \prod_{i=1}^N \text{range}(S_i),$$

and accordingly operator $\mathcal{R}_{BNN}$ is replaced by the linear operator $\mathcal{R}_{BNNG}$ defined as

$$\begin{aligned}\mathcal{R}_{BNNG} &: H_D \longrightarrow H,\\ (\mathbf{U}_i)_{0\le i\le N} &\longmapsto U_0 + (I_d - P_g)\sum_{i=1}^{N} R_{\Gamma_i}^T D_{\Gamma_i}\mathbf{U}_i\,.\end{aligned} \tag{7.93}$$

It can easily be checked from section 7.8.2 that the surjectivity of $\mathcal{R}_{BNNG}$ and the stable decomposition are unchanged since $W_0 \subset W_{GenEO}$.

Lemma 7.40 (Surjectivity of $\mathcal{R}_{BNNG}$). *Operator $\mathcal{R}_{BNNG}$ is surjective.*

Proof. Similarly to (7.83), we have

$$\begin{aligned}\mathbf{U} &= P_g\mathbf{U} + (I_d - P_g)\mathbf{U} = P_g\mathbf{U} + (I_d - P_g)(I_d - P_g)\mathbf{U}\\ &= P_g\mathbf{U} + (I_d - P_g)\sum_{i=1}^{N} R_{\Gamma_i}^T D_{\Gamma_i} R_{\Gamma_i}(I_d - P_g)\mathbf{U}\\ &= P_g\mathbf{U} + (I_d - P_g)\sum_{i=1}^{N} R_{\Gamma_i}^T D_{\Gamma_i}(I_d - P_i) R_{\Gamma_i}(I_d - P_g)\mathbf{U}\\ &\quad + (I_d - P_g)\sum_{i=1}^{N} R_{\Gamma_i}^T D_{\Gamma_i} P_i R_{\Gamma_i}(I_d - P_g)\mathbf{U}.\end{aligned} \tag{7.94}$$

The last term of this equation is zero since for all subdomains i, $P_i R_{\Gamma_i}(I_d - P_g)\mathbf{U} \in \ker S_i$, and thus

$$\sum_{i=1}^{N} R_{\Gamma_i}^T D_{\Gamma_i} P_i R_{\Gamma_i}(I_d - P_g)\mathbf{U} \in W_0 \subset W_{GenEO}$$

and $I_d - P_g$ is a projection parallel to W_{GenEO}. Finally we have proved that for all $\mathbf{U} \in H$, if we define $\mathcal{U} \in H_D$ by

$$\mathcal{U} := (P_g\mathbf{U}, ((I_d - P_i)R_{\Gamma_i}(I_d - P_g)\mathbf{U})_{1\le i\le N}), \tag{7.95}$$

we have $\mathcal{R}_{BNNG}(\mathcal{U}) = \mathbf{U}$. □

As for the stable decomposition estimate with a stability constant $c_T = 1$, since projection P_g is $\mathbb{S}$-orthogonal as is P_0, it suffices to replace P_0 by P_g in (7.85). The change in coarse space will improve the stability of the linear operator $\mathcal{R}_{BNNG}$.

Lemma 7.41 (Continuity of $\mathcal{R}_{BNN}$). *Let k_2 be defined as in (7.75). Then the stability property (7.9) is verified with the constant $c_R = \max(1, k_2\tau)$.*

Proof. Let $\mathcal{U} \in H_D$; since P_g is an $\mathbb{S}$-orthogonal projection on W_{GenEO}, we have

$$\begin{aligned}&(\mathbb{S}\mathcal{R}_{BNNG}(\mathcal{U}), \mathcal{R}_{BNNG}(\mathcal{U}))\\ &= \left(\mathbb{S}(\mathbf{U}_0 + (I_d - P_g)\sum_{i=1}^{N} R_{\Gamma_i}^T D_{\Gamma_i}\mathbf{U}_i), \mathbf{U}_0 + (I_d - P_g)\sum_{i=1}^{N} R_{\Gamma_i}^T D_{\Gamma_i}\mathbf{U}_i\right)\\ &= (\mathbb{S}\mathbf{U}_0, \mathbf{U}_0) + \left(\mathbb{S}(I_d - P_g)\sum_{i=1}^{N} R_{\Gamma_i}^T D_{\Gamma_i}\mathbf{U}_i, (I_d - P_g)\sum_{i=1}^{N} R_{\Gamma_i}^T D_{\Gamma_i}\mathbf{U}_i\right).\end{aligned} \tag{7.96}$$

Since P_g is the $\mathbb{S}$-orthogonal projection on W_{GenEO} and

$$\sum_{i=1}^{N} R_{\Gamma_i}^T D_{\Gamma_i} \xi_i \mathbf{U}_i \in W_{GenEO},$$

we have

$$\left(\mathbb{S}(I_d - P_g)\sum_{i=1}^{N} R_{\Gamma_i}^T D_{\Gamma_i} \mathbf{U}_i, (I_d - P_g)\sum_{i=1}^{N} R_{\Gamma_i}^T D_{\Gamma_i} \mathbf{U}_i\right)$$
$$= \left(\mathbb{S}(I_d - P_g)\sum_{i=1}^{N} R_{\Gamma_i}^T D_{\Gamma_i}(I_d - \xi_i)\mathbf{U}_i, (I_d - P_g)\sum_{i=1}^{N} R_{\Gamma_i}^T D_{\Gamma_i}(I_d - \xi_i)\mathbf{U}_i\right).$$

Thus using equality (7.96), Lemma 7.29, and then Lemma 7.37 we have

$$\begin{aligned}
&(\mathbb{S}\mathcal{R}_{BNNG}(\mathcal{U}), \mathcal{R}_{BNNG}(\mathcal{U}))\\
&\quad = (\mathbb{S}\mathbf{U}_0, \mathbf{U}_0)\\
&\quad\quad + \left(\mathbb{S}(I_d - P_g)\sum_{i=1}^{N} R_{\Gamma_i}^T D_{\Gamma_i}(I_d - \xi_i)\mathbf{U}_i, (I_d - P_g)\sum_{i=1}^{N} R_{\Gamma_i}^T D_{\Gamma_i}(I_d - \xi_i)\mathbf{U}_i\right)\\
&\quad \le (\mathbb{S}\mathbf{U}_0, \mathbf{U}_0) + k_2\tau\sum_{i=1}^{N}\left(S_i R_{\Gamma_i}^T D_{\Gamma_i}(I_d - \xi_i)\mathbf{U}_i, R_{\Gamma_i}^T D_{\Gamma_i}(I_d - \xi_i)\mathbf{U}_i\right)\\
&\quad \le \max(1, k_2\tau)\left((\mathbb{S}\mathbf{U}_0, \mathbf{U}_0) + \sum_{i=1}^{N}(S_i\mathbf{U}_i, \mathbf{U}_i)\right)\\
&\quad \le \max(1, k_2\tau)\, b(\mathbf{U}, \mathbf{U}),
\end{aligned} \tag{7.97}$$

where τ is the user-defined threshold. □

To sum up, by applying the fictitious space lemma (Lemma 7.4) we have the following.

Theorem 7.42 (Spectral estimate for BNNG). *Let τ be a user-defined threshold to build the GenEO coarse space W_{GenEO} (Definition 7.38), let P_g be defined by (7.91), and let k_2 be defined by (7.75). Then, the eigenvalues of the Neumann–Neumann preconditioned (7.82) system satisfy the following estimate:*

$$\boxed{1 \le \lambda(M_{BNNG}^{-1}\mathbb{S}) \le \max(1, k_2\tau).}$$

7.8.4 ▪ Efficient implementation of the BNNG method

As for the hybrid Schwarz method in section 7.5.1, when M_{BNNG}^{-1} is used as a preconditioner in the PCG method to solve the linear system (7.69)

$$\mathbb{S}\, U_\Gamma = G_\Gamma,$$

an efficient implementation can be done if the initial guess is such that the initial residual is $\mathbb{S}$-orthogonal to W_{GenEO}. It can be achieved simply by taking as the initial guess

$$U_{\Gamma 0} := Z_{GenEO}(Z_{GenEO}^T \mathbb{S} Z_{GenEO})^{-1} Z_{GenEO}^T G_\Gamma .$$

In this case the application of the preconditioner M_{BNNG}^{-1} in the PCG algorithm can be replaced by the application of

$$(I_d - P_g) \sum_{i=1}^{N} R_{\Gamma_i}^T D_{\Gamma_i} S_i^\dagger (I_d - P_i) D_{\Gamma_i} R_{\Gamma_i} .$$

Chapter 8

Parallel implementation of Schwarz methods

As depicted in previous chapters, domain decomposition methods can be used to design extremely robust parallel preconditioners. The purpose of this chapter is to give an introduction to their use on parallel computers through their implementation in the free finite element package FreeFem++ [108]. We start with a self-contained script of a three-dimensional elasticity problem; see section 8.1. The solver can be either one of the domain decomposition methods introduced in the book, a direct solver, or an algebraic multigrid method. The last two solvers are available via the PETSc [8, 7] interface. In section 8.2, we show strong and weak scalability results for various problems on both small and large numbers of cores. In the last section, section 8.3, we explain the algebraic formalism used for domain decomposition methods that bypasses a global numbering of the unknowns.

8.1 ▪ A parallel FreeFem++ script

Section 8.1.1 is devoted to the formulation of the elasticity problem and its geometry. Such an equation typically arises in computational solid mechanics for modeling small deformations of compressible bodies. The solver of the corresponding linear system will be chosen in section 8.1.2. In section 8.1.3, we explain how to visualize the solution. The complete script `elasticity-3d.edp` is given at the end of this section in section 8.1.4.

8.1.1 ▪ Three-dimensional elasticity problem

The mechanical properties of a solid can be characterized by its Young modulus E and Poisson ratio ν or alternatively by its Lamé coefficients λ and μ. These coefficients relate to each other by the following formulas:

$$\lambda = \frac{E\nu}{(1+\nu)(1-2\nu)} \quad \text{and} \quad \mu = \frac{E}{2(1+\nu)}. \tag{8.1}$$

The reference geometry of the solid is the open set $\Omega \subset \mathbb{R}^3$. The solid is subjected to a volume force $\mathbf{f}$ and on part of its boundary (Γ_3) to a normal stress $\mathbf{g}$. It is clamped on the other part of its boundary (Γ_1). Let $\mathbf{u} : \Omega \to \mathbb{R}^3$ denote the displacement of the solid. In Cartesian coordinates (x_1, x_2, x_3), the linearized strain tensor $\varepsilon(\mathbf{u})$ is given by

$$\epsilon_{ij}(\mathbf{u}) := \frac{1}{2}\left(\frac{\partial u_i}{\partial x_j} + \frac{\partial u_j}{\partial x_i}\right) \text{ for } 1 \leq i, j \leq 3.$$

The inner product of two strain tensors is defined by

$$\epsilon(\mathbf{u}):\epsilon(\mathbf{v}) := \sum_{i=1}^{3}\sum_{j=1}^{3}\epsilon_{ij}(\mathbf{u})\epsilon_{ij}(\mathbf{v}).$$

The divergence of the displacement is defined by

$$\nabla\cdot\mathbf{u} := \sum_{i=1}^{3}\frac{\partial u_i}{\partial x_i}.$$

The finite element approximation of the system of linear elasticity is defined by a variational formulation: Find a displacement $\mathbf{u}:\Omega\to\mathbb{R}^3$ in $\mathbb{P}_2$ so that for any test function $\mathbf{v}:\Omega\to\mathbb{R}^3$ in $\mathbb{P}_2$, we have

$$a(\mathbf{u},\mathbf{v}) := \int_{\Omega}\lambda\nabla\cdot\mathbf{u}\,\nabla\cdot\mathbf{v}+\mu\,\varepsilon(\mathbf{u}):\varepsilon(\mathbf{v})-\int_{\Omega}\mathbf{f}\cdot\mathbf{v}=0, \tag{8.2}$$

where

- λ and ν are defined by formula (8.1), and Young's modulus E and Poisson's ratio ν vary between two sets of values, $(E_1,\nu_1)=(2\cdot 10^{11}, 0.25)$ and $(E_2,\nu_2)=(10^7, 0.45)$, and
- $\mathbf{f}$ are the body forces, in this case only the gravity.

In the FreeFem++ script, (8.2) corresponds to three parametrized macros: `epsilon`, `div`, and `Varf`.

```
real Sqrt = sqrt(2.);
macro epsilon(u)[dx(u), dy(u#B), dz(u#C), (dz(u#B) + dy(u#C)) / Sqrt, (dz(u) + dx(u#C)) / Sqrt, (dy(u) + dx(u#B)) / Sqrt]// EOM
macro div(u)(dx(u) + dy(u#B) + dz(u#C))// EOM

macro Varf(varfName, meshName, PhName)
    coefficients(meshName, PhName)
    varf varfName(def(u), def(v)) = intN(meshName)(lambda * div(u) * div(v) + 2. * mu * (epsilon(u)' * epsilon(v))) + intN(meshName)(f * vB) + on(1, BC(u, 0));
// EOM
```

Listing 8.1. `./PARALLEL/FreefemProgram/elasticity-3d.edp`

A first geometry is defined via a three-dimensional mesh of a (simplified) bridge that is built in two steps:

- The mesh of a 2D bridge is built: `ThGlobal2d`.
- A third dimension is added to the previous 2D mesh using the `buildlayers` function, see Figure 8.1.

Note that the three-dimensional mesh is not actually built but a macro is defined to do this afterwards.

Figure 8.1. *Partitioning into eight subdomains.*

```
real depth = 0.25;
int discrZ = getARGV("−discrZ", 1);
real L = 2.5;
real H = 0.71;
real Hsupp = 0.61;
real r = 0.05;
real l = 0.35;
real h = 0.02;
real width = 2.5*L/4.;
real alpha = asin(h/(2.*r))/2;
/*# twoDsequentialMesh #*/
border a0a(t=0, 1){x=0; y=−t*Hsupp; label=2;};
border a0(t=0, (L − width)/2.){x=t; y=−Hsupp; label=1;};
border a0b(t=0, 1){x=(L − width)/2.; y=−(1−t)*Hsupp; label=2;};
border aa(t=0, 0.5^(1/0.75)){x=L/2. − width/2.*cos(pi*t^0.75); y=sin(pi*t^0.75)/4.; label=2;};
border ab(t=0, 0.5^(1/0.75)){x=L/2. + width/2.*cos(pi*t^0.75); y=sin(pi*t^0.75)/4.; label=2;};
border a2a(t=0, 1){x=(L + width)/2.; y=−t*Hsupp; label=2;};
border a2(t=(L + width)/2., L){x=t; y=−Hsupp; label=1;};
border a2b(t=0, 1){x=L; y=−(1−t)*Hsupp; label=2;};
border e(t=0, 1){x=L; y=t*H; label=2;};
border c(t=0, 1){x=(1−t)*L; y=H; label=3;};
border d(t=0, 1){x=0; y=(1−t)*H; label=2;};
mesh ThGlobal2d = buildmesh(a0(global * (L − width)/(2.*L)) + a0a(global*Hsupp/L) +
    a0b(global*Hsupp/L) + a2(global * (L − width)/(2*L)) + a2a(global*Hsupp/L) +
    a2b(global*Hsupp/L) + aa(global * width/(2*L)) + ab(−global * width/(2*L)) + e(global*H/L)
    + c(global) + d(global*H/L));
ThGlobal2d = adaptmesh(ThGlobal2d, 1/200., IsMetric=1, nbvx=100000);
/*# twoDsequentialMeshEnd #*/
macro minimalMesh()Cube(CC, BB, LL)// EOM
macro generateTh(name)name = buildlayers(ThGlobal2d, discrZ, zbound=[0, depth])// EOM
int[int, int] LL = [[1,3], [2,2], [2,2]];
real[int, int] BB = [[0,10], [0,1], [0,1]];
int[int] CC = [1, 1, 1];
```

Listing 8.2. `./PARALLEL/FreefemProgram/elasticity-3d.edp`

This geometry is partitioned and distributed among the MPI processes.

From now on, all the tasks can be computed concurrently, meaning that each MPI process is in charge of only one subdomain and variables are local to each process. Then a parallel mesh refinement is made by cutting each tetrahedron into 8 tetrahedra. This corresponds to a mesh refinement factor s equal to 2. Note also that at this point, the displacement field $\mathbf{u}$ is approximated by $\mathbb{P}_2$ continuous piecewise quadratic functions.

```
func Pk = [P2, P2, P2];

build(generateTh, Th, ThBorder, ThOverlap, D, numberIntersection, arrayIntersection,
    restrictionIntersection, Wh, Pk, mpiCommWorld, s)
ThGlobal2d = square(1, 1);
```

Listing 8.3. `./PARALLEL/FreefemProgram/elasticity-3d.edp`

The Young modulus and Poisson's ratio are heterogeneous, and the exterior forces are constant.

Remark 8.1.1. *In our numerical experiments, Poisson's ratio being relatively far from the incompressible limit of* 0.5, *it is not necessary to switch to a mixed finite element formulation since there is no locking effect.*

```
real f = -900000.;
func real stripes(real a, real b, real paramA, real paramB) {
    int da = int(a * 10);
    return (da == (int(da / 2) * 2) ? paramB : paramA);
}

macro coefficients(meshName, PhName)
    fespace PhName(meshName, P0);
    PhName Young = stripes(y, x, 2e11, 1e7);
    PhName poisson = stripes(y, x, 0.25, 0.45);
    PhName tmp = 1. + poisson;
    PhName mu = Young / (2. * tmp);
    PhName lambda = Young * poisson / (tmp * (1. - 2. * poisson));// EOM
```

Listing 8.4. `./PARALLEL/FreefemProgram/elasticity-3d.edp`

8.1.2 ▪ Native DDM solvers and the PETSc interface

At this point, the physics, the geometry, and the discretization of the three-dimensional elasticity problem have been given in the script. In order to find the finite element approximation of the displacement field, we have to solve the corresponding global algebraic system which is actually never assembled. Its distributed storage among the processes depends on whether we use native FreeFem++ solvers or other solvers via the PETSc interface. The native FreeFem++ solvers are

- a parallel GMRES solver which is the default parallel solver in FreeFem++;
- a one-level Schwarz method, either ASM or RAS; and
- the two-level Schwarz method with a GenEO coarse space; see Chapter 7.

They are implemented inside HPDDM, a framework for high-performance domain decomposition methods, available at `https://github.com/hpddm/hpddm`. Its interface with FreeFem++ also includes substructuring methods, like the FETI and BDD methods, as described in Chapter 6. The solvers interfaced with PETSc are

- the PETSc parallel GMRES solver,
- the multifrontal parallel solver MUMPS [5], and
- GAMG: an algebraic multigrid solver [2].

```
solver = getARGV("--solver", 0);
if(solver == 0)
{
  if(mpirank == 0) {
     cout << "What kind of solver would you like to use ?" << endl;
     cout << " [1] PETSc GMRES" << endl;
     cout << " [2] GAMG" << endl;
     cout << " [3] MUMPS" << endl;
     cout << " [10] ASM" << endl;
     cout << " [11] RAS" << endl;
     cout << " [12] Schwarz GenEO" << endl;
     cout << " [13] GMRES" << endl;
     cout << "Please type in a number: ";
     cin >> solver;
     if(solver != 1 && solver != 2 && solver != 3 && solver != 4 && solver != 10 && solver != 11 && solver != 12) {
        cout << "Wrong choice, using GMRES instead !" << endl;
        solver = 10;
     }
  }
}
broadcast(processor(0), solver);
```

Listing 8.5. `./PARALLEL/FreefemProgram/elasticity-3d.edp`

FreeFem++ interface

If the native FreeFem++ solvers are used, the local stiffness matrices are assembled concurrently:

```
assemble(A, rhs, Wh, Th, ThBorder, Varf)
```

Listing 8.6. `./PARALLEL/FreefemProgram/elasticity-3d.edp`

The local operator is attached to a distributed structure `dschwarz` which is a FreeFem++ type. If necessary, the GenEO coarse space is built:

```
matrix N;
if(mpisize > 1 && solver == 12) {
   EVproblem(vPbNoPen, Th, Ph)
   matrix noPen = vPbNoPen(Wh, Wh, solver = CG);
   attachCoarseOperator(mpiCommWorld, Aglob, A = noPen,);
}
```

Listing 8.7. `./PARALLEL/FreefemProgram/elasticity-3d.edp`

The distributed linear system is solved by a call that includes additional command line arguments that are automatically passed to the solvers.

```
DDM(Aglob, u[], rhs, dim = getARGV("-gmres_restart", 60), iter = getARGV("-iter", 100), eps =
    getARGV("-eps", 1e-8), solver = solver - 9);
```

Listing 8.8. `./PARALLEL/FreefemProgram/elasticity-3d.edp`

PETSc interface

If the PETSc interface is used, the local stiffness matrix K and the local load vector `rhs` are built concurrently from the variational forms for all subdomains $1 \leq i \leq N$.

```
Varf(vPb, Th, Ph)
matrix A = vPb(Wh, Wh);
rhs = vPb(0, Wh);
dmatrix Mat(A, arrayIntersection, restrictionIntersection, D, bs = 3);
```

Listing 8.9. `./PARALLEL/FreefemProgram/elasticity-3d.edp`

`dmatrix` is a FreeFem++ type.

If an algebraic multigrid method is used via the PETSc interface, the near null space must be provided in order to enhance the convergence; see, e.g., [109]. For an elasticity problem, it is made of the rigid body motions which are three translations and three rotations, here along the axis.

```
    Wh[int] def(Rb)(6);
    [Rb[0], RbB[0], RbC[0]] = [ 1, 0, 0];
    [Rb[1], RbB[1], RbC[1]] = [ 0, 1, 0];
    [Rb[2], RbB[2], RbC[2]] = [ 0, 0, 1];
    [Rb[3], RbB[3], RbC[3]] = [ y, -x, 0];
    [Rb[4], RbB[4], RbC[4]] = [-z, 0, x];
    [Rb[5], RbB[5], RbC[5]] = [ 0, z, -y];
```

Listing 8.10. `./PARALLEL/FreefemProgram/elasticity-3d.edp`

Eventually, the solver may be called by passing command line parameters to PETSc:

```
    set(Mat, sparams = "-pc_type gamg -ksp_type gmres -pc_gamg_threshold 0.05
        -ksp_monitor", nearnullspace = Rb);
}
else if(solver == 3)
    set(Mat, sparams = "-pc_type lu -pc_factor_mat_solver_package mumps
        -mat_mumps_icntl_7 2 -ksp_monitor");
mpiBarrier(mpiCommWorld);
timing = mpiWtime();
u[] = Mat^-1 * rhs;
```

Listing 8.11. `./PARALLEL/FreefemProgram/elasticity-3d.edp`

8.1.3 ▪ Validation of the computation

The true residual is computed by first performing a parallel matrix vector product either via matrix `Mat` that is interfaced with the PETSc interface

```
res = Mat * u[];
```

Listing 8.12. `./PARALLEL/FreefemProgram/elasticity-3d.edp`

or via the matrix `Aglob` in the "native" FreeFem++ Schwarz format.

```
res = Aglob * u[];
```

Listing 8.13. `./PARALLEL/FreefemProgram/elasticity-3d.edp`

Once the linear system is solved and the residual is computed, the rest of the program does not depend on whether a FreeFem++ or a PETSc solver was used. The residual is subtracted from the distributed right-hand side `rhs` and the distributed scalar product is performed based on Lemma 8.2:

```
res -= rhs;
real rhsnorm = dscalprod(D, rhs, rhs);
real dist = dscalprod(D, res, res);
if(mpirank == 0)
    cout << " --- normalized L^2 norm of the true residual: " << sqrt(dist / rhsnorm) << endl;
```

Listing 8.14. `./PARALLEL/FreefemProgram/elasticity-3d.edp`

The three-dimensional solution can be visualized by the `plotMPI` command.

```
plotMPI(Th, u, "Global solution", Pk, def, 3, 1);
meshN ThMoved = movemesh3(Th, transfo = [x + u, y + uB, z + uC]);
plotMPI(ThMoved, u, "Moved mesh", Pk, def, 3, 1);
```

Listing 8.15. `./PARALLEL/FreefemProgram/elasticity-3d.edp`

Timings for the solvers are given in section 8.2.

8.1.4 ▪ Parallel script

For the sake of completeness, the full script is given here, split into five parts.

```
macro K()real// EOM
macro def()def3// EOM
macro init()init3// EOM
macro BC()BC3// EOM
macro meshN()mesh3// EOM
macro intN()int3d// EOM
macro measureN()volume// EOM
macro bbN()bb3// EOM
include "Schwarz/macro.idp"

/*# problemPhysics #*/
real Sqrt = sqrt(2.);
macro epsilon(u)[dx(u), dy(u#B), dz(u#C), (dz(u#B) + dy(u#C)) / Sqrt, (dz(u) + dx(u#C)) / Sqrt, (dy(u) + ↲
```

```
    dx(u#B)) / Sqrt]// EOM
macro div(u)(dx(u) + dy(u#B) + dz(u#C))// EOM

macro Varf(varfName, meshName, PhName)
    coefficients(meshName, PhName)
    varf varfName(def(u), def(v)) = intN(meshName)(lambda * div(u) * div(v) + 2. * mu * (epsilon(u)' * epsilon(v))) + intN(meshName)(f * vB) + on(1, BC(u, 0));
// EOM
/*# problemPhysicsEnd #*/
/*# vfGENEO #*/
macro EVproblem(varfName, meshName, PhName)
    coefficients(meshName, PhName)
    varf varfName(def(u), def(v)) = intN(meshName)(lambda * div(u) * div(v) + 2. * mu * (epsilon(u)' * epsilon(v))) + on(1, BC(u, 0));
// EOM
/*# vfGENEOEnd #*/

/*# sequentialMesh #*/
real depth = 0.25;
int discrZ = getARGV("-discrZ", 1);
real L = 2.5;
real H = 0.71;
real Hsupp = 0.61;
real r = 0.05;
real l = 0.35;
real h = 0.02;
real width = 2.5*L/4.;
real alpha = asin(h/(2.*r))/2;
```

Listing 8.16. ./PARALLEL/FreefemProgram/elasticity-3d.edp

```
border a0a(t=0, 1){x=0; y=-t*Hsupp; label=2;};
border a0(t=0, (L - width)/2.){x=t; y=-Hsupp; label=1;};
border a0b(t=0, 1){x=(L - width)/2.; y=-(1-t)*Hsupp; label=2;};
border aa(t=0, 0.5^(1/0.75)){x=L/2. - width/2.*cos(pi*t^0.75); y=sin(pi*t^0.75)/4.; label=2;};
border ab(t=0, 0.5^(1/0.75)){x=L/2. + width/2.*cos(pi*t^0.75); y=sin(pi*t^0.75)/4.; label=2;};
border a2a(t=0, 1){x=(L + width)/2.; y=-t*Hsupp; label=2;};
border a2(t=(L + width)/2., L){x=t; y=-Hsupp; label=1;};
border a2b(t=0, 1){x=L; y=-(1-t)*Hsupp; label=2;};
border e(t=0, 1){x=L; y=t*H; label=2;};
border c(t=0, 1){x=(1-t)*L; y=H; label=3;};
border d(t=0, 1){x=0; y=(1-t)*H; label=2;};
mesh ThGlobal2d = buildmesh(a0(global * (L - width)/(2.*L)) + a0a(global*Hsupp/L) + a0b(global*Hsupp/L) + a2(global * (L - width)/(2*L)) + a2a(global*Hsupp/L) + a2b(global*Hsupp/L) + aa(global * width/(2*L)) + ab(-global * width/(2*L)) + e(global*H/L) + c(global) + d(global*H/L));
ThGlobal2d = adaptmesh(ThGlobal2d, 1/200., IsMetric=1, nbvx=100000);
/*# twoDsequentialMeshEnd #*/
macro minimalMesh()Cube(CC, BB, LL)// EOM
macro generateTh(name)name = buildlayers(ThGlobal2d, discrZ, zbound=[0, depth])// EOM
int[int, int] LL = [[1,3], [2,2], [2,2]];
real[int, int] BB = [[0,10], [0,1], [0,1]];
int[int] CC = [1, 1, 1];
/*# sequentialMeshEnd #*/

include "Schwarz/additional_macro.idp"

int overlap = getARGV("-overlap", 1);

if(mpirank == 0) {
```

```
    cout << " —— " << mpirank << "/" << mpisize;
    cout << " - input parameters: global size = " << global << " - refinement factor = " << s << " - precision = " << getARGV("-eps", 1e-8) << " - overlap = " << overlap << " - with partitioner? = " << partitioner << endl;
}

/*# parallelMesh #*/
func Pk = [P2, P2, P2];

build(generateTh, Th, ThBorder, ThOverlap, D, numberIntersection, arrayIntersection, restrictionIntersection, Wh, Pk, mpiCommWorld, s)
ThGlobal2d = square(1, 1);
```

Listing 8.17. `./PARALLEL/FreefemProgram/elasticity-3d.edp`

```
Wh def(u);
/*# chooseSolver #*/
solver = getARGV("-solver", 0);
if(solver == 0)
{
  if(mpirank == 0) {
      cout << "What kind of solver would you like to use ?" << endl;
      cout << " [1] PETSc GMRES" << endl;
      cout << " [2] GAMG" << endl;
      cout << " [3] MUMPS" << endl;
      cout << " [10] ASM" << endl;
      cout << " [11] RAS" << endl;
      cout << " [12] Schwarz GenEO" << endl;
      cout << " [13] GMRES" << endl;
      cout << "Please type in a number: ";
      cin >> solver;
      if(solver != 1 && solver != 2 && solver != 3 && solver != 4 && solver != 10 && solver != 11 && solver != 12) {
        cout << "Wrong choice, using GMRES instead !" << endl;
        solver = 10;
      }
  }
}
broadcast(processor(0), solver);
/*# chooseSolverEnd #*/

/*# physicalParameters #*/
real f = -900000.;
func real stripes(real a, real b, real paramA, real paramB) {
    int da = int(a * 10);
    return (da == (int(da / 2) * 2) ? paramB : paramA);
}

macro coefficients(meshName, PhName)
    fespace PhName(meshName, P0);
    PhName Young = stripes(y, x, 2e11, 1e7);
    PhName poisson = stripes(y, x, 0.25, 0.45);
    PhName tmp = 1. + poisson;
    PhName mu = Young / (2. * tmp);
    PhName lambda = Young * poisson / (tmp * (1. - 2. * poisson));// EOM
```

Listing 8.18. `./PARALLEL/FreefemProgram/elasticity-3d.edp`

```
real[int] res(Wh.ndof);
real[int] rhs(Wh.ndof);

if(solver == 1 || solver == 2 || solver == 3) {
/*# StiffnessRhsMatrix #*/
    Varf(vPb, Th, Ph)
    matrix A = vPb(Wh, Wh);
    rhs = vPb(0, Wh);
    dmatrix Mat(A, arrayIntersection, restrictionIntersection, D, bs = 3);
/*# StiffnessRhsMatrixEnd #*/
    // sparams will override command line arguments !
    if(solver == 2) {
/*# rigidBodyMotion #*/
        Wh[int] def(Rb)(6);
        [Rb[0], RbB[0], RbC[0]] = [ 1, 0, 0];
        [Rb[1], RbB[1], RbC[1]] = [ 0, 1, 0];
        [Rb[2], RbB[2], RbC[2]] = [ 0, 0, 1];
        [Rb[3], RbB[3], RbC[3]] = [ y, -x, 0];
        [Rb[4], RbB[4], RbC[4]] = [-z, 0, x];
        [Rb[5], RbB[5], RbC[5]] = [ 0, z, -y];
/*# rigidBodyMotionEnd #*/
/*# SolverPETSc #*/
        set(Mat, sparams = "-pc_type gamg -ksp_type gmres -pc_gamg_threshold 0.05 
            -ksp_monitor", nearnullspace = Rb);
    }
    else if(solver == 3)
        set(Mat, sparams = "-pc_type lu -pc_factor_mat_solver_package mumps 
            -mat_mumps_icntl_7 2 -ksp_monitor");
    mpiBarrier(mpiCommWorld);
    timing = mpiWtime();
    u[] = Mat^-1 * rhs;
/*# SolverPETScEnd #*/
    timing = mpiWtime() - timing;
/*# matrixVectorPETSc #*/
    res = Mat * u[];
/*# matrixVectorPETScEnd #*/
}
else {
    /*# localMatrix #*/
    assemble(A, rhs, Wh, Th, ThBorder, Varf)
```

Listing 8.19. ./PARALLEL/FreefemProgram/elasticity-3d.edp

```
    dschwarz Aglob(A, arrayIntersection, restrictionIntersection, scaling = D);

    mpiBarrier(mpiCommWorld);
    timing = mpiWtime();
    /*# coarseSpace #*/
    matrix N;
    if(mpisize > 1 && solver == 12) {
        EVproblem(vPbNoPen, Th, Ph)
        matrix noPen = vPbNoPen(Wh, Wh, solver = CG);
        attachCoarseOperator(mpiCommWorld, Aglob, A = noPen,);
    }
    /*# coarseSpaceEnd #*/
    /*# SolverDDM #*/
    DDM(Aglob, u[], rhs, dim = getARGV("-gmres_restart", 60), iter = getARGV("-iter", 100), eps = 
        getARGV("-eps", 1e-8), solver = solver - 9);
```

```
    /*# SolverDDMEnd #*/
    timing = mpiWtime() - timing;
    /*# matrixVectorFFpp #*/
    res = Aglob * u[];
    /*# matrixVectorFFppEnd #*/
}
/*# trueResidual #*/
res -= rhs;
real rhsnorm = dscalprod(D, rhs, rhs);
real dist = dscalprod(D, res, res);
if(mpirank == 0)
    cout << " --- normalized L^2 norm of the true residual: " << sqrt(dist / rhsnorm) << endl;
/*# trueResidualEnd #*/

mpiBarrier(mpiCommWorld);
if(mpirank == 0)
    cout << " --- time to solution: " << timing << endl;
/*# Visualization #*/
plotMPI(Th, u, "Global solution", Pk, def, 3, 1);
meshN ThMoved = movemesh3(Th, transfo = [x + u, y + uB, z + uC]);
plotMPI(ThMoved, u, "Moved mesh", Pk, def, 3, 1);
/*# VisualizationEnd #*/
```

Listing 8.20. `../PARALLEL/FreefemProgram/elasticity-3d.edp`

8.2 ▪ Numerical experiments

In section 8.2.1, we present results obtained on a few cores with the above script from section 8.1. Section 8.2.2 shows the scalability of the method with a large number of cores solving both the system of linear elasticity and the problem of scalar diffusion.

8.2.1 ▪ Small-scale computations

Results and timings for solving this problem with 263,000 unknowns on 8 cores running at 1.6 GHz are given in Table 8.1. The parameter τ is the relative threshold used for dropping edges in the aggregation graphs of the multigrid preconditioner, while the parameter ν is the number of local deflation vectors computed per subdomain in the GenEO coarse space. The multigrid implementation is based on GAMG [2], which is bundled into PETSc [8, 7]. The results for exactly the same problem as before on 64 cores are given in Table 8.2.

For the GenEO method, the computational times vary slightly when the parameter ν varies around its optimal value. The iteration count decreases when the parameter ν increases. On the other hand, when ν is increased the cost of the factorization of the coarse operator increases. For the multigrid method, the computational times vary rapidly with the parameter τ.

8.2.2 ▪ Large-scale computations

Results were obtained on Curie, a Tier-0 system for PRACE;[2] with a peak performance of 1.7 PFLOP/s. They were first published in the article [118] nominated for the best paper award at SC13[3] and were also disseminated in the PRACE Annual Report 2013 [153, pp. 22–23] as a "success story." Curie is composed of 5040 nodes made of two eight-core

[2]Partnership for Advanced Computing in Europe `http://www.prace-ri.eu/`.

[3]Among five other papers out of 90 accepted papers out of 457 submissions.

Table 8.1. 263,000 *unknowns with* 8 *cores (timings are in seconds).*

Solver	Time to solution	Number of iterations
GAMG—$\tau = 0.01$	211.5	383
GAMG—$\tau = 0.025$	131.7	195
GAMG—$\tau = 0.05$	138.8	172
GAMG—$\tau = 0.1$	not available	> 400 (in 333.3)
GAMG—$\tau = 0.2$	not available	> 400 (in 1395.8)
GenEO—$\nu = 20$	135.2	192
GenEO—$\nu = 25$	112.1	107
GenEO—$\nu = 30$	103.5	69
GenEO—$\nu = 35$	96.1	38
GenEO—$\nu = 40$	103.4	34
MUMPS	68.1	3

Table 8.2. 263,000 *unknowns with* 64 *cores (timings are in seconds).*

Solver	Time to solution	Number of iterations
GAMG—$\tau = 0.01$	30.6	310
GAMG—$\tau = 0.025$	26.6	232
GAMG—$\tau = 0.05$	25.7	179
GAMG—$\tau = 0.1$	not available	> 400 (in 62.5 sec.)
GAMG—$\tau = 0.2$	not available	> 400 (in 263.7 sec.)
GenEO—$\nu = 20$	25.3	200
GenEO—$\nu = 25$	15.1	32
GenEO—$\nu = 30$	15.5	26
MUMPS	not available	failed to setup

Intel Sandy Bridge processors clocked at 2.7 GHz. Its interconnect is an InfiniBand QDR full fat tree, and the MPI implementation was BullxMPI version 1.1.16.5. Intel compilers and MKL in their version 13.1.0.146 were used for all binaries and shared libraries, and as the linear algebra backend for both dense and sparse computations in the framework. Finite element matrices were obtained with FreeFem++. The speedup and efficiency are displayed in terms of the number of MPI processes. In these experiments, each MPI process is assigned a single subdomain and two OpenMP threads. Since the preconditioner is not symmetric, we use GMRES. The computation is stopped when the relative residual is less than 10^{-6}.

Strong scaling experiments

We modified the script of section 8.1 to handle the two-dimensional elasticity case in addition to the three-dimensional case. In 2D, piecewise cubic basis functions are used and the system of equations has approximately 33 nonzero entries per row. It is of constant size equal to almost two billion unknowns. In 3D, piecewise quadratic basis functions are used and the system of equations has approximately 83 nonzero entries per row. The system is of constant size equal to almost 300 million unknowns. These are so-called strong scaling experiments. Geometries are partitioned with METIS. After the partitioning step, each local mesh is refined concurrently by splitting each triangle or tetrahedron into multiple smaller elements. This means that the simulation starts with a relatively poor global mesh (26 million triangles in 2D, 10 million tetrahedra in 3D), which is then refined in parallel

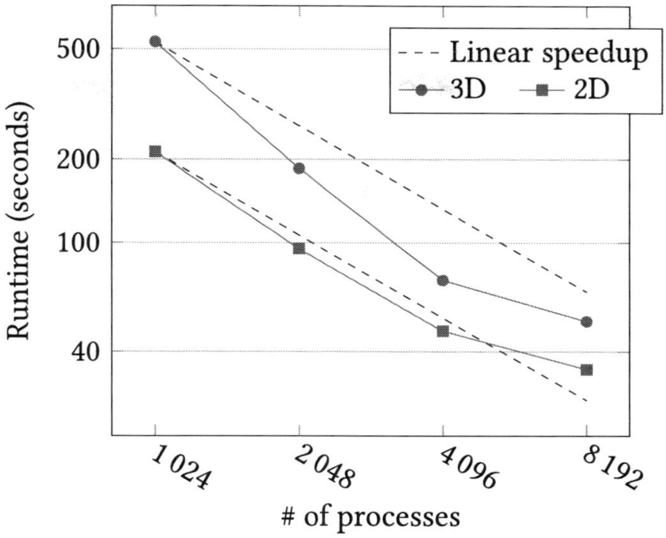

Figure 8.2. *Timings of various simulations.*

Table 8.3. *Breakdown of the timings used for Figure* 8.3.

	N	Factorization	Deflation	Solution	# of it.	Total	# of d.o.f.
3D	1 024	177.9 s	264.0 s	77.4 s	28	530.6 s	$293.98 \cdot 10^6$
	2 048	62.7 s	97.3 s	20.4 s	23	186.0 s	
	4 096	19.6 s	35.7 s	9.7 s	20	73.1 s	
	8 192	6.3 s	22.1 s	6.0 s	27	51.8 s	
2D	1 024	37.0 s	131.8 s	34.3 s	28	213.2 s	$2.14 \cdot 10^9$
	2 048	17.5 s	53.8 s	17.5 s	28	95.1 s	
	4 096	6.9 s	27.1 s	8.6 s	23	47.7 s	
	8 192	2.0 s	20.8 s	4.8 s	23	34.5 s	

(thrice in 2D, twice in 3D). A nice speedup is obtained from $1024 \times 2 = 2048$ to $8192 \times 2 = 16384$ threads as shown in Figure 8.2.

In 3D, we achieve superlinear speedups when going from one thousand to four thousand cores as shown in Figure 8.2. From four thousand to eight thousand cores, the speedup is sublinear. At peak performance, on 16384 threads, the speedup relative to 2048 threads equals $\frac{530.56}{51.76}$, which is approximately a tenfold decrease in runtime. In 2D, the speedup is linear at the beginning and then sublinear. At peak performance, on 16384 threads, the speedup relative to 2048 threads equals $\frac{213.20}{34.54}$, which is approximately a sixfold decrease in runtime.

According to Table 8.3, the costly operations in the construction of the two-level preconditioner are the solution of each local eigenvalue problem (column *Deflation*) and the factorization of each local matrix $\{A_i\}_{i=1}^N$ (column *Factorization*). In 3D, the complexity of the factorization phase typically grows superlinearly with respect to the number of unknowns. In both 2D and 3D, the solution of the eigenproblem is the limiting factor for achieving better speedups. This can be explained by the fact that the Lanczos method,

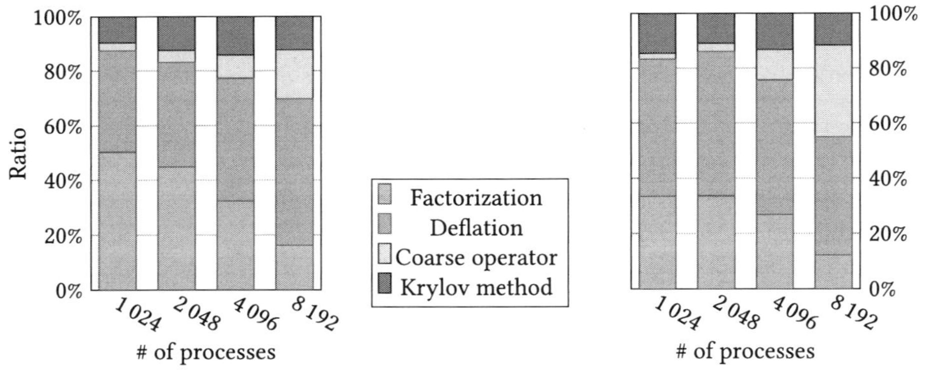

Figure 8.3. *Comparison of the time spent in various steps for building and using the preconditioner in 2D (left) and 3D (right).*

on which ARPACK [127] is based, tends to perform better for larger ratios $\left\{\frac{n_i}{\nu_i}\right\}_{1\le i\le N}$, but these values decrease as the subdomains get smaller. The local number of deflation vectors, ν_i, is uniform across subdomains and ranges from 15 to 20. For fewer but larger subdomains, the time to compute the solution (column *Solution*), i.e., the time for the GMRES to converge, is almost equal to that spent in local forward eliminations and back substitutions and the communication times are negligible. When the decompositions become bigger, subdomains are smaller, and hence each local solution is computed faster and global communications have to be taken into account.

To infer a more precise idea of the communication-to-computation ratio, Figure 8.3 is quite useful. It shows for each computation the relative cost of each phase in percentages. The first two steps for computing the local factorizations and deflation vectors are purely concurrent and do not involve any communication. Thus, it is straightforward to get a lower bound of the aforementioned ratio. The time spent assembling the coarse operator and for the Krylov method to converge is comprised of both communications and computations.

To assess the need for such a sophisticated preconditioner, the convergence histogram of a simple one-level method versus this two-level method is displayed in Figure 8.4. One can easily understand that, while the cost of building the preconditioner cannot be neglected, it is necessary to ensure the convergence of the Krylov method: after more than 10 minutes, the one-level preconditioner barely decreases the relative error to 2×10^{-5}, while it takes 214 seconds for the two-level method to converge to the desired tolerance; see row 5 of Table 8.3. That is at least a threefold speedup. For larger decompositions, the need for two-level methods is even more obvious.

Weak scaling experiments

Moving on to the weak scaling properties (see Definition 4.2), the problem now being solved is a scalar equation of diffusivity,

$$\begin{aligned} -\nabla\cdot(\kappa\nabla u) &= 1 \quad \text{in } \Omega,\\ u &= 0 \quad \text{on } [0;1]\times\{0\}, \end{aligned} \tag{8.3}$$

with highly heterogeneous coefficients. Parameter κ varies from 1 to $3\cdot 10^6$ as displayed in Figure 8.5.The partitioned domain is $\Omega=[0;1]^d$ ($d=2$ or 3) with piecewise quartic basis functions in 2D yielding linear systems with approximately 23 nonzero entries per row

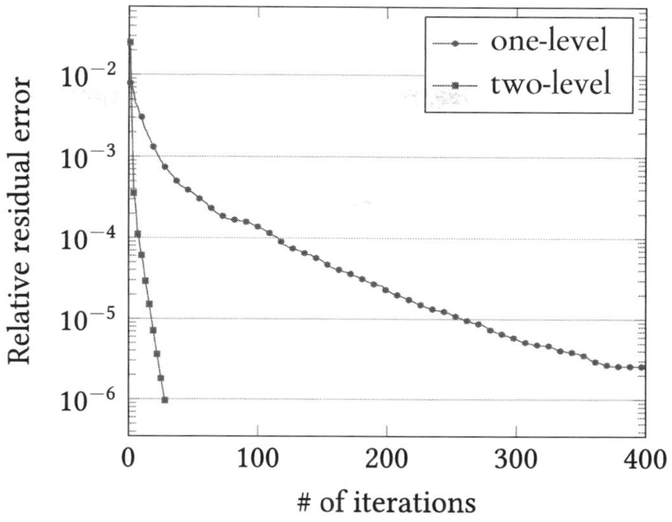

Figure 8.4. *Convergence of the restarted GMRES*(40) *for a 2D problem of linear elasticity using* 1024 *subdomains. Timings for the setup and solution phases are available in Table* 8.3*, using a one-level method, the convergence is not reached after* 10 *minutes.*

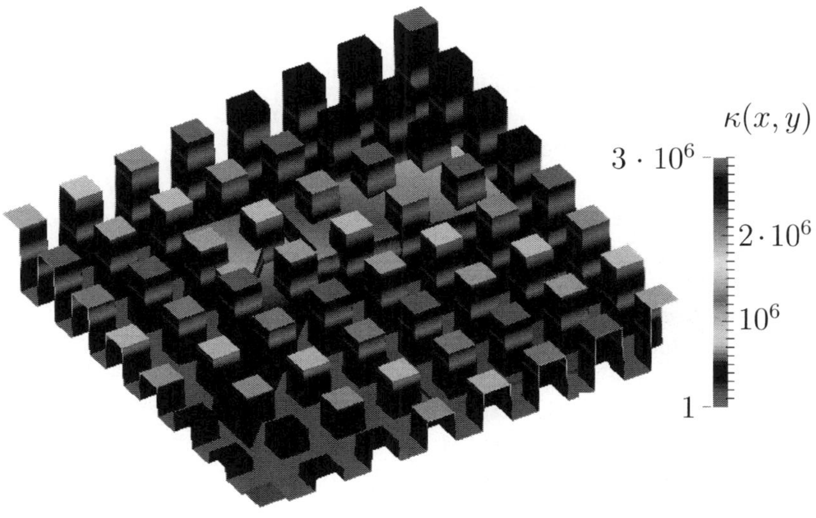

Figure 8.5. *Diffusivity* κ *used for the two-dimensional weak scaling experiments with channels and inclusions.*

and piecewise quadratic basis functions in 3D yielding linear systems with approximately 27 nonzero entries per row. After using Green's formula, its variational formulation is, for all test functions $v \in H_0^1(\Omega)$

$$a(u,v) := \int_\Omega \kappa \nabla u \cdot \nabla v - \int_\Omega f \cdot v = 0,$$

where f is a source term. Such an equation typically arises for modeling flows in porous media or in computational fluid dynamics. The only work needed is changing the mesh used for computations, as well as the variational formulation of the problem in FreeFem++. On average, there is a constant number of degrees of freedom per subdomain equal to roughly 280 thousand in 3D and nearly 2.7 million in 2D. As for the strong scaling experiment, after building and partitioning a global coarse mesh with a few million elements, each local mesh is refined independently to ensure a constant size per subdomain as the decomposition gets bigger. The efficiency remains at almost 90% thanks to only a slight change in the factorization time for the local problems and in the construction of the deflation vectors. In 3D, the initial problem of 74 million unknowns is solved in 200 seconds on 512 threads. When using 16384 threads, the size of the problem is approximately 2.3 billion unknowns. It is solved in 215 seconds for an efficiency of approximately 90%. In 2D, the initial problem of 695 million unknowns is solved in 175 seconds on 512 threads. When using 16384 threads, the problem size is approximately 22.3 billion unknowns. It is solved in 187 seconds for an efficiency of approximately 96%. For these kinds of scales, the most penalizing step in the algorithm is the construction of the coarse operator, especially in 3D, see Figure 8.6. A non-negligible increase in the time spent to assemble and factorize the Galerkin correction matrix is responsible for the slight loss of efficiency. As a result, a multilevel (more than two-level) approach would be necessary.

8.3 ▪ FreeFem++ algebraic formulation

We present here our method for computing a global matrix vector product or the action of a global one-level preconditioner, using only their local components plus point-to-point (between a pair of processes) communications. The special feature of this implementation is that no global numbering of the unknowns is needed. This is the way domain decomposition methods have been implemented in FreeFem++.

Recall that the set of indices $\mathcal{N}$ is decomposed into N sets $(\mathcal{N}_i)_{1\le i\le N}$. An MPI process is attached to each subset. Let $n := \#\mathcal{N}$ be the number of degrees of freedom of the global finite element space. A global vector $\mathbf{U} \in \mathbb{R}^n$ is stored in a distributed way. Each process $i, 1 \le i \le N$, stores the local vector $\mathbf{U}_i := R_i \mathbf{U} \in \mathbb{R}^{\#\mathcal{N}_i}$, where R_i is the restriction operator introduced in Definition 1.11 in Chapter 1. Total storage of $\sum_{i=1}^N \#\mathcal{N}_i$ scalars must be allocated, which is larger than the size n of the original system. The extra cost in memory is not a problem since it is distributed among the N MPI processes. The unknowns in the intersection of two subsets of degrees of freedom are duplicated. It is important to ensure that the result $(v_i)_{1\le i\le N}$ of all linear algebra operators applied to this representation will preserve its coherence, that is, the duplicated degrees of freedom should share the same values across the subdomains $\mathcal{N}_i$, $1 \le i \le N$.

Definition 8.1. *A sequence of local vectors* $(\mathbf{V}_i)_{1\le i\le N} \in \Pi_{i=1}^N \mathbb{R}^{\#\mathcal{N}_i}$ *is coherent if there exists a vector* $\mathbf{V} \in \mathbb{R}^n$ *such that for all* $1 \le i \le N$*, we have*

$$\mathbf{V}_i = R_i \mathbf{V}.$$

Another equivalent definition could be that for all $1 \le i, j \le N$, we have

$$R_j^T R_j R_i^T \mathbf{U}_i = R_i^T R_i R_j^T \mathbf{U}_j .$$

We have to define basic linear algebra operations based on this distributed storage and ensure that all linear algebra operations are performed so that the coherence is enforced, up to roundoff errors.

We start with the scalar product of two distributed vectors.

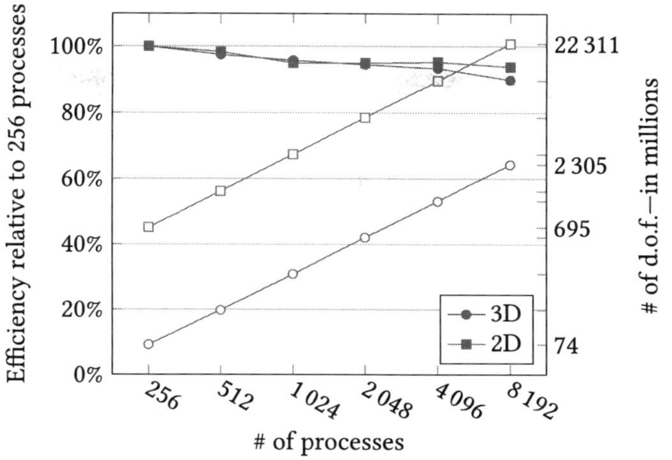

(a) *Timings of various simulations*

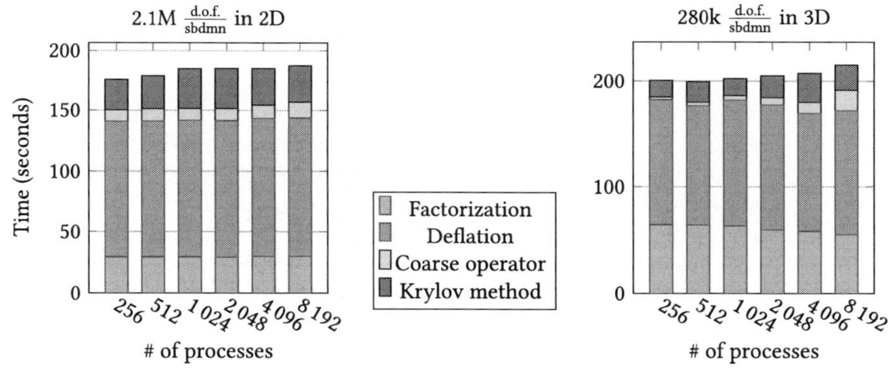

(b) *Comparison of the time spent in various steps for building and using the preconditioner*

	N	Factorization	Deflation	Solution	# of it.	Total	# of d.o.f.
3D	256	64.2 s	117.7 s	15.8 s	13	200.6 s	$74.6 \cdot 10^6$
	512	64.0 s	112.2 s	19.9 s	18	199.4 s	$144.7 \cdot 10^6$
	1 024	63.2 s	118.6 s	16.2 s	14	202.4 s	$288.8 \cdot 10^6$
	2 048	59.4 s	117.6 s	21.3 s	17	205.3 s	$578.0 \cdot 10^6$
	4 096	58.1 s	110.7 s	27.9 s	20	207.5 s	$1.2 \cdot 10^9$
	8 192	55.0 s	116.6 s	23.6 s	17	215.2 s	$2.3 \cdot 10^9$
2D	256	29.4 s	111.3 s	25.7 s	29	175.8 s	$696.0 \cdot 10^6$
	512	29.6 s	111.5 s	28.0 s	28	179.1 s	$1.4 \cdot 10^9$
	1 024	29.4 s	112.2 s	33.6 s	28	185.2 s	$2.8 \cdot 10^9$
	2 048	29.2 s	112.2 s	33.7 s	28	185.2 s	$5.6 \cdot 10^9$
	4 096	29.8 s	113.7 s	31.0 s	26	185.4 s	$11.2 \cdot 10^9$
	8 192	29.8 s	113.8 s	30.7 s	25	187.6 s	$22.3 \cdot 10^9$

(c) *Breakdown of the timings used for the top figure*

Figure 8.6. *Weak scaling experiments.*

Lemma 8.2. *Let* $\mathbf{U}, \mathbf{V} \in \mathbb{R}^n$. *We have the following formula for their scalar product* $(\mathbf{U}, \mathbf{V})$:

$$(\mathbf{U},\mathbf{V}) = \sum_{i=1}^{N} (R_i \mathbf{U}, D_i R_i \mathbf{V}).$$

Proof. Using the partition of unity (1.25)

$$\mathrm{Id} = \sum_{i=1}^{N} R_i^T D_i R_i ,$$

we have

$$(\mathbf{U},\mathbf{V}) = \left(\mathbf{U}, \sum_{i=1}^{N} R_i^T D_i R_i \mathbf{V}\right) = \sum_{i=1}^{N} (R_i \mathbf{U}, D_i R_i \mathbf{V}) = \sum_{i=1}^{N} (\mathbf{U}_i, D_i \mathbf{V}_i). \qquad \square$$

Local scalar products are performed concurrently. Thus, the implementation is parallel except for the sum which corresponds to an `MPI_AllReduce` call across the N MPI processes. Note also that the implementation relies on the knowledge of a partition of unity so that the FreeFem++ syntax is `dscalprod(D,u,v)`.

An `axpy` procedure $y \leftarrow \alpha x + y$ for $x, y \in \mathbb{R}^n$ and $\alpha \in \mathbb{R}$ is easily implemented concurrently for distributed vectors in the form

$$y_i \leftarrow \alpha x_i + y_i \quad \forall 1 \le i \le N .$$

The matrix-vector product is more intricate. Let $A \in \mathbb{R}^{n\times n}$ be a matrix and let $\mathbf{U}, \mathbf{V} \in \mathbb{R}^n$ be two global vectors. We want to compute $\mathbf{V} := A\mathbf{U}$ based on a distributed storage of $\mathbf{U}$ in the form $(\mathbf{U}_i)_{1\le i \le N}$. Using the partition of unity identity (1.25), we have for all $1 \le i \le N$

$$\mathbf{V}_i := R_i A \mathbf{U} = \sum_{j=1}^{N} R_i A R_j^T D_j R_j \mathbf{U} = \sum_{j=1}^{N} R_i A R_j^T D_j \mathbf{U}_j .$$

If matrix A arises from a finite element discretization of a partial differential equation, as is the case here when using FreeFem++, it is possible to simplify the matrix-vector product. The sparsity pattern of matrix A can be inferred from the graph of the underlying mesh. Let k, l be two degrees of freedom associated to the basis functions ϕ_k and ϕ_l. If their supports have a zero-measure intersection, then

$$A_{kl} = 0 .$$

We can take advantage of this sparsity pattern in the following way. A degree of freedom $k \in \mathcal{N}_j$ is interior to $\mathcal{N}_j$ if for all $i \ne j$ and all $l \in \mathcal{N}_i \setminus \mathcal{N}_j$, $A_{kl} = 0$. Otherwise, it is said to be a boundary degree of freedom. If the overlap is sufficient, it is possible to choose diagonal matrix D_j with zero entries for the boundary degrees of freedom. Then all nonzero rows of matrix $A R_j^T D_j$ have indices in $\mathcal{N}_j$, that is,

$$R_i A R_j^T D_j = R_i R_j^T R_j A R_j^T D_j .$$

Moreover, for non-neighboring subdomains indexed i and j, the interaction matrix $R_i A R_j^T$ is zero. Denoting $\mathcal{O}_i$ the neighbors of a subdomain i,

$$\mathcal{O}_i := \{j \in [\![1;N]\!] \, : \, j \neq i \text{ and } R_i A R_j^T \neq 0\},$$

we have

$$\mathbf{V}_i := R_i A \mathbf{U} = \sum_{j=1}^{N} R_i A R_j^T D_j R_j \mathbf{U} = \underbrace{(R_i A R_i^T)}_{A_{ii}} D_i \mathbf{U}_i + \sum_{j \in \mathcal{O}_i} R_i R_j^T \underbrace{(R_j A R_j^T)}_{A_{jj}} D_j \mathbf{U}_j .$$

Matrix $A_{jj} := R_j A R_j^T$ is local to domain j. The matrix operation $R_i R_j^T$ corresponds to copying data on the overlap from subdomain j to subdomain i. Therefore, the matrix-vector product is computed in three steps:

- concurrent computing of $(R_j A R_j^T) D_j \mathbf{U}_j$ for all $1 \leq j \leq N$;
- neighbor-to-neighbor MPI communications;
- concurrent sum of neighbor contributions.

More details can be found in the PhD thesis of P. Jolivet, written in English [116].

Since we have basic linear algebra subroutines, we have all the necessary ingredients for solving concurrently the linear system $A x = b$ by a Krylov method such as CG (conjugate gradient) or GMRES. We now turn our attention to domain decomposition methods. The ASM preconditioner reads as

$$M_{ASM}^{-1} := \sum_{j=1}^{N} R_j^T A_{jj}^{-1} R_j .$$

Let $\mathbf{R}$ be a distributed residual so that each MPI process stores $\mathbf{R}_j := R_j \mathbf{R}$, $1 \leq j \leq N$. We want to compute for each subdomain $1 \leq i \leq N$ the restriction of the application of the ASM preconditioner to $\mathbf{R}$:

$$R_i M_{ASM}^{-1} \mathbf{R} = R_i \sum_{j=1}^{N} R_j^T A_{jj}^{-1} R_j \mathbf{R} = A_{ii}^{-1} \mathbf{R}_i + \sum_{j \in \mathcal{O}_i} (R_i R_j^T) A_{jj}^{-1} \mathbf{R}_j .$$

This task is performed by first solving concurrently on all subdomains a linear system:

$$A_{jj} \mathbf{Z}_j = \mathbf{R}_j \, \forall 1 \leq j \leq n . \tag{8.4}$$

Then data transfers between neighboring subdomains implement the $R_i R_j^T \mathbf{Z}_j$ formula. The contributions from neighboring subdomains are summed locally. This pattern is very similar to that of the matrix-vector product.

The RAS preconditioner reads as

$$M_{RAS}^{-1} := \sum_{j=1}^{N} R_j^T D_j A_{jj}^{-1} R_j .$$

Its application to a distributed residual r consists in computing

$$R_i M_{RAS}^{-1} \mathbf{R} = R_i \sum_{j=1}^{N} R_j^T D_j A_{jj}^{-1} R_j \mathbf{R} = D_i A_{ii}^{-1} \mathbf{R}_i + \sum_{j \in \mathcal{O}_i} (R_i R_j^T) D_j A_{jj}^{-1} \mathbf{R}_j .$$

The only difference in the implementation of RAS compared to ASM lies in the concurrent multiplication of the local solution to (8.4) by the partition of unity matrix D_j before data transfer between neighboring subdomains.

Bibliography

[1] Y. Achdou, P. Le Tallec, F. Nataf, and M. Vidrascu. A domain decomposition preconditioner for an advection-diffusion problem. *Comput. Methods Appl. Mech. Engrg.*, 184:145–170, 2000. (Cited on p. 131)

[2] M. Adams, H. Bayraktar, T. Keaveny, and P. Papadopoulos. Ultrascalable implicit finite element analyses in solid mechanics with over a half a billion degrees of freedom. In *Proceedings of the 2004 ACM/IEEE Conference on Supercomputing, (SC04)*, IEEE Computer Society, 2004, 34:1–15, 2004. (Cited on pp. 207, 213)

[3] H. Alcin, B. Koobus, O. Allain, and A. Dervieux. Efficiency and scalability of a two-level Schwarz algorithm for incompressible and compressible flows. *Internat. J. Numer. Methods Fluids*, 72(1):69–89, 2013. ISSN: 0271-2091. DOI: `10.1002/fld.3733`. URL: `http://dx.doi.org/10.1002/fld.3733` (Cited on p. 115)

[4] A. Alonso-Rodríguez and L. Gerardo-Giorda. New nonoverlapping domain decomposition methods for the harmonic Maxwell system. *SIAM J. Sci. Comput.*, 28(1):102–122, 2006. (Cited on p. 72)

[5] P. R. Amestoy, I. S. Duff, J.-Y. L'Excellent, and J. Koster. A fully asynchronous multifrontal solver using distributed dynamic scheduling. *SIAM J. Matrix Anal. and Appl.*, 23(1):15–41, 2001. (Cited on p. 207)

[6] X. Antoine, Y. Boubendir, and C. Geuzaine. A quasi-optimal non-overlapping domain decomposition algorithm for the Helmholtz equation. *J. Comput. Phys.*, 231(2):262–280, 2012. (Cited on p. 72)

[7] S. Balay, S. Abhyankar, M.F. Adams, J. Brown, P. Brune, K. Buschelman, V. Eijkhout, W. D. Gropp, D. Kaushik, M. G. Knepley, L. Curfman McInnes, K. Rupp, B. F. Smith, and H. Zhang. PETSc Web Page 2014. URL: `http://www.mcs.anl.gov/petsc` (Cited on pp. viii, 203, 213)

[8] S. Balay, W. D. Gropp, L. Curfman McInnes, and B. F. Smith. Efficient management of parallelism in object oriented numerical software libraries. In *Modern Software Tools in Scientific Computing*. E. Arge, A. M. Bruaset, and H. P. Langtangen, editors. Birkhäuser Press, Basel, 1997, pp. 163–202. (Cited on pp. viii, 203, 213)

[9] R. Barrett, M. W. Berry, T. F. Chan, J. Demmel, J. Donato, J. Dongarra, V. Eijkhout, R. Pozo, C. Romine, and H. V. der Vorst. *Templates for the Solution of Linear Systems: Building Blocks for Iterative Methods*, SIAM, Philadelphia, 1993. (Cited on pp. 84, 93)

[10] H. Barucq, J. Diaz, and M. Tlemcani. New absorbing layers conditions for short water waves. *J. Comput. Phys.*, 229(1):58–72. ISSN: 0021-9991. DOI: `10.1016/j.jcp.2009.08.033`. URL: `http://dx.doi.org/10.1016/j.jcp.2009.08.033` (Cited on p. 154)

[11] G. K. Batchelor. *An Introduction to Fluid Dynamics*. Cambridge University Press, Cambridge, 1999. ISBN: 0-521-66396-2 (Cited on p. 154)

[12] M. Benzi. Preconditioning techniques for large linear systems: a survey. *J. Comput. Phys.*, 182(2):418–477, 2002. (Cited on p. viii)

[13] J. P. Berenger. A perfectly matched layer for the absorption of electromagnetic waves. *J. Comput. Phys.*, 114:185–200, 1994. (Cited on pp. 54, 72)

[14] J.-F. Bourgat, R. Glowinski, P. Le Tallec, and M. Vidrascu. Variational formulation and algorithm for trace operator in domain decomposition calculations. In *Domain Decomposition Methods*. Tony Chan, Roland Glowinski, Jacques Périaux, and Olof Widlund, editors. SIAM, Philadelphia, 1989, pp. 3–16. (Cited on pp. 131, 137, 191)

[15] H. Brézis. *Analyse Fonctionnelle: Théorie et Applications*. Dunod, Paris, 1983. (Cited on p. 139)

[16] P. N. Brown and Homer F. Walker. GMRES on (nearly) singular systems. *SIAM J. Matrix Anal. Appl.*, 18(1):37–51, 1997. (Cited on p. 97)

[17] X.-C. Cai, W. D. Gropp, D. E. Keyes, R. G. Melvin, and D. P. Young. Parallel Newton–Krylov–Schwarz algorithms for the transonic full potential equation. *SIAM J. Sci. Comput.*, 19:246–265, 1998. (Cited on p. viii)

[18] X.-C. Cai and D. E. Keyes. Nonlinearly preconditioned inexact Newton algorithms. *SIAM J. Sci. Comput.*, 24:183–200, 2002. (Cited on p. viii)

[19] X.-C. Cai, M. A. Casarin, F. W. Elliott Jr., and O. B. Widlund. Overlapping Schwarz algorithms for solving Helmholtz's equation. *Contemporary Mathematics*, vol. 218. Amer. Math. Soc., Providence, RI, 1998, pp. 391–399. (Cited on p. 62)

[20] X.-C. Cai, C. Farhat, and M. Sarkis. A minimum overlap restricted additive Schwarz preconditioner and applications to 3D flow simulations. *Contemporary Mathematics*, vol. 218. Amer. Math. Soc., Providence, RI, 1998, pp. 479–485. (Cited on p. 5)

[21] X.-C. Cai and M. Sarkis. A restricted additive Schwarz preconditioner for general sparse linear systems. *SIAM J. Sci. Comput.*, 21:792–797, 1999. (Cited on pp. 5, 187)

[22] X.-C. Cai and O. B. Widlund. Domain decomposition algorithms for indefinite elliptic problems. SIAM J. Sci. Statist. Comput., 13(1):243–258, 1992. (Cited on p. 62)

[23] T. F. Chan and T. P. Mathew. Domain decomposition algorithms. In *Acta Numerica 1994*, pp. 61–143. Cambridge University Press, Cambridge, 1994. (Cited on p. vii)

[24] A. Chapman and Y. Saad. Deflated and augmented Krylov subspace techniques. *Numer. Linear Algebra Appl.*, 4(1):43–66, 1997. ISSN: 1070-5325. DOI: `10.1002/(SICI)1099-1506(199701/02)4:1<43::AID-NLA99>3.0.CO;2-Q`. (Cited on p. 106)

[25] T. Chartier, R. D. Falgout, V. E. Henson, J. Jones, T. Manteuffel, S. McCormick, J. Ruge, and P. S. Vassilevski. Spectral AMGe (ρAMGe). *SIAM J. Sci. Comput.*, 25(1):1–26, 2003. ISSN: 1064-8275. DOI: `10.1137/S106482750139892X`. URL: `http://dx.doi.org/10.1137/S106482750139892X` (Cited on p. 129)

[26] P. L. Chebyshev. *Théorie des mécanismes connus sous le nom de parallélogrammes*. Imprimerie de l'Académie impériale des Sciences, 1853. (Cited on p. 85)

[27] C. Chevalier and F. Pellegrini. PT-SCOTCH: A tool for efficient parallel graph ordering. *Parallel Computing*, 6–8(34):318–331, 2008. (Cited on pp. 11, 23, 113, 117)

[28] P. Chevalier. Méthodes numériques pour les tubes hyperfréquences. Résolution par décomposition de domaine. PhD thesis, Université Paris VI, 1998. (Cited on p. 72)

[29] P. Chevalier and F. Nataf. Symmetrized method with optimized second-order conditions for the Helmholtz equation. *Contemporary Mathematics*, vol. 218. Amer. Math. Soc., Providence, RI, 1998, pp. 400–407. (Cited on p. 72)

[30] W. C. Chew and W. H. Weedon. A 3D perfectly matched medium from modified Maxwell's equations with stretched coordinates. Microwave Opt. Technol. Lett., 7:599–604, 1994. (Cited on pp. 54, 72)

[31] P. G. Ciarlet. *The Finite Element Method for Elliptic Problems*. North–Holland, Amsterdam, 1978. (Cited on p. 139)

[32] T. Cluzeau, V. Dolean, F. Nataf, and A. Quadrat. Symbolic preconditioning techniques for linear systems of partial differential equations. Technical report (7953), INRIA, 2012. URL: `http://hal.inria.fr/hal-00694468`. (Cited on pp. 154, 158)

[33] T. Cluzeau, V. Dolean, F. Nataf, and A. Quadrat. Symbolic techniques for domain decomposition methods. In *Domain Decomposition Methods in Science and Engineering XX, Lecture Notes in Computational Science and Engineering*, vol. 91. Springer, Berlin, 2013, pp. 27–38. URL: `http://hal.archives-ouvertes.fr/hal-00664092`. (Cited on pp. 154, 158)

[34] Fr. Collino, G. Delbue, P. Joly, and A. Piacentini. A new interface condition in the non-overlapping domain decomposition for the Maxwell equations. *Comput. Methods Appl. Mech. Engrg.*, 148:195–207, 1997. (Cited on p. 72)

[35] F. Collino, S. Ghanemi, and P. Joly. Domain decomposition method for harmonic wave propagation: a general presentation. *Comput. Methods Appl. Mech. Engrg.*, 184(2–4):171–211, 2000. Vistas in domain decomposition and parallel processing in computational mechanics. ISSN: 0045-7825. DOI: `10.1016/S0045-7825(99)00228-5`. URL: `http://dx.doi.org/10.1016/S0045-7825(99)00228-5` (Cited on p. 39)

[36] L. Conen, V. Dolean, R. Krause, and F. Nataf. A coarse space for heterogeneous Helmholtz problems based on the Dirichlet-to-Neumann operator. *J. Comput. Appl. Math.*, 271:83–99, 2014. ISSN: 0377-0427. DOI: `10.1016/j.cam.2014.03.031`. URL: `http://dx.doi.org/10.1016/j.cam.2014.03.031` (Cited on pp. 75, 115)

[37] L. C. Cowsar, J. Mandel, and M. F. Wheeler. Balancing domain decomposition for mixed finite elements. *Math. Comp.*, 64(211):989–1015, 1995. (Cited on pp. 138, 142)

[38] T. A. Davis and I. S. Duff. A combined unifrontal-multifrontal method for unsymmetric sparse matrices. *ACM Trans. Math. Software*, 25(1):1–19, 1999. (Cited on p. 21)

[39] Y.-H. De Roeck and P. Le Tallec. Analysis and test of a local domain decomposition preconditioner. In *Fourth International Symposium on Domain Decomposition Methods for Partial Differential Equations*. Roland Glowinski, Yuri Kuznetsov, Gérard Meurant, Jacques Périaux, and Olof Widlund, editors. SIAM, Philadelphia, 1991, pages 112–128. (Cited on p. 131)

[40] E. de Sturler. Lecture notes on iterative methods. 2014. URL: `http://www.math.vt.edu/people/sturler/LectureNotes/IterativeMethods.html` (Cited on p. 77)

[41] Q. Deng, An analysis for a nonoverlapping domain decomposition iterative procedure, *SIAM J. Sci. Comput.*, 18:1517–1525, 1997. (Cited on p. 72)

[42] B. Després. Décomposition de domaine et problème de Helmholtz. *C.R. Acad. Sci. Paris*, 1(6):313–316, 1990. (Cited on p. 59)

[43] B. Després. Domain decomposition method and the Helmholtz problem. In *Mathematical and Numerical Aspects of Wave Propagation Phenomena (Strasbourg, 1991)*, Philadelphia, 1991, SIAM, pages 44–52. (Cited on pp. 45, 64)

[44] B. Després. Domain decomposition method and the Helmholtz problem. II. In *Second International Conference on Mathematical and Numerical Aspects of Wave Propagation (Newark, DE, 1993)*. SIAM, Philadelphia, 1993, pages 197–206. (Cited on pp. 17, 35, 38, 41, 62)

[45] B. Després. Méthodes de décomposition de domaine pour les problèmes de propagation d'ondes en régimes harmoniques. PhD thesis, Paris IX, 1991. (Cited on p. 45)

[46] B. Després, P. Joly, and J. E. Roberts. A domain decomposition method for the harmonic Maxwell equations. In *Iterative Methods in Linear Algebra (Brussels, 1991)*. North-Holland, Amsterdam, 1992, pp. 475–484. (Cited on pp. 38, 41, 46, 59)

[47] C. R. Dohrmann. A preconditioner for substructuring based on constrained energy minimization. *SIAM J. Sci. Comput.*, 25(1):246–258, 2003. ISSN: 1064-8275. DOI: `10.1137/S1064827502412887`. URL: `http://dx.doi.org/10.1137/S1064827502412887` (Cited on pp. 131, 147)

[48] C. R. Dohrmann and Olof B. Widlund. An overlapping Schwarz algorithm for almost incompressible elasticity. *SIAM J. Numer. Anal.*, 47(4):2897–2923, 2009. (Cited on p. 113)

[49] C. R. Dohrmann and Olof B. Widlund. Hybrid domain decomposition algorithms for compressible and almost incompressible elasticity. *Internat. J. Numer. Methods Engrg.*, 82(2):157–183, 2010. (Cited on p. 113)

[50] V. Dolean, M. J. Gander, and L.G. Giorda. Optimized Schwarz methods for Maxwell equations. *SIAM J. Sci. Comput.*, 31(3):2193–2213, 2009. (Cited on p. 72)

[51] V. Dolean, M. J. Gander, S. Lanteri, J.-F. Lee, and Z. Peng. Effective transmission conditions for domain decomposition methods applied to the time-harmonic curl–curl Maxwell's equations. *J. Comput. Phys.*, 280:232–247, 2015. (Cited on p. 72)

[52] V. Dolean, M. J. Gander, S. Lanteri, J.-F. Lee, and Z. Peng. Optimized Schwarz methods for curl-curl time-harmonic Maxwell's equations. In *Proceedings of the 21st International Domain Decomposition Conference. Lecture Notes in Computational Science and Engineering*, vol. 98, Jocelyne Erhel, Martin J. Gander, Laurence Halpern, Taoufik Sassi, and Olof Widlund, editors. Springer, Berlin, 2014, pp. 587–595. (Cited on p. 72)

[53] V. Dolean, S. Lanteri, and F. Nataf. Construction of interface conditions for solving compressible Euler equations by non-overlapping domain decomposition methods. *Internat. J. Numer. Meth. Fluids*, 40:1485–1492, 2002. (Cited on p. 72)

[54] V. Dolean, S. Lanteri, and F. Nataf. Convergence analysis of a Schwarz type domain decomposition method for the solution of the Euler equations. *Appl. Numer. Math.*, 49:153–186, 2004. (Cited on p. 72)

[55] V. Dolean, S. Lanteri, and F. Nataf. Optimized interface conditions for domain decomposition methods in fluid dynamics. *Internat. J. Numer. Methods Fluids*, 40:1539–1550, 2002. (Cited on p. 72)

[56] V. Dolean, S. Lanteri, and R. Perrussel. A domain decomposition method for solving the three-dimensional time-harmonic Maxwell equations discretized by discontinuous Galerkin methods. *J. Comput. Phys.*, 227(3):2044–2072, 2008. (Cited on p. 72)

[57] V. Dolean, S. Lanteri, and R. Perrussel. Optimized Schwarz algorithms for solving time-harmonic Maxwell's equations discretized by a discontinuous Galerkin method. *IEEE Trans. Magn.*, 44(6):954–957, 2008. (Cited on p. 72)

[58] V. Dolean and F. Nataf. A new domain decomposition method for the compressible Euler equations. *ESAIM Math. Model. Numer. Anal.*, 40(4):689–703, 2006. ISSN: 0764-583X. DOI: 10.1051/m2an:2006026. URL: http://dx.doi.org/10.1051/m2an:2006026 (Cited on pp. 154, 158)

[59] V. Dolean and F. Nataf. A new domain decomposition method for the compressible Euler equations using Smith factorization. In *Domain Decomposition Methods in Science and Engineering XVII. Lecture Notes in Computational Science and Engineering* vol. 60. Springer, Berlin, 2008, pp. 331–338. DOI: 10.1007/978-3-540-75199-1_40. URL: http://dx.doi.org/10.1007/978-3-540-75199-1_40 (Cited on p. 154)

[60] V. Dolean, F. Nataf, and G. Rapin. Deriving a new domain decomposition method for the Stokes equations using the Smith factorization. *Math. Comp.*, 78(266):789–814, 2009. ISSN: 0025-5718. DOI: 10.1090/S0025-5718-08-02172-8. URL: http://dx.doi.org/10.1090/S0025-5718-08-02172-8 (Cited on pp. 154, 158, 159)

[61] V. Dolean, F. Nataf, and G. Rapin. How to use the Smith factorization for domain decomposition methods applied to the Stokes equations. In *Domain Decomposition Methods in Science and Engineering XVII. Lecture Notes in Computational Science and Engineering* vol. 60. Springer, Berlin, 2008, pp. 477–484. DOI: 10.1007/978-3-540-75199-1_60. URL: http://dx.doi.org/10.1007/978-3-540-75199-1_60 (Cited on p. 154)

[62] V. Dolean, F. Nataf, R. Scheichl, and N. Spillane. Analysis of a two-level Schwarz method with coarse spaces based on local Dirichlet-to-Neumann maps. *Comput. Methods Appl. Math.*, 12(4):391–414, 2012. ISSN: 1609-4840. DOI: 10.2478/cmam-2012-0027. URL: http://dx.doi.org/10.2478/cmam-2012-0027 (Cited on pp. 122, 129, 161)

[63] Z. Dostál, David Horák, and Dan Stefanica. A scalable feti-dp algorithm for a coercive variational inequality. *Applied Numerical Mathematics*, 54(3–4):378 – 390, 2005. Selected papers from the 16th Chemnitz Finite Element Symposium 2003. (Cited on p. viii)

[64] M. Dryja, M. V. Sarkis, and O. B. Widlund. Multilevel Schwarz methods for elliptic problems with discontinuous coefficients in three dimensions. *Numer. Math.*, 72(3):313–348, 1996. (Cited on p. 113)

[65] M. Dryja and O. B. Widlund. Schwarz methods of Neumann-Neumann type for three-dimensional elliptic finite element problems. *Comm. Pure Appl. Math.*, 48(2):121–155, 1995. (Cited on p. 195)

[66] M. Dryja and O. B. Widlund. Some domain decomposition algorithms for elliptic problems. In *Iterative Methods for Large Linear Systems, Proceedings of the Conference on Iterative Methods for Large Linear Systems* (Austin, TX, 1988). Linda Mayes and David Kincaid, editors. Academic Press, San Diego, CA, 1989, pages 273–291. (Cited on p. 5)

[67] O. Dubois. Optimized Schwarz methods for the advection-diffusion equation and for problems with discontinuous coefficients. PhD thesis. McGill University, 2007. (Cited on p. 72)

[68] O. Dubois, M. J. Gander, S. Loisel, A. St-Cyr, and D. B. Szyld. The optimized Schwarz method with a coarse grid correction. *SIAM J. Sci. Comput.*, 34(1):A421–A458, 2012. ISSN: 1064-8275. DOI: 10.1137/090774434. URL: http://dx.doi.org/10.1137/090774434. (Cited on p. 72)

[69] I. S. Duff and J. K. Reid. The multifrontal solution of indefinite sparse symmetric linear equations. *ACM Trans. Math. Software*, 9(3):302–325, 1983. ISSN: 0098-3500. DOI: 10.1145/356044.356047. URL: http://dx.doi.org/10.1145/356044.356047. (Cited on p. 131)

[70] Y. Efendiev, J. Galvis, R. Lazarov, and J. Willems. Robust domain decomposition preconditioners for abstract symmetric positive definite bilinear forms. *ESAIM Math. Model. Numer. Anal.*, 46(5):1175–1199, 2012. ISSN: 0764-583X. DOI: `10.1051/m2an/2011073`. URL: `http://dx.doi.org/10.1051/m2an/2011073`. (Cited on pp. 129, 161, 162)

[71] E. Efstathiou and M. J. Gander. RAS: Understanding restricted additive Schwarz. Technical report 06, McGill University, 2002. (Cited on p. 3)

[72] M. El Bouajaji, V. Dolean, M. J. Gander, and S. Lanteri. Comparison of a one and two parameter family of transmission conditions for Maxwell's equations with damping. In *Domain Decomposition Methods in Science and Engineering XX*, Springer, Berlin, 2012. (Cited on p. 72)

[73] M. El Bouajaji, V. Dolean, M. J. Gander, and S. Lanteri. Optimized Schwarz methods for the time-harmonic Maxwell equations with dampimg. *SIAM J. Sci. Comput.*, 34(4):A2048–A2071, 2012. (Cited on p. 72)

[74] M. El Bouajaji, V. Dolean, M. J. Gander, S. Lanteri, and R. Perrussel. Domain decomposition methods for electromagnetic wave propagation problems in heterogeneous media and complex domains. In *Domain Decomposition Methods in Science and Engineering XIX*, Springer, Berlin, 2011, pp. 5–16. (Cited on p. 72)

[75] B. Engquist and A. Majda. Absorbing boundary conditions for the numerical simulation of waves. *Math. Comp.*, 31(139):629–651, 1977. (Cited on pp. 54, 57, 64)

[76] J. Erhel and F. Guyomarc'h. An augmented conjugate gradient method for solving consecutive symmetric positive definite linear systems. *SIAM J. Matrix Anal. Appl.*, 21(4):1279–1299, 2000. (Cited on p. 106)

[77] C. Farhat, P. Avery, R. Tezaur, and J. Li. FETI-DPH: a dual-primal domain decomposition method for acoustic scattering. *J. Comput. Acoust.*, 13(3):499–524, 2005. ISSN: 0218-396X. DOI: `10.1142/S0218396X05002761`. URL: `http://dx.doi.org/10.1142/S0218396X05002761`. (Cited on p. 115)

[78] C. Farhat, M. Lesoinne, P. LeTallec, K. Pierson, and D. Rixen. FETI-DP: A dual–primal unified FETI method. I. A faster alternative to the two-level FETI method. *Internat. J. Numer. Methods Engrg.*, 50(7):1523–1544, 2001. ISSN: 0029-5981. DOI: `10.1002/nme.76`. URL: `http://dx.doi.org/10.1002/nme.76`. (Cited on pp. 131, 147)

[79] C. Farhat, A. Macedo, and M. Lesoinne. A two-level domain decomposition method for the iterative solution of high-frequency exterior Helmholtz problems. *Numer. Math.*, 85(2):283–303, 2000. (Cited on p. 115)

[80] C. Farhat and F.-X. Roux. A method of finite element tearing and interconnecting and its parallel solution algorithm. *Internat. J. Numer. Methods Engrg.*, 32:1205–1227, 1991. (Cited on pp. 35, 138, 142)

[81] C. Farhat and F.-X. Roux. An unconventional domain decomposition method for an efficient parallel solution of large-scale finite element systems. *SIAM J. Sci. Statist. Comput.*, 13:379–396, 1992. (Cited on p. 131)

[82] E. Flauraud. Méthode de décomposition de domaine pour des milieux poreux faillés. PhD thesis, Paris VI, 2001. (Cited on p. 72)

[83] J. Galvis and Y. Efendiev. Domain decomposition preconditioners for multiscale flows in high contrast media: Reduced dimension coarse spaces. *Multiscale Model. Simul.*, 8(5):1621–1644, 2010. ISSN: 1540-3459. DOI: `10.1137/100790112`. URL: `http://dx.doi.org/10.1137/100790112`. (Cited on pp. 129, 161)

[84] J. Galvis and Y. Efendiev. Domain decomposition preconditioners for multiscale flows in high-contrast media. *Multiscale Model. Simul.*, 8(4):1461–1483, 2010. ISSN: 1540-3459. DOI: 10.1137/090751190. URL: http://dx.doi.org/10.1137/090751190. (Cited on pp. 129, 161)

[85] M. J. Gander. On the influence of geometry on optimized Schwarz methods. *SēMA J.*, 53(1):71–78, 2011. ISSN: 1575-9822 (Cited on p. 72)

[86] M. J. Gander. Optimized Schwarz methods. *SIAM J. Numer. Anal.*, 44(2):699–731, 2006. (Cited on pp. 59, 72)

[87] M. J. Gander. Optimized Schwarz methods for Helmholtz problems. In *Thirteenth International Conference on Domain Decomposition Methods*, 2001, pp. 245–252. (Cited on p. 67)

[88] M. J. Gander. Schwarz methods over the course of time. *Electron. Trans. Numer. Anal.*, 31:228–255, 2008. (Cited on p. 1)

[89] M. J. Gander, L. Halpern, and F. Magoulès. An optimized Schwarz method with two-sided Robin transmission conditions for the Helmholtz equation. *Internat. J. Numer. Methods Fluids*, 55(2):163–175, 2007. (Cited on p. 72)

[90] M. J. Gander and F. Kwok. Best Robin parameters for optimized Schwarz methods at cross points. *SIAM J. Sci. Comput.*, 34(4):A1849–A1879, 2012. ISSN: 1064-8275. DOI: 10.1137/110837218. URL: http://dx.doi.org/10.1137/110837218. (Cited on p. 72)

[91] M. J. Gander, F. Magoulès, and F. Nataf. Optimized Schwarz methods without overlap for the Helmholtz equation. *SIAM J. Sci. Comput.*, 24(1):38–60, 2002. (Cited on pp. 51, 62, 65, 72)

[92] M. J. Gander and F. Nataf. AILU: A preconditioner based on the analytic factorization of the elliptic operator. *Numer. Linear Algebra Appl.*, 7(7–8):505–526, 2000. Preconditioning techniques for large sparse matrix problems in industrial applications (Minneapolis, MN, 1999). ISSN: 1070-5325. DOI: 10.1002/1099-1506(200010/12)7:7/8<505::AID-NLA210>3.0.CO;2-Z. (Cited on p. 72)

[93] M. J. Gander and Y. Xu. Optimized Schwarz methods for circular domain decompositions with overlap. *SIAM J. Numer. Anal.*, 52(4):1981–2004, 2014. ISSN: 0036-1429. DOI: 10.1137/130946125. URL: http://dx.doi.org/10.1137/130946125. (Cited on p. 72)

[94] F. R. Gantmacher. *Theorie des matrices*, Dunod, Paris, 1966. (Cited on p. 155)

[95] L. Gerardo-Giorda, P. Le Tallec, and F. Nataf. A Robin–Robin preconditioner for advection–diffusion equations with discontinuous coefficients. *Comput. Methods Appl. Mech. Engrg.*, 193:745–764, 2004. (Cited on p. 131)

[96] L. Gerardo-Giorda and F. Nataf. Optimized Schwarz methods for unsymmetric layered problems with strongly discontinuous and anisotropic coefficients. Technical report (561). submitted. Ecole Polytechnique, CMAP, CNRS UMR 7641, Paris, France, 2004. URL: http://www.cmap.polytechnique.fr/preprint/repository/561.pdf (Cited on p. 72)

[97] L. Giraud and A. Haidar. Parallel algebraic hybrid solvers for large 3D convection-diffusion problems. *Numer. Algorithms*, 51(2):151–177, 2009. ISSN: 1017-1398. DOI: 10.1007/s11075-008-9248-x. URL: http://dx.doi.org/10.1007/s11075-008-9248-x. (Cited on p. 190)

[98] L. Giraud, A. Haidar, and L. T. Watson. Parallel scalability study of hybrid preconditioners in three dimensions. *Parallel Comput.*, 34(6–8):363–379, 2008. ISSN: 0167-8191. DOI: 10.1016/j.parco.2008.01.006. URL: http://dx.doi.org/10.1016/j.parco.2008.01.006. (Cited on p. 190)

[99] D. Givoli. *Numerical Methods for Problems in Infinite Domains*, Elsevier, Amsterdam, 1992. (Cited on p. 54)

[100] G. H. Golub and C. F. Van Loan. *Matrix Computations, 2nd edition*, Johns Hopkins University Press, Baltimore, MD, 1989. (Cited on pp. 85, 93)

[101] P. Gosselet and C. Rey. Non-overlapping domain decomposition methods in structural mechanics. *Arch. Comput. Methods Engrg.*, 13(4):515–572, 2006. ISSN: 1134-3060. DOI: 10.1007/BF02905857. URL: http://dx.doi.org/10.1007/BF02905857. (Cited on p. 154)

[102] I. Graham, P. O. Lechner, and R. Scheichl. Domain decomposition for multiscale PDEs. *Numer. Math.*, 106(4):589–626, 2007. (Cited on p. 161)

[103] A. Greenbaum. *Iterative Methods for Solving Linear Systems. Frontiers in Applied Mathematics* vol. 17, SIAM, Philadelphia, 1997. ISBN: 0-89871-396-X (Cited on pp. 78, 93)

[104] A. Greenbaum, V. Pták, and Z. Strakoš. Any nonincreasing convergence curve is possible for GMRES. *SIAM J. Matrix Anal. Appl.*, 17(3):465–469, 1996. (Cited on p. 93)

[105] M. Griebel and P. Oswald. On the abstract theory of additive and multiplicative Schwarz algorithms. *Numer. Math.*, 70(2):163–180, 1995. ISSN: 0029-599X. DOI: 10.1007/s002110050115. URL: http://dx.doi.org/10.1007/s002110050115. (Cited on pp. 164, 165)

[106] R. Haferssas, P. Jolivet, and F. Nataf. A robust coarse space for optimized Schwarz methods SORAS-GenEO-2. 2015. URL: https://hal.archives-ouvertes.fr/hal-01100926 (Cited on p. 186)

[107] T. Hagstrom, R. P. Tewarson, and A. Jazcilevich. Numerical experiments on a domain decomposition algorithm for nonlinear elliptic boundary value problems. *Appl. Math. Lett.*, 1(3):299–302, 1988. (Cited on p. 56)

[108] F. Hecht. New development in freefem++. *J. Numer. Math.*, 20(3–4):251–265, 2012. ISSN: 1570-2820 (Cited on pp. viii, 18, 189, 203)

[109] Van Emden Henson and Ulrike Meier Yang. BoomerAMG: A parallel algebraic multigrid solver and preconditioner. *Appl. Numer. Math.*, 41(1):155–177, 2002. (Cited on p. 208)

[110] R. L. Higdon. Absorbing boundary conditions for difference approximations to the multidimensional wave equations. *Math. Comp.*, 47(176):437–459, 1986. (Cited on p. 66)

[111] R. A. Horn and C. R. Johnson. *Matrix Analysis, 2nd edition*, Cambridge University Press, Cambridge, 2013. ISBN: 978-0-521-54823-6 (Cited on p. 93)

[112] C. Japhet. Optimized Krylov-Ventcell method. Application to convection-diffusion problems. In *Proceedings of the 9th International Conference on Domain Decomposition Methods*, Petter E. Bjørstad, Magne S. Espedal, and David E. Keyes, editors. 1998, pp. 382–389. (Cited on p. 65)

[113] C. Japhet and F. Nataf. The best interface conditions for domain decomposition methods: Absorbing boundary conditions. In *Absorbing Boundaries and Layers, Domain Decomposition Methods: Applications to Large Scale Computations*. L. Tourrette and L. Halpern, editors. Nova Science, New York. 2001. (Cited on pp. 59, 68, 72)

[114] C. Japhet, F. Nataf, and F. Rogier. The optimized order 2 method. Application to convection-diffusion problems. *Future Gener. Comput. Syst.*, 18(1):17–30, 2001. (Cited on pp. 59, 65, 72)

[115] C. Japhet, F. Nataf, and F.-X. Roux. The optimized order 2 method with a coarse grid preconditioner. Application to convection-diffusion problems. In *Ninth International Conference on Domain Decompositon Methods in Science and Engineering*, P. Bjorstad, M. Espedal, and D. Keyes, editors. John Wiley & Sons, New York, 1998, pages 382–389. (Cited on pp. 72, 104)

[116] P. Jolivet. Méthodes de décomposition de domaine. Application au calcul haute performance. PhD thesis, Université de Grenoble, 2014. URL: `http://people.inf.ethz.ch/pjolivet/thesis.pdf`. (Cited on p. 221)

[117] P. Jolivet, V. Dolean, F. Hecht, F. Nataf, C. Prud'homme, and N. Spillane. High performance domain decomposition methods on massively parallel architectures with freefem++. *J. Numer. Math.*, 20(3–4):287–302, 2012. ISSN: 1570-2820. (Cited on p. 113)

[118] P. Jolivet, F. Hecht, F. Nataf, and C. Prud'homme. Scalable domain decomposition preconditioners for heterogeneous elliptic problems. In *Proceedings of the 2013 ACM/IEEE Conference on Supercomputing (SC13)*. Best paper finalist. ACM, 2013, 80:1–11. (Cited on p. 213)

[119] G. Karypis and V. Kumar. METIS: A software package for partitioning unstructured graphs, partitioning meshes, and computing fill-reducing orderings of sparse matrices. Technical report, Department of Computer Science, University of Minnesota, Minneapolis, MN, 1998. URL: `http://glaros.dtc.umn.edu/gkhome/views/metis`. (Cited on pp. 113, 117)

[120] G. Karypis and V. Kumar. METIS—unstructured graph partitioning and sparse matrix ordering system, version 2.0. Technical report, Department of Computer Science, University of Minnesota, Minneapolis, MN, 1995. (Cited on pp. 11, 23)

[121] J.-H. Kimn and M. Sarkis. Restricted overlapping balancing domain decomposition methods and restricted coarse problems for the Helmholtz problem. *Comput. Methods Appl. Mech. Engrg.*, 196(8):1507–1514, 2007. ISSN: 0045-7825. DOI: `10.1016/j.cma.2006.03.016`. URL: `http://dx.doi.org/10.1016/j.cma.2006.03.016`. (Cited on p. 115)

[122] A. Klawonn, M. Lanser, and O. Rheinbach. Nonlinear FETI-DP and BDDC methods. *SIAM J. Sci. Comput.*, 36(2):A737–A765, 2014. (Cited on p. viii)

[123] A. Klawonn and O. B. Widlund. FETI and Neumann–Neumann iterative substructuring methods: Connections and new results. *Comm. Pure Appl. Math.*, 54:57–90, 2001. (Cited on pp. 138, 142)

[124] A. Le Hyaric. Freefem++-cs: An integrated environment for freefem++. URL: `https://www.ljll.math.upmc.fr/lehyaric/ffcs/doc.php` (Cited on p. 18)

[125] P. Le Tallec. Domain decomposition methods in computational mechanics. In *Computational Mechanics Advances, vol. 1 (2)*, J. Tinsley Oden, editor. pp. 121–220. North-Holland, Amsterdam, 1994. (Cited on pp. 131, 195)

[126] S.-C. Lee, Marinos Vouvakis, and Jin-Fa Lee. A non-overlapping domain decomposition method with non-matching grids for modeling large finite antenna arrays. *J. Comput. Phys.*, 203(1):1–21, 2005. (Cited on p. 72)

[127] R. B. Lehoucq, D. C. Sorensen, and C. Yang. ARPACK users' guide: soluton of large-scale eigenvalue problems with implicitly restarted Arnoldi methods. SIAM, Philadelpha, 1998. (Cited on p. 216)

[128] P.-L. Lions. On the Schwarz alternating method. II. In *Domain Decomposition Methods*, Tony Chan, Roland Glowinski, Jacques Périaux, and Olof Widlund, editors. SIAM, Philadelphia, 1989, pp. 47–70. (Cited on p. 1)

[129] P.-L. Lions. On the Schwarz alternating method. III. In *Third International Symposium on Domain Decomposition Methods for Partial Differential Equations*, Tony F. Chan, Roland Glowinski, Jacques Périaux, and Olof Widlund, editors. SIAM, Philadelphia, 1990. (Cited on p. 1)

[130] P.-L. Lions. On the Schwarz alternating method. In *First International Symposium on Domain Decomposition Methods for Partial Differential Equations*, Tony F. Chan, Roland Glowinski, Jacques Périaux, and Olof Widlund, editors. SIAM, Philadelphia, 1990. (Cited on pp. 17, 35, 38, 55, 59)

[131] G. Lube, L. Mueller, and F.-C. Otto. A non-overlapping domain decomposition method for the advection-diffusion problem. *Computing*, 64:49–68, 2000. (Cited on p. 72)

[132] J. Mandel. Balancing domain decomposition. *Comm. Appl. Numer. Methods*, 9:233–241, 1992. (Cited on pp. 35, 131, 178, 179, 181, 188)

[133] J. Mandel and M. Brezina. Balancing domain decomposition for problems with large jumps in coefficients. *Math. Comp.*, 65(216):1387–1401, 1996. (Cited on pp. 113, 195)

[134] J. Mandel and M. Brezina. Balancing domain decomposition: Theory and computations in two and three dimensions. Technical report (UCD/CCM 2). Center for Computational Mathematics, University of Colorado at Denver, 1993. (Cited on p. 131)

[135] J. Mandel, C. R. Dohrmann, and Radek Tezaur. An algebraic theory for primal and dual substructuring methods by constraints. *Appl. Numer. Math.*, 54:167–193, 2005. (Cited on p. 131)

[136] J. Mandel and B. Sousedík. BDDC and FETI-DP under minimalist assumptions. *Computing*, 81(4):269–280, 2007. ISSN: 0010-485X. DOI: `10.1007/s00607-007-0254-y`. URL: `http://dx.doi.org/10.1007/s00607-007-0254-y`. (Cited on p. 147)

[137] J. Mandel and B. Sousedík. Coarse spaces over the ages. In *Domain Decomposition Methods in Science and Engineering XIX, Lecture Notes in Computational Science and Engineering*, vol. 78. Springer, Berlin, 2011, pp. 213–220. DOI: `10.1007/978-3-642-11304-8_23`. URL: `http://dx.doi.org/10.1007/978-3-642-11304-8_23`. (Cited on p. 106)

[138] T. P. A. Mathew (ed). Domain decomposition methods for the numerical solution of partial differential equations. *Lecture Notes in Computational Science and Engineering*, vol. 6. Springer, Berlin, 2008. ISBN: 978-3-540-77205-7. DOI: `10.1007/978-3-540-77209-5`. URL: `http://dx.doi.org/10.1007/978-3-540-77209-5`. (Cited on p. vii)

[139] A. M. Matsokin and S. V. Nepomnyaschikh. A Schwarz alternating method in a subspace. *Soviet Math.*, 29(10):78–84, 1985. (Cited on p. 5)

[140] G. Meurant. *Computer solution of large linear systems*, volume 28 of *Studies in Mathematics and its Applications*. North-Holland Publishing Co., Amsterdam, 1999. (Cited on p. viii)

[141] G. Meurant. *The Lanczos and Conjugate Gradient Algorithms: From Theory to Finite Precision Computations. Software, Environments, and Tools*, vol. 19. SIAM, Philadelphia, 2006. ISBN: 978-0-898716-16-0. DOI: `10.1137/1.9780898718140`. URL: `http://dx.doi.org/10.1137/1.9780898718140`. (Cited on p. 93)

[142] F. Nataf. A new approach to perfectly matched layers for the linearized Euler system. *J. Comput. Phys.*, 214(2):757–772, 2006. ISSN: 0021-9991. DOI: 10.1016/j.jcp.2005.10.014. URL: http://dx.doi.org/10.1016/j.jcp.2005.10.014. (Cited on p. 154)

[143] F. Nataf. Absorbing boundary conditions in block Gauss–Seidel methods for convection problems. *Math. Models Methods Appl. Sci.*, 6(4):481–502, 1996. (Cited on pp. 60, 72)

[144] F. Nataf and F. Nier. Convergence rate of some domain decomposition methods for overlapping and nonoverlapping subdomains. *Numer. Math.*, 75(3):357–77, 1997. (Cited on p. 72)

[145] F. Nataf and F. Rogier. Factorization of the convection-diffusion operator and the Schwarz algorithm. *Math. Models Methods Appl. Sci.*, 5(1):67–93, 1995. (Cited on p. 41)

[146] F. Nataf, F. Rogier, and E. de Sturler. Optimal interface conditions for domain decomposition methods. Technical report (301). Ecole Polytechnique, CMAP, Paris, France, 1994. (Cited on p. 56)

[147] F. Nataf, H. Xiang, and V. Dolean. A two level domain decomposition preconditioner based on local Dirichlet-to-Neumann maps. *C. R. Mathématique*, 348(21–22):1163–1167, 2010. (Cited on pp. 113, 129)

[148] F. Nataf, H. Xiang, V. Dolean, and N. Spillane. A coarse space construction based on local Dirichlet-to-Neumann maps. *SIAM J. Sci Comput.*, 33(4):1623–1642, 2011. (Cited on pp. 113, 129, 161)

[149] S. V. Nepomnyaschikh. Decomposition and fictitious domains for elliptic boundary value problems. In *Fifth International Symposium on Domain Decomposition Methods for Partial Differential Equations*. David E. Keyes, Tony F. Chan, Gérard A. Meurant, Jeffrey S. Scroggs, and Robert G. Voigt, editors. SIAM, Philadelphia, 1992, pp. 62–72. (Cited on pp. 164, 165)

[150] S. V. Nepomnyaschikh. Mesh theorems of traces, normalizations of function traces and their inversions. *Sov. J. Numer. Anal. Math. Modeling*, 6:1–25, 1991. (Cited on pp. 164, 165, 188)

[151] R. A. Nicolaides. Deflation of conjugate gradients with applications to boundary value problems. *SIAM J. Numer. Anal.*, 24(2):355–365, 1987. ISSN: 0036-1429. DOI: 10.1137/0724027. URL: http://dx.doi.org/10.1137/0724027. (Cited on pp. 103, 107, 129)

[152] F. Nier. Remarques sur les algorithmes de décomposition de domaines. In *Seminaire: Équations aux Dérivées Partielles*, 1998–1999, École Polytechnique, Paris, 1999. (Cited on p. 56)

[153] M. Oorsprong, F. Berberich, V. Teodor, T. Downes, S. Erotokritou, S. Requena, E. Hogan, M. Peters, S. Wong, A. Gerber, E. Emeriau, R. Guichard, G. Yepes, K. Ruud, et al., editors. *PRACE Annual Report 2013*. Insight Publishers, Bristol, UK, 2014. (Cited on p. 213)

[154] M. L. Parks, E. de Sturler, G. Mackey, D. D. Johnson, and S. Maiti. Recycling Krylov subspaces for sequences of linear systems. *SIAM J. Sci. Comput.*, 28(5):1651–1674, 200. ISSN: 1064-8275. DOI: 10.1137/040607277. URL: http://dx.doi.org/10.1137/040607277. (Cited on p. 106)

[155] L. F. Pavarino and O. B. Widlund. Balancing Neumann-Neumann methods for incompressible Stokes equations. *Comm. Pure Appl. Math.*, 55(3):302–335, 2002. (Cited on p. 131)

[156] A. Pechstein and C. Pechstein. A FETI method for a TDNNS discretization of plane elasticity. Technical report (2013-05), Johannes Kepler University Linz, Austria, 2013. URL: http://www.numa.uni-linz.ac.at/publications/List/2013/2013-05.pdf. (Cited on p. 159)

[157] A. Pechstein and J. Schöberl. Anisotropic mixed finite elements for elasticity. *Internat. J. Numer. Methods Engrg.*, 90(2):196–217, 2012. ISSN: 0029-5981. DOI: 10.1002/nme.3319. URL: http://dx.doi.org/10.1002/nme.3319. (Cited on p. 159)

[158] C. Pechstein. *Finite and Boundary Element Tearing and Interconnecting Solvers for Multiscale Problems*, Springer, Berlin, 2013. (Cited on p. 147)

[159] C. Pechstein and R. Scheichl. Analysis of FETI methods for multiscale PDEs. *Numer. Math.*, 111(2):293–333, 2008. (Cited on p. 113)

[160] C. Pechstein and R. Scheichl. Scaling up through domain decomposition. *Appl. Anal.*, 88(10–11):1589–1608, 2009. (Cited on pp. 113, 161)

[161] C. Pechstein and R. Scheichl. Weighted Poincaré inequalities. Technical report (NuMa-Report 2010-10). Institute of Computational Mathematics, Johannes Kepler University Linz, Austria, 2010. (Cited on p. 128)

[162] Z. Peng and J.-F. Lee. Non-conformal domain decomposition method with second-order transmission conditions for time-harmonic electromagnetics. *J. Comput. Phys.*, 229(16):5615–5629, 2010. (Cited on p. 72)

[163] Z. Peng, V. Rawat, and J.-F. Lee. One way domain decomposition method with second order transmission conditions for solving electromagnetic wave problems. *J. Comput. Phys.*, 229(4):1181–1197, 2010. (Cited on p. 72)

[164] J. Poulson, B. Engquist, S. Li, and L. Ying. A parallel sweeping preconditioner for heterogeneous 3D Helmholtz equations. *SIAM J. Sci. Comput.*, 35(3):C194–C212, 2013. ISSN: 1064-8275. DOI: 10.1137/120871985. URL: http://dx.doi.org/10.1137/120871985. (Cited on p. 72)

[165] A. Quarteroni and A. Valli. *Domain Decomposition Methods for Partial Differential Equations*. Oxford, 1999. (Cited on p. vii)

[166] V. Rawat and J.-F. Lee. Nonoverlapping domain decomposition with second order transmission condition for the time-harmonic Maxwell's equations. *SIAM J. Sci. Comput.*, 32(6):3584–3603, 2010. ISSN: 1064-8275 (Cited on p. 72)

[167] F.-X. Roux, F. Magoulès, L. Series, and Y. Boubendir. Approximation of optimal interface boundary conditons for two-Lagrange multiplier FETI method. In *Domain Decomposition Methods in Science and Engineering. Lecture Notes in Computational Science and Engineering*, vol. 40. Springer, Berlin, 2005, pp. 283–290. (Cited on p. 51)

[168] Y. Saad. Analysis of augmented Krylov subspace methods. *SIAM J. Matrix Anal. Appl.*, 18(2):435–449, 1997. ISSN: 0895-4798. DOI: 10.1137/S0895479895294289. URL: http://dx.doi.org/10.1137/S0895479895294289. (Cited on p. 106)

[169] Y. Saad. *Iterative Methods for Sparse Linear Systems*. PWS Publishing Company, Boston, MA, 1996; reprinted, SIAM, Philadelphia, 2003. (Cited on pp. 78, 85, 88, 92, 93, 106)

[170] Y. Saad and M. H. Schultz. GMRES: A generalized minimal residual algorithm for solving nonsymmetric linear systems. *SIAM J. Sci. Statist. Comput.*, 7:856–869, 1986. (Cited on p. 88)

[171] A. Schädle, L. Zschiedrich, S. Burger, R. Klose, and F. Schmidt. Domain decomposition method for Maxwell's equations: scattering off periodic structures. *J. Comput. Phys.*, 226(1):477–493, 2007. ISSN: 0021-9991. DOI: 10.1016/j.jcp.2007.04.017. URL: http://dx.doi.org/10.1016/j.jcp.2007.04.017. (Cited on p. 72)

[172] R. cheichl and E. Vainikko. Additive Schwarz with aggregation-based coarsening for elliptic problems with highly variable coefficients. *Computing*, 80(4):319–343, 2007. (Cited on p. 161)

[173] R. Scheichl, P. S. Vassilevski, and L. T. Zikatanov. Weak approximation properties of elliptic projections with functional constraints. *Multiscale Model. Simul.*, 9(4):1677–1699, 2011. ISSN: 1540-3459. DOI: `10.1137/110821639`. URL: `http://dx.doi.org/10.1137/110821639`. (Cited on pp. 129, 161)

[174] H. A. Schwarz. Über einen Grenzübergang durch alternierendes Verfahren. *Vierteljahrsschrift der Naturforschenden Gesellschaft in Zürich*, 15:272–286, 1870. (Cited on pp. viii, 1)

[175] B. F. Smith, P. E. Bjørstad, and W. Gropp. Domain Decomposition: Parallel Multilevel Methods for Elliptic Partial Differential Equations. Cambridge University Press, Cambridge, 1996. (Cited on pp. vii, 142, 143)

[176] H. J. Stephen Smith. On systems of linear indeterminate equations and congruences. *Phil. Trans. R. Soc. Lond.*, 151:293–326, 1861. (Cited on p. 155)

[177] N. Spillane, V. Dolean, P. Hauret, F. Nataf, C. Pechstein, and R. Scheichl. A robust two-level domain decomposition preconditioner for systems of PDEs. *C. R. Math. Acad. Sci. Paris*, 349(23–24):1255–1259, 2011. ISSN: 1631-073X. DOI: `10.1016/j.crma.2011.10.021`. URL: `http://dx.doi.org/10.1016/j.crma.2011.10.021`. (Cited on p. 161)

[178] N. Spillane, V. Dolean, P. Hauret, F. Nataf, C. Pechstein, and R. Scheichl. Abstract robust coarse spaces for systems of PDEs via generalized eigenproblems in the overlaps. *Numer. Math.*, 126(4):741–770, 2014. ISSN: 0029-599X. DOI: `10.1007/s00211-013-0576-y`. URL: `http://dx.doi.org/10.1007/s00211-013-0576-y`. (Cited on pp. 161, 186, 188)

[179] N. Spillane, V. Dolean, P. Hauret, F. Nataf, and D. Rixen. Solving generalized eigenvalue problems on the interfaces to build a robust two-level FETI method. *C. R. Math. Acad. Sci. Paris*, 351(5–6):197–201, 2013. ISSN: 1631.073X. DOI: `10.1016/j.crma.2013.03.010`. URL: `http://dx.doi.org/10.1016/j.crma.2013.03.010`. (Cited on pp. 186, 188, 190)

[180] N. Spillane and D. Rixen. Automatic spectral coarse spaces for robust finite element tearing and interconnecting and balanced domain decomposition algorithms. *Internat. J. Numer. Methods Engrg.*, 95(11):953–990, 2013. ISSN: 0029-5981. DOI: `10.1002/nme.4534`. URL: `http://dx.doi.org/10.1002/nme.4534`. (Cited on p. 190)

[181] A. St-Cyr, M. J. Gander, and S. J. Thomas. Optimized multiplicative, additive, and restricted additive Schwarz preconditioning. *SIAM J. Sci. Comput.*, 29(6):2402–2425, 2007. ISSN: 1064-8275. DOI: `10.1137/060652610`. URL: `http://dx.doi.org/10.1137/060652610` (Cited on pp. 51, 53, 187)

[182] P. Le Tallec, J. Mandel, and M. Vidrascu. A Neumann–Neumann domain decomposition algorithm for solving plate and shell problems. *SIAM J. Numer. Anal.*, 35:836–867, 1998. (Cited on p. 131)

[183] P. Le Tallec and A. Patra. Non-overlapping domain decomposition methods for adaptive hp approximations of the Stokes problem with discontinuous pressure fields. *Comput. Methods Appl. Mech. Engrg.*, 145:361–379, 1997. (Cited on p. 131)

[184] J. M. Tang, R. Nabben, C. Vuik, and Y. A. Erlangga. Comparison of two-level preconditioners derived from deflation, domain decomposition and multigrid methods. *J. Sci. Comput.*, 39(3):340–370, 2009. (Cited on pp. 106, 178)

[185] A. Toselli and O. Widlund. *Domain Decomposition Methods - Algorithms and Theory. Springer Series in Computational Mathematics*, vol. 34. Springer, Berlin, 2005. (Cited on pp. vii, 5, 106, 118, 120, 124, 129, 147, 166)

[186] H. A. van der Vorst. Bi-CGSTAB: A fast and smoothly converging variant of Bi-CG for the solution of nonsymmetric linear systems. *SIAM J. Sci. Statist. Comput.*, 13(2):631–644, 1992. ISSN: 0196-5204. DOI: `10.1137/0913035`. URL: `http://dx.doi.org/10.1137/0913035`. (Cited on pp. 88, 93)

[187] J. Willems. Robust multilevel methods for general symmetric positive definite operators. *SIAM J. Numer. Anal.*, 52(1):103–124, 2014. ISSN: 0036-1429. DOI: `10.1137/120865872`. URL: `http://dx.doi.org/10.1137/120865872`. (Cited on p. 162)

[188] F. Willien, I. Faille, F. Nataf, and F. Schneider. Domain decomposition methods for fluid flow in porous medium. In *Sixth European Conference on the Mathematics of Oil Recovery*, 1998. (Cited on pp. 68, 72)

[189] J. T. Wloka, B. Rowley, and B. Lawruk. *Boundary Value Problems for Elliptic Systems*. Cambridge University Press, Cambridge, 1995. (Cited on p. 155)

[190] J. Xu. Theory of Multilevel Methods. PhD thesis, Cornell University, Ithaca, NY, 1989. (Cited on p. 166)

Index